Climate Change and Global Health

Climate Change and Global Health

———————

Edited by

Professor Colin D. Butler

Faculty of Health
The University of Canberra
University Drive
Bruce
ACT 2617
Australia

www.cabi.org

CABI is a trading name of CAB International

CABI	CABI
Nosworthy Way	38 Chauncy Street
Wallingford	Suite 1002
Oxfordshire OX10 8DE	Boston, MA 02111
UK	USA

Tel: +44 (0)1491 832111	Tel: +1 800 552 3083 (toll free)
Fax: +44 (0)1491 833508	E-mail: cabi-nao@cabi.org
E-mail: info@cabi.org	
Website: www.cabi.org	

A catalogue record for this book is available from the British Library, London, UK.

Library of Congress Cataloging-in-Publication Data

Climate change and global health / edited by Professor Colin D. Butler, Faculty of Health, The University of Canberra.
    p. cm.
  Includes bibliographical references and index.
  ISBN 978-1-78064-265-9 (hbk : alk. paper)
  1. Climatic changes--Health aspects. 2. Medical climatology. 3. World health.
I. Butler, Colin D.

  RA793.C5766 2014
  616.9'88--dc23

ISBN-13: 978 1 78064 265 9

Commissioning editor: Rachel Cutts
Editorial assistant: Emma McCann
Production editor: James Bishop

Typeset by SPi, Pondicherry, India
Printed and bound by CPI Group (UK) Ltd, Croydon, CR0 4YY

Contents

——————————

Contributors

Adeladza K. Amegah, Center for Environmental and Respiratory Health Research, Faculty of Medicine, University of Oulu, P.O. Box 5000, FI-90014 Oulu, Finland. E-mail: Adeladza. Amegah@oulu.fi

Dr Godson Ana, Department of Environmental Health Sciences, Faculty of Public Health, College of Medicine, University of Ibadan, PMB 5017, Ibadan, Nigeria. E-mail: anagrow@yahoo.com

Associate Professor Hilary Bambrick, Centre for Health Research, School of Medicine, Centre for Health Research Building 3, Campbelltown Campus, University of Western Sydney, Locked Bag 1797, Penrith, NSW 2751, Australia. E-mail: H.Bambrick@uws.edu.au

Associate Professor Paul J. Beggs, Department of Environment and Geography, Faculty of Science, Macquarie University, NSW 2109, Australia. E-mail: paul.beggs@mq.edu.au

Bernard Bett, International Livestock Research Institute, P.O. Box 30709, Nairobi 00100, Kenya.

Dr Menno Bouma, London School of Hygiene and Tropical Medicine, London, WC1E 7HT, UK. E-mail: Menno.Bouma@lshtm.ac.uk

Kathryn Bowen, National Centre for Epidemiology and Population Health, The Australian National University, Canberra, ACT 0200, Australia. E-mail: kathrynjbowen@gmail.com

Devin C. Bowles, National Centre for Epidemiology and Population Health, The Australian National University, Canberra, ACT 0200, Australia. E-mail: devin.bowles@anu.edu.au

Dr Mark Braidwood, School of Public Health & Community Medicine, UNSW Medicine, Room 314B, Samuels Building, UNSW Sydney NSW 2052, Australia. E-mail: m.braidwood@unsw.edu.au

Dr Bianca Brijnath, Department of General Practice, School of Primary Health Care, Monash University, Building 1, 270 Ferntree Gully Road, Notting Hill, Victoria 3168, Australia. E-mail: bianca.brijnath@monash.edu

Professor Colin D. Butler, Faculty of Health, The University of Canberra, University Drive, Bruce, ACT 2617, Australia, and Visiting Fellow, National Centre for Epidemiology and Population Health, The Australian National University, Canberra, ACT 0200, Australia, and Medical Director, Benevolent Organisation for Development, Health & Insight (BODHI). E-mail: colin.butler@canberra.edu.au

Professor Guéladio Cissé, Swiss Tropical and Public Health Institute, University of Basel, Socinstrasse 57, CH-4002 Basel, Switzerland. E-mail: Gueladio.Cisse@unibas.ch

Professor Ulisses Confalonieri, CPqRR – Fundação Oswaldo Cruz (FIOCRUZ), LAESA, Av. Augusto de Lima, 1715, Barro Preto - Belo Horizonte, MG, 30190-002 Brazil. E-mail: uconfalonieri@gmail.com

Dr Kristie L. Ebi, ClimAdapt, LLC, 13212 42nd Ave NE, Seattle, WA 98125 USA. E-mail: krisebi@essllc.org

Professor Marilyn Aparicio Effen, Climate Change, Environmental and Health Unit, Instituto Boliviano de Biologia de Altura (IBBA), Facultad de Medicina – Universidad Mayor de San Andrés, Bolivia. E-mail: marilyneffen@gmail.com

Professor Birgitta Evengård, Arctic Research Center (ARCUM), Division of Infectious Diseases, Umeå University, 90187 Sweden. E-mail: birgitta.evengard@umu.se

Emeritus Professor Ole Faergeman, Emeritus Professor of Preventive Cardiology, Department of Cardiology B, Aarhus University Hospital, Tage Hansens Gade 2, 8000 Aarhus C, Denmark. E-mail: ferryman@mail.tele.dk

Dr Delia Grace, International Livestock Research Institute, PO Box 30709, Nairobi 00100, Kenya. E-mail: D.GRACE@CGIAR.ORG

Dr Matthew Haigh, School of Oriental and African Studies, University of London, Russell Square, London, WC1H 0XG, UK. E-mail: haighmatthew@yahoo.com.au

Professor Sir Andy Haines, Department of Social & Environmental Health Research (SEHR), Faculty of Public Health and Policy, London School of Hygiene and Tropical Medicine, Keppel Street, London, WC1E 7HT, UK. E-mail: Andy.Haines@lshtm.ac.uk

Dr Shakoor Hajat, London School of Hygiene and Tropical Medicine, 15–17 Tavistock Place, London, WC1H 9SH, UK. E-mail: shakoor.hajat@lshtm.ac.uk

Simon Hales, Research Associate Professor, University of Otago, Wellington, PO Box 7343, Mein Street, Wellington, New Zealand. E-mail: haknabus@gmail.com

Dr Elizabeth G. Hanna, National Centre for Epidemiology and Population Health, Australian National University, Canberra, ACT 0200, Australia. E-mail: Liz.Hanna@anu.edu.au

Professor Yasushi Honda, Faculty of Health and Sport Sciences, The University of Tsukuba, 1-1-1 Tennodai (Comprehensive Res Build D), Tsukuba 305-8577, Japan. E-mail: honda@taiiku.tsukuba.ac.jp

Professor Jouni J.K. Jaakkola, Professor and Director, Center for Environmental and Respiratory Health Research, University of Oulu, PO Box 5000, FI-90014 Oulu, Finland. E-mail: jouni.jaakkola@oulu.fi

Professor Shinjiro Kanae, Department of Civil Engineering, Tokyo Institute of Technology Japan.

Professor Patrick Kinney, Department of Environmental Health Sciences, Mailman School of Public Health, Columbia University, 722 W 168th St, 11th floor, New York, NY 10032, USA. E-mail: plk3@columbia.edu

Professor Tord Kjellstrom, Umeå Centre for Global Health Research, Division of Epidemiology and Global Health, Department of Public Health and Clinical Medicine, Umeå University, SE-901 85 Umeå, Sweden, and National Centre for Epidemiology and Population Health, Australian National University, Canberra, ACT 0200, Australia, and Institute of International Health, University College London, UK. E-mail: tord.kjellstrom@epiph.umu.se; tord.kjellstrom@anu.edu.au

Dr Bruno Lemke, School of Health, Nelson-Marlborough Institute of Technology, Nelson, New Zealand. E-mail: Bruno.Lemke@nmit.ac.nz

Dr. Rebekah Lucas, Umeå Centre for Global Health Research, Division of Epidemiology and Global Health, Department of Public Health and Clinical Medicine, Umeå University, SE-901 85 Umeå, Sweden. E-mail: rebekah.lucas@epiph.umu.se

Dr Lachlan McIver, National Centre for Epidemiology and Population Health, The Australian National University, Canberra, ACT 0200, Australia. E-mail: lachlan.mciver@gmail.com

Professor A.J. (Tony) McMichael, Emeritus Professor (Population Health), The Australian National University, Canberra, ACT 0200, Australia. E-mail: tony.mcmichael@anu.edu.au

Dr Andrew Mathieson, The Australian National University, Canberra, ACT 0200, Australia. E-mail: andrewmathieson7@hotmail.com

Ms Haruka Morita, Department of Environmental Health Sciences, Mailman School of Public Health, Columbia University, 722 W 168th St, 11th floor, New York, NY 10032, USA. E-mail: hm2487@columbia.edu

Dr Nick H. Ogden, Senior Research Scientist, Zoonoses Division, Centre for Food-borne, Environmental and Zoonotic Infectious Diseases, Public Health Agency of Canada, 3200

Sicotte, CP 5000, Saint-Hyacinthe, QC, J2S 7C6, Canada. E-mail: nicholas_ogden@phac-aspc.gc.ca

Professor Taken Oki, Institute of Industrial Science, The University of Tokyo, Japan.

Matthias Otto, School of Health, Nelson-Marlborough Institute of Technology, Nelson, New Zealand.

Dr Lisa Page, Consultant Liaison Psychiatrist, Sussex Partnership NHS Foundation Trust, and Honorary Clinical Senior Lecturer, Brighton and Sussex Medical School; and Mental Health Liaison Team, Level 5, Royal Sussex County Hospital, Eastern Road, Brighton, East Sussex, BN2 5BR UK. E-mail: lisa.2.page@kcl.ac.uk

Dr Alan J. Parkinson, Deputy Director, Arctic Investigations Program, Centers for Disease Control and Prevention, 4055 Tudor Centre Drive, Anchorage, Alaska 99508, USA. E-mail: ajp1@cdc.gov

Professor Mercedes Pascual, Rosemary Grant Collegiate Professor of Ecology and Evolutionary Biology, Department of Ecology and Evolutionary Biology, University of Michigan, 2019 Kraus Natural Sciences Bldg, 830 North University, Ann Arbor, MI 48109-1048, USA; and Howard Hughes Medical Institute, Chevy Chase, Maryland 20815-6789, USA. E-mail: pascual@umich.edu

Professor Germán Poveda, Department of Geosciences and Environment, Facultad de Minas, Carrera 80 x Calle 65, Bloque M2-315, Universidad Nacional de Colombia, Medellín, Colombia. E-mail: gpoveda@unal.edu.co

Dr Ana Flávia Quintão, Escola de Saúde Pública do Estado de Minas Gerais (ESP-MG), avenue Augusto de Lima, 2,061, Barro Preto, Belo Horizonte, Minas Gerais, 30.190-002, Brazil. E-mail: ana.fonseca@esp.mg.gov.br; anaflaviaqf@gmail.com

Professor Mala Rao OBE, Professor of International Health, Institute for Health and Human Development, University of East London, Stratford Campus, Water Lane, London, E15 4LZ, UK. E-mail: mala.rao@doctors.org.uk

Rafael Reuveny, Professor, School of Public and Environmental Affairs (SPEA), Indiana University 10th Street and Fee Lane, Bloomington, IN 47405, USA. E-mail: rreuveny@indiana.edu

Dr Horacio Riojas, Director de Salud Ambiental, Instituto Nacional de Salud Pública, Av. Universidad 655 Santa María Ahuacatitlán, CP 62100, Cuernavaca Morelos, México. E-mail: hriojas@insp.mx

Dr Subhashis Sahu, Ergonomics and Occupational Physiology Laboratory, Department of Physiology, University of Kalyani, Kalyani-741235, West Bengal, India. E-mail: skcsahu@yahoo.co.in; sahuphysiolku@gmail.com

Professor Jan C. Semenza, Head of Health Determinants Programme, Office of the Chief Scientist, European Centre for Disease Prevention and Control, SE-171 83 Stockholm, Sweden. E-mail: Jan.Semenza@ecdc.europa.eu

Dr Manpreet Singh, Dalberg Global Development Advisors, Nairobi, Kenya. E-mail: manpreet1@gmail.com

Byron Smith, School of Divinity, University of Edinburgh, New College, Mound Place, Edinburgh, EH1 2LX, Scotland, UK; and 398 Bronte Road, Waverley, NSW 2024, Australia. E-mail: byron@thesmiths.id.au

Professor Will Steffen, Adjunct Professor, Fenner School of Environment and Society, The Australian National University, Canberra, ACT 0200, Australia; and Guest Researcher, The Stockholm Resilience Centre, Sweden. E-mail: will.steffen@anu.edu.au

Dr Robin Stott, Climate and Health Council, UK, and 15 Egerton Drive, Greenwich, London, SE10 8JS, UK. E-mail: stott@dircon.co.uk

Dr Lucy Telfar Barnard, Postdoctoral Fellow, He Kainga Oranga/Housing and Health Research Programme, Department of Public Health, School of Medicine and Health Sciences, University of Otago, Wellington, Box 7343, Wellington South, New Zealand. E-mail: lucy.telfar-barnard@otago.ac.nz

Professor Vidhya Venugopal, Professor and Head (Industrial Hygiene Services), Department of Environmental Health Engineering, Sri Ramachandra University, Porur, Chennai-600 116, India. E-mail: vvidhya@ehe.org.in; rvidhyaa@yahoo.com

Professor Mark L. Wahlqvist AO, Emeritus Professor of Medicine, Monash University, Australia; and Visiting Professor, Institute of Population Health Science, National Health Research Institute, Taiwan; and Director, Fuli Institute of Food Science, Zhejiang University, China. E-mail: profmlw@ nhri.org.tw

Dr Sue Wareham OAM, Vice President, Medical Association for Prevention of War (Australia), 22 Wybalena Grove, Cook, ACT 2614, Australia. E-mail: warehams@ozemail.com.au

Dr Haylee J. Weaver, School of Science, Education and Engineering (Building J), University of the Sunshine Coast, Locked Bag 4, Maroochydore DC, QLD 4558, Australia. E-mail: hweaver@usc. edu.au

Acronyms

_____

AC	air conditioning
ACTs	artemisinin combination therapies
AI	aerosol index
AIMS	Atlantic, Indian Ocean, Mediterranean and South China Sea (islands)
ALRI	acute lower respiratory tract infection
AOD	aerosol optical depth
AOSIS	Association of Small Island States
AOT	aerosol optical thickness
AR	allergic rhinitis
ASEAN	Association of South-east Asian Nations
ASF	animal source food
BOD	burden of disease
BRIC	Brazil, Russia, India and China
C	Celsius
CARICOM	Caribbean Commission
CCRIS	Caribbean Catastrophe Risk Insurance Scheme
CDC	Centers for Disease Control and Prevention
CE	Common Era
CFE	carbon fertilization effect
CH_4	methane
CI	confidence interval
CL	cutaneous leishmaniasis
CMIP5	Climate Model Intercomparison Project 5
CO	carbon monoxide
CO_2	carbon dioxide
CO_2e	carbon dioxide equivalents
CO_{TC}	total column CO
CRED	Centre for Research on the Epidemiology of Disasters
CSIRO	Commonwealth Scientific and Industrial Research Organisation
DALYs	disability-adjusted life years
DDT	dichlorodiphenyltrichloroethane
DF	dengue fever

DHF	dengue haemorrhagic fever
DOTS	directly observed treatment, short-course
DPR	Democratic Peoples' Republic
DSR	daily severity rating
ECDC	European Centre for Disease Prevention and Control
EIP	extrinsic incubation period
ENSO	El Niño–Southern Oscillation
EP	epidemic potential
EPA	Environmental Protection Agency
EU	European Union
EWE	extreme weather events
EWM	excess winter mortality
EWMI	excess winter mortality index
EWS	early warning system
FAO	Food and Agriculture Organization of the United Nations
FBP	fire behaviour prediction
FCS	Falciparum Climate Suitability
FWI	fire weather index
GCM	global climate model or general circulation model
GDP	gross domestic product
GHGs	greenhouse gases
GSOD	global summary of the day
HAT	human African trypanosomiasis
HH	household
HIAs	health impact assessments
HIV/AIDS	human immunodeficiency virus/acquired immunodeficiency syndrome
Hothaps	high occupational temperature health and productivity suppression
hpA	hectopascals (a unit of pressure equal to a millibar: 1 hPa = 1 mb)
HPAI	highly pathogenic avian influenza
IAM	integrated assessment modelling
ICT	information and communication technologies
IDB	Inter-American Development Bank
IEQ	indoor environmental quality
IIASA	International Institute for Applied Systems Analysis
IOC	Indian Ocean Commission
IOM	Institute of Medicine
IPCC	Intergovernmental Panel on Climate Change
IPPNW	International Physicians for the Prevention of Nuclear War
JGCRI-PNNL	Joint Global Change Research Institute Pacific Northwest National Laboratory
LA	Latin America
LDC	less developed country
MDG	Millennium Development Goal
Mha	million hectares
MPC	maximum permissible concentration
MRSA	methicillin-resistant *Staphylococcus aureus*
MSR	monthly severity rating
Mt	million tonnes
NASA	National Aeronautics and Space Administration
NCCHAP	National Climate Change and Health Action Plan
NCDs	non-communicable diseases
NIES	National Institute for Environmental Studies
NOAA	National Oceanographic and Atmospheric Administration

NO_x	nitrogen oxides
NO_2	nitrogen dioxide
N_2O	nitrous oxide
NSW	New South Wales
NTD	neglected tropical disease
O_3	ozone
OR	odds ratio
PAH	polycyclic aromatic hydrocarbon
PAHO	Pan American Health Organization
PAR	population at risk
PBL	Netherlands Environmental Assessment Agency (http://www.pbl.nl/en)
PCB	polychlorinated biphenyl
PCR	polymerase chain reaction
PDR	Peoples' Democratic Republic
PIC	Pacific Island country
PM	particulate matter
POPs	persistent organic pollutants
ppm	parts per million
PTSD	post-traumatic stress disorder
RC	rhinoconjunctivitis
RCP	representative concentration pathways
RF	Russian Federation
Rh	rhinitis
RR	relative risk
RSV	respiratory syncytial virus
RT-PCR	reverse transcriptase polymerase chain reaction
RVF	Rift Valley fever
SARS	Severe Acute Respiratory Syndrome
SDG	Sustainable Development Goal
SEAR-D	South-east Asian Region-D (Bangladesh, Bhutan, Democratic People's Republic of Korea, India, Maldives, Myanmar, Nepal)
SES	socio-economic status
SIDS	small island developing states
SIV	simian immunodeficiency virus
SPA	shared climate policy assumption
SRES	Special Report on Emission Scenarios
SREX	Special Report on Extreme Events
SSA	single-scattering albedo
SSP	shared socio-economic pathways
SSR	seasonal severity rating
SST	sea surface temperature
STH	soil-transmitted helminths
Ta	air temperature
tg	terragram
Tg	globe temperature
Tmax	maximum temperature
Tmin	minimum temperature
Tnwb	natural wet bulb temperature
TBE	tick-borne encephalitis
UN	United Nations
UNDP	United Nations Development Programme
UNFCCC	UN Framework Convention on Climate Change

UNHCR	United Nations High Commissioner for Refugees
UN-OHRLLS	UN Office of the High Representative for the Least Developed Countries, Landlocked Countries and Small Island Developing States
USA	United States of America
UV-B	ultraviolet B
VC	vectorial capacity
VIA	vulnerability, impacts and adaptation
VOCs	volatile organic compounds
WBGT	wet bulb globe temperature
WEMA	water-efficient maize for Africa
WHO	World Health Organization

Acknowledgements

_____

This book, almost three years in preparation, has involved numerous people in its various stages, from conception to completion. I particularly want to thank Prof Colin Soskolne, Fiona Armstrong, Prof Val Brown, Dr Susie Burke, Susan Woldenberg Butler, Prof Tony Capon, Prof Eric Chivian, Rachel Cutts, Winfried K. Dallmann, Dr Kris Ebi, Andy Epstein, A/Prof David Harley, Prof Pat Kinney, Prof John Last, Prof Tony McMichael, Luiz Ribiero, Prof David Shearman, Dr Atanu Sarkar, Prof Adrian Sleigh, Dr Peter Tait, Dr Elvina Viennet, Louise Whelan and all 56 contributing authors (some of whom are also mentioned above).

Dedication

Colin D. Butler

_____

This book is dedicated to the late Dr Paul Epstein (1943–2011), who died when the book was being planned. I first became aware of Paul's work in 1992. At that time, I was working as a solo general practitioner in a small Tasmanian country town. Before the Internet, one of my few connections with international health was to skim _The Lancet_, delivered weekly by the postman. In one such issue, soon after the Rio Earth Summit, I read Paul's report of a symposium held in Rio, which led to the Heidelberg Appeal and reminded me that Gross Domestic Product needed adjustment for loss of 'natural' capital.

The Appeal, signed by more than 50 Nobel Prize laureates, stressed the central role of science in measuring and preserving resources and in motivating the public and politicians to help protect them. Paul wrote:

> the Appeal concludes: 'The greatest evils which stalk our earth are ignorance and oppression, and not science, technology, and industry, whose instruments, when adequately managed, are indispensible tools of a future shaped by humanity, by itself and for itself, overcoming major problems like overpopulation, starvation, and world-wide disease.'
>
> (Epstein, 1992)

These stirring words, reported by someone from the Harvard Medical School, were immensely encouraging to me. Tasmania, though far from Brazil, was the site of the world's first Green political party (founded 1974) and in its capital city, Hobart, I had already heard lectures by Paul Ehrlich and David Suzuki. It seemed clear to me that adverse global environmental change, together with nuclear weapons, represented the largest threats to human well-being on the planet. While studying tropical medicine at the London School of Hygiene in 1990, I had met Maurice King, a guest lecturer with whom I had talked non-stop for three hours, and who also perceived climate change (together with high population growth rates in Africa) as extremely important (King, 1990). People like Maurice were rare among doctors. Dr Bob Brown, a former GP and then leader of the Tasmanian Green Party, was another – but, by then, Bob was no longer working directly in health.

Paul was probably the first US-based health worker that I knew of who so clearly recognized the links between adverse planetary change and the risk to future human well-being, connections stressed by Ehrlich and Suzuki. These changes include the relentless conversion of forests and wetlands to fields and farms, with the resultant loss of biodiversity, and the inexorable rise in climate-changing gases. Both phenomena are underpinned by an ongoing rise in levels of human population, aspiration and consumption (Butler, 1994), issues that the Australian epidemiologist, Tony McMichael, calls 'planetary overload' (McMichael, 1993). These risks accumulate, driven by high-level denial of 'limits to growth' (Meadows _et al._, 1972; Butler, 2004).

Epstein became one of the leading academics active in alerting those concerned with public health to the risks of climate change. Highlights include his key role (with Andy Haines and Tony McMichael) in an early series of articles on climate change and health in *The Lancet* (Epstein and Sharp, 1993; Epstein *et al.*, 1993; Haines *et al.*, 1993), his work with the reinsurance industry (Epstein *et al.*, 2005) and his co-authored critique of the cost of coal, including its health hazards (Epstein *et al.*, 2011). Before his premature death, Paul and the science writer, Dan Ferber, also co-wrote *Changing Planet, Changing Health: How the Climate Crisis Threatens Our Health and What We Can Do about It* (Epstein and Ferber, 2011).

I was fortunate to meet Paul twice, though we corresponded occasionally from 1994 until 2011. In 1994, he sent me collected papers used to teach at the then newly formed Center for Health and the Global Environment at Harvard, which he established with Eric Chivian, a founder of the International Physicians for the Prevention of Nuclear War.

I first met Paul in 1997, in London, with a group of physicians active in the International Society of Doctors for the Environment. Paul was active in the US affiliate, at that time, Physicians for Human Rights. The other encounter, six years later, was in New York City, for an inaugural meeting of the work that led to the report on climate change and disasters, led by Paul and funded by the United Nations Development Programme (UNDP) and Swiss Re (Epstein *et al.*, 2005). Paul had invited me because of my involvement with the Millennium Ecosystem Assessment, the conceptual framework of which linked human well-being to ecosystem services (Millennium Ecosystem Assessment, 2003). At that meeting, Paul had arranged for Mark Malloch-Brown, then head of the UNDP (and substantially responsible for the text of the Millennium Development Goals, including its hasty phrases about sustainability) (Tran, 2012), to talk to us, and the next day we met at the UN Headquarters building. Paul's connection with extremely influential people also included Al Gore, who consulted with Paul when preparing his Nobel Prize-winning movie, *An Inconvenient Truth* (Vitello, 2011). My final contact with Paul was when I invited him to write a chapter for this book. Sadly, he had to decline, due to his illness.

There is no space here to provide more than these glimpses of Paul, but accessible articles about his life include an obituary in the *New York Times* (Vitello, 2011) and an open access article in *PLoS Biology* (Clapp, 2012). These reveal a dedication to issues of social justice and health care, evident from his student years in the USA, later extending to Mozambique, Nicaragua and El Salvador. Paul also linked the breakdown of sanitation in parts of Latin America to structural adjustment and the onerous conditions imposed by international lending agencies such as the International Monetary Fund (IMF) on national governments in developing countries. The resulting disinvestment in public health infrastructure led to unsafe sewage management, contaminated water and, ultimately, the cholera outbreak in Latin America in the early 1990s (Clapp, 2012). This opportunity for cholera was also facilitated by warmer sea surface temperatures associated with El Niño, as well as the unregulated transfer of bilge water from Asia to the Peruvian coast (Epstein *et al.*, 1993).

I have a sense that Paul was considered by some, at times, to have crossed from science to activism, to eschew the most conservative interpretation of evidence. Perhaps, in part, this reflects his extensive clinical experience, including in low-income settings, where prognoses are often based on limited data, leavened by experience. Doctors have a duty of care to think ahead, to think in a precautionary manner and to sound warnings in the interests of their patient, whether an individual, a population or the whole planet. The world needs far more Paul Epsteins, and it needs far more people, individually and collectively, to respond responsibly to the sober cautions that he and his colleagues, through their scientific assessments and prognostications, have sounded, if not for their own futures, then for the generations to follow.

References

Butler, C.D. (1994) Overpopulation, overconsumption, and economics. *The Lancet* 343, 582–584.

Butler, C.D. (2004) Human carrying capacity and human health. *PLoS Medicine* 1, 192–194.

Clapp, R. (2012) Paul Epstein (1943–2011): a life of commitment to health and social justice. *PLoS Biology* 10, e1001284.

Epstein, P.R. (1992) Brazil: climate of change? *The Lancet* 339, 1529.

Epstein, P.R. and Ferber, D. (2011) *Changing Planet, Changing Health: How the Climate Crisis Threatens Our Health and What We Can Do about It.* University of California Press, Berkeley, California.

Epstein, P.R. and Sharp, D. (1993) Medicine in a warmer world. *The Lancet* 342, 1003–1004.

Epstein, P.R., Ford, T.E. and Colwell, R.R. (1993) Marine ecosystems. *The Lancet* 342, 1216–1219.

Epstein, P.R., Mills, E., Frith, K., Linden, E., Thomas, B. and Weireter, R. (eds) (2005) *Climate Change Futures. Health, Ecological and Economic Dimensions Summary for Policymakers.* The Center for Health and the Global Environment, Swiss Re, UNDP, Harvard Medical School, Boston, Massachusetts.

Epstein, P.R., Buonocore, J.J., Eckerle, K., Hendryx, M., Stout, B.M. III, Heinberg, R., *et al.* (2011) Full cost accounting for the life cycle of coal. *Annals of the New York Academy of Science* 1219, 73–98.

Haines, A., Epstein, P.R. and McMichael, A.J. (1993) Global health watch: monitoring impacts of environmental change. *The Lancet* 342, 1464–1469.

King, M. (1990) Health is a sustainable state. *The Lancet* 336, 664–667.

McMichael, A.J. (1993) *Planetary Overload. Global Environmental Change and the Health of the Human Species.* Cambridge University Press, Cambridge, UK.

Meadows, D., Meadows, D., Randers., J. and Behrens, W. III (1972) *The Limits to Growth.* Universe Books, New York.

Millennium Ecosystem Assessment (2003) *Ecosystems and Human Well-being. A Framework for Assessment.* Island Press, Washington, DC.

Tran, M. (2012) Mark Malloch-Brown: developing the MDGs was a bit like nuclear fusion. *The Guardian*, 16 November 2012 (http://www.guardian.co.uk/global-development/2012/nov/16/mark-malloch-brown-mdgs-nuclear, accessed 3 March 2014).

Vitello, P. (2011) Dr Paul Epstein, expert in public health, is dead at 67. *New York Times*, 14 November 2011 (http://www.nytimes.com/2011/11/15/health/dr-paul-epstein-public-health-expert-dies-at-67.html?_r=0, accessed 3 March 2014).

Foreword

Climate change is arguably the most pressing example of policy failure on a global scale that confronts humanity. The evidence that the world's climate is changing, and that human activities are responsible, has gathered strength in recent years, and yet policy makers have shown themselves unable to act decisively to address the causes and plan for the consequences. The reasons for this failure include the opposition of powerful vested interests to change, the perception that such change is difficult and expensive, particularly in a time of financial austerity, and lack of public understanding of the science underpinning climate change.

Climate change will have wide-ranging effects on human health, although there are clearly uncertainties, not just about the rate and magnitude of future climatic change but also about medium- to long-term socio-economic trends, which are strong determinants of health in their own right, as well as the capacity of populations to adapt to climate change. There are also questions about the appropriate counterfactual scenario against which to compare the health effects of climate change. The heaviest burden of impacts will be experienced in the second half of this century and beyond by a population of perhaps 10 billion or so, mostly living in what are today's low- and middle-income countries, but the effects will be felt worldwide. Many scientists consider that the probability of 'dangerous' climate change, causing major adverse consequences for humanity, increases substantially when the rise in global mean temperature exceeds 2°C, and this now appears very likely to occur.

In the light of the threats to health posed by climate change, it is important to advance understanding of the potential mechanisms by which it can impact on human health, as well as the potential role of adaptation and mitigation actions in reducing the adverse effects of climate change.

This book provides invaluable evidence for public health professionals about the effects of climate, their vulnerability to climate change and how appropriate policies can enhance the resilience of communities to climatic change. It provides a comprehensive overview of a rapidly evolving field of research, which will be invaluable to those with responsibility for protecting populations from the effects of extreme events (e.g. heatwaves, floods, droughts and intense tropical cyclones) that are likely to increase in frequency with climate change. It will also be of interest to the wider public health and development communities, who need to be aware of how our changing climate could cause shifts in disease risk, including changes in the distribution of vector-borne and other climate-sensitive communicable diseases and a range of other health outcomes. The book demonstrates why it is imperative to move towards 'low-carbon' development, which achieves improvements in health and well-being without undermining the world life-support systems on which humanity ultimately depends.

Although the growth in interest in the public health implications of climatic change is relatively recent, we owe much to the public health pioneers who first pointed to the potential health impacts some 20 years ago, notably Paul Epstein and Tony McMichael. Sadly, Paul's recent death has robbed us of his vision and energy, but this book acknowledges his and Tony's leadership, which was crucial in raising the profile of the topic before it became widely accepted as a legitimate, and indeed an increasingly vital, focus of concern for public health.

Sir Andy Haines
London, UK
November 2013

1 The Anthropocene: A Planet Under Pressure

Will Steffen

Fenner School of Environment and Society, The Australian National University, Canberra, Australia, and The Stockholm Resilience Centre, Sweden

1.1 The Nature of the Earth System

The advent of climate change, and the broader issue of global change (Steffen *et al.*, 2004), has begun to alter humanity's perspective on the planet we evolved in and is now beginning to influence at the worldwide scale. To understand how humanity's role in the functioning of the Earth System (the term I use here to refer to the dynamic and systemic features of the global environment), a basic understanding of Earth history is essential.

For most of its 4.6 billion-year history, life has been a feature of the Earth System and has wrought vast and significant changes to the planet's environment, including the atmosphere. But, for contemporary issues, the past 60 million years or so are most relevant (Zalasiewicz *et al.*, 2012). During that period, the Earth has cooled slowly from a much warmer state, the continents have migrated into their current positions and the great polar ice sheets have appeared from the earlier ice-free world – first the Antarctic ice sheets about 34 million years ago and then the northern hemisphere ice sheets, only about 10–12 million years ago.

Fully modern humans – *Homo sapiens* – are a much more recent phenomenon; we are only about 200,000–250,000 years old as a species (Oppenheimer, 2004), and our hominid ancestors are only a few million years old. The period of the planet during which humans evolved is called the late Quaternary; it is characterized by a rich biodiversity dominated by mammals and a climate that oscillates in a strikingly regular pattern between long, cold ice ages and shorter, intervening warm periods (Petit *et al.*, 1999; EPICA Community Members, 2004).

The regular climatic pattern of the late Quaternary provides strong evidence that the Earth as a whole operates as a single complex system, albeit a highly complex one. Many specific features of the late Quaternary environment support this perspective (Scheffer, 2009):

- The two states of the late Quaternary are well defined, with the ice ages apparently the more stable state.
- The oscillations between these states are exceptionally regular, with a periodicity of about 100,000 years through the past million years, and a periodicity of about 40,000 years before that. This is an example of 'phase locking', in which the internal dynamics of the Earth System are in synchronization with a regular, but weak, external forcing agent. In this case, regular variations in the Earth's orbit are the weak external forcing and the waxing and waning of the northern hemisphere ice sheets are probably the feature of the Earth's internal dynamics that locks into various modes of the Earth's orbit.
- Perhaps the most important feature of the Earth System is the 'limit cycles' – that is, as

the Earth cycle oscillates between ice ages and warm periods, the variation in temperature, greenhouse gas (GHG) concentrations and other features of the environment are tightly constrained between well-defined limits.

- Feedbacks – both reinforcing (positive) and damping (negative) – are essential features of the complex Earth System in the late Quaternary. The two most important of these are: (i) the waxing and waning of the great northern hemisphere ice sheets, which change the reflectivity of the Earth's surface; and (ii) the concentration of atmospheric GHGs, the most important of which is carbon dioxide (CO_2). Both of these feedbacks significantly alter the energy balance at the Earth's surface and modulate the rather modest variations in incoming solar radiation.

1.2 The Expanding Human Enterprise

For nearly all of our existence, the relationship between humans and the Earth System has essentially been one way: the features of the Earth System described above, and many modes of natural variability on finer spatial and temporal scales, have had significant influences, sometimes devastating ones, on the viability of human societies. But humans have operated under the planetary radar screen in terms of their influences back on to the functioning of the Earth System. Our activities may have had environmental impacts at local and sometimes even regional levels, but we did not have any appreciable influence on the functioning of the Earth System on the global level (Steffen et al., 2007, and references therein).

That benign relationship changed with the advent of the Industrial Revolution late in the 18th century, first in England and then elsewhere in Europe and the New World. The most striking feature of the Industrial Revolution was the widespread use of fossil fuels, which broke through a long-standing bottleneck in the development of the human enterprise. Until then, the extent and rate of activity of civilization was constrained tightly by the limits of animal muscle power – that is, the power of humans and their domesticated animals, augmented to a small extent by water power. Fossil fuels shattered that bottleneck.

The impacts of fossil fuel use were not confined to their direct impact on energy production and transport. Arguably more important were the flow-on effects that a cheap and plentiful new energy source facilitated, together with science, governance and popular acceptance. These effects included: (i) the capability to 'fix' unreactive nitrogen from the atmosphere and to make it available to support food production; (ii) the provision of fresh water in reliable ways to most human settlements; and (iii) the design and construction of much more effective sewage systems, leading to a great improvement in human health. These factors led to an explosion of the human population, and to their capability to produce and consume goods and services (McNeill, 2000; Hibbard et al., 2006).

The vast changes in the human enterprise since the Industrial Revolution are presented in Fig. 1.1. The timescale begins at 1750, to capture the beginning of the Industrial Revolution, and continues to the beginning of the new millennium. The graphs encompass indicators not only of population and economic activity but also of resource use, communication, transport and globalization. One feature stands out strongly in the figures – the year 1950. In each of the panels, the rate of increase rises sharply about that year, and there is no appreciable value at all for several of the indicators before 1950.

This remarkable explosion of the human enterprise soon after the Second World War is sometimes called the 'Great Acceleration' (Hibbard et al., 2006). Many excellent narratives have been created to explain this 20th century phenomenon (e.g. McNeill, 2000), and out of these, a few important drivers of the Great Acceleration can be distilled (Hibbard et al., 2006):

- Global connectivity – people, finance and goods and services became much more connected around the globe as new technologies in communications and transport built the foundation for what is now known as globalization.
- Neo-liberal economics – a free market economic system became dominant and, by the end of the 20th century, had achieved

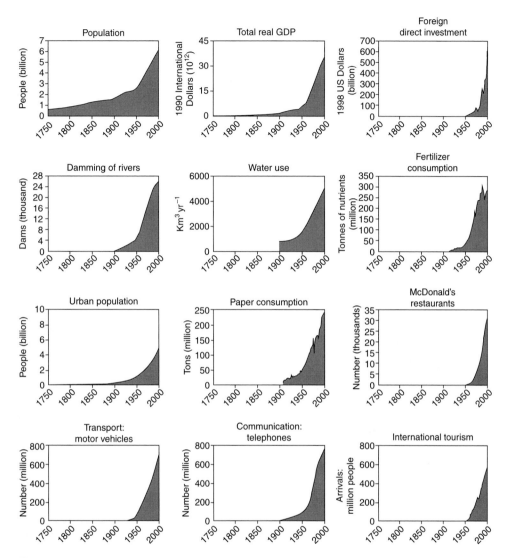

Fig. 1.1. Increasing rates of change in several indicators of human activity since the beginning of the Industrial Revolution. Significant increases in the rates of change of all indicators occur simultaneously around 1950, ushering in a period of dramatic and unprecedented change, often called the Great Acceleration (Steffen *et al.*, 2004) (which includes references to the original data for each of the 12 panels). (Reprinted with kind permission of Springer Science+Business Media.)

virtually complete coverage of the entire global economic system. The Bretton Woods institutions, created following the Second World War, such as the World Bank and the International Monetary Fund, played a key role in this trend.

• Armies of scientists and engineers – the war effort generated a large cadre of scientists and engineers, and this significant technical

capacity was tapped by the civilian economy soon after hostilities had ceased.

• The economic growth imperative – the need to rebuild economies after the devastation of the Depression and Second World War led to a period of rapid growth and the embedding of continuous, strong economic growth as a core value of late 20th century societies.

The post-1950 Great Acceleration was supported by an abundance of natural resources, especially cheap fossil fuels such as petroleum and coal, and was driven almost entirely by the wealthy countries of the world, which represented no more than 20% of the human population. The first decade of the 21st century brought some changes to this pattern. The environmental contribution from wealthy countries is now amplified by several rapidly industrializing economies, such as those of Brazil, Russia, India and China (the so-called BRIC countries), as well as South Africa and Indonesia. Furthermore, the era of abundant natural resources, especially cheap fossil fuels, seems to be coming rapidly to an end, with the concept of 'peak oil' (e.g. Sorrell *et al.*, 2009) becoming the poster child for a growing array of resource constraints.

1.3 Global Change

The Great Acceleration is an important phenomenon in terms of human history, but it has perhaps even greater significance for the history of planet Earth. The imprint of the human enterprise on the environment can be seen at remarkably early times in our planetary existence, but almost always at local and regional levels only. As shown in Fig. 1.2, human activities have recently become so pervasive that their imprint on the environment at the global level is obvious. The intertwining planetary-level changes in human activities and the environment is now often called 'global change' (e.g. Steffen *et al.*, 2004).

Undoubtedly, the most well known of the changes depicted in Fig. 1.2 is that of the energy balance at the Earth's surface, resulting in what is commonly known as 'climate change' or 'global warming'. The warming of the Earth's surface is unequivocal, and there are many other associated changes in the climate system, ranging from alteration in the patterns of precipitation to a slowly but inexorably rising sea level (IPCC, 2007). Although natural variability makes some aspects of climate change difficult to interpret, there is a high degree of scientific consensus that human activities, most notably the increase of GHGs in the atmosphere, are the primary cause of the temperature increase and associated climate changes we have experienced

since the mid-20th century (IPCC, 2007). Perhaps most worrying is that the influence of human-driven climate change can already be seen in the characteristics of some extreme climatic events (IPCC, 2012), the aspect of climate that most directly and obviously affects human well-being.

While much of the media attention has been focused on climate change, largely missing from the public discourse is the fact that many other human-driven changes in the global environment are perhaps even more pronounced than climate change. A good example is the nitrogen cycle, which has been even more modified by human action than the carbon cycle (Galloway and Cowling, 2002). Currently, human activities – the direct fixation of unreactive atmospheric nitrogen (N_2) into reactive forms used primarily in fertilizers and the planting of leguminous crops – fixes more nitrogen from the atmosphere than all of the natural terrestrial fixation processes combined. In essence, human activities are inserting into the Earth System about 135 million tonnes (Mt) per annum of additional reactive nitrogen compounds (Galloway and Cowling, 2002). These then cascade through the environment as additional flows from the land to freshwater systems and the coastal seas, as a range of gaseous compounds that contribute to local air pollution and add to the GHG burden in the atmosphere, and as reactive compounds that slowly accumulate in the soil.

The loss of biodiversity (bottom right panel of Fig. 1.2) is another change to the planetary environment that can be attributed unequivocally to human activities. This impact goes far beyond the loss of charismatic species, but also has implications for the provision of ecosystem services on which human well-being depends (Cardinale, 2011). It is estimated that the rate of species extinctions, as measured by extinctions of mammal, bird and amphibian species, is now 100–1000 times greater than the natural background level, and is expected to rise by another factor of 10, at least by 2100, if present trends continue (Millennium Ecosystem Assessment, 2005). It is even possible that there is a planetary-level tipping point beyond which biodiversity loss will become so rapid and uncontrollable that the Earth may experience the sixth great extinction event in its history (Barnosky *et al.*, 2011).

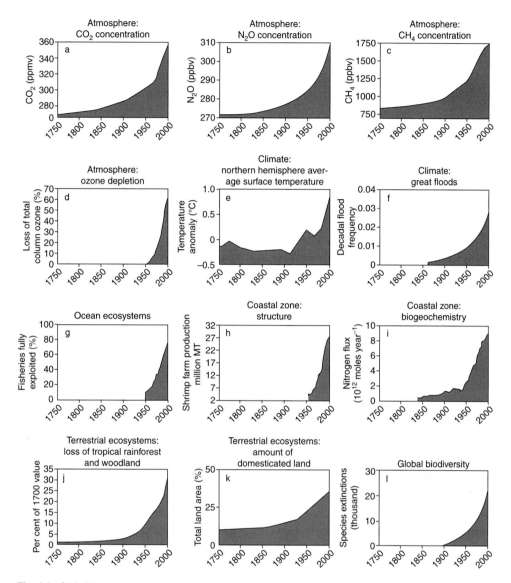

Fig. 1.2. Globally aggregated changes in the Earth System: 1750–2000. These result from the dramatic increase in human activity shown in Fig. 1.1 (Steffen *et al.*, 2004) (which includes references to the original data for each of the 12 panels). (Reprinted with kind permission of Springer Science+Business Media.)

Nowhere more than in land systems has the direct impact of human activities been apparent, as shown in the bottom left two panels of Fig. 1.2. This is not surprising as humans, being terrestrial creatures, obtain most of their ecosystem services from terrestrial ecosystems. The loss of tropical forests and woodlands has been a dominant feature of land-system change in the 20th century and has continued through the first decade of the 21st century. Overall, the human imprint on the Earth's land surface is perhaps best summarized as the increase in 'domesticated land', that is in croplands and pastures, at the expense of forests and other more natural land covers (bottom middle panel of Fig. 1.2).

Methods have been developed to integrate the overall human impact on the global environment. One of them is the application of the I = PAT identity (Holdren and Ehrlich, 1974), in which the overall impact (I) is a function of the aggregation of the global population (P); their affluence (A), which is used as a proxy for consumption of goods and services; and the technology (T) used to produce the goods and services. Not surprisingly, the global impact of the human enterprise shows a dramatic rise since 1950, in the Great Acceleration (Kolbert, 2011). While population increase is, indeed, a factor in growing environmental impact, it is not nearly as important as the increase in consumption per capita.

A second method of estimating the aggregate global impact of humanity is the so-called 'global footprint analysis', which estimates the area of land and ocean per person required to provide all of the resources needed to support each person's consumption and to absorb and process the wastes that are generated (Global Footprint Network, 2011). The country-by-country results are aggregated to give a global total, which can be expressed as the number of planet Earths required to support the human enterprise. At present, that total is about 1.35 planets and rising, which implies that the human enterprise is now in overshoot, consuming more of the Earth's natural resources than can be replenished annually – that is, we are eating into the natural capital of the planet and thus reducing its capacity to support future generations.

1.4 The Anthropocene: A Planetary Game Changer

The ultimate consequence of the rapidly accelerating suite of global changes was summarized by Nobel Laureate, Paul Crutzen, in the concept of the Anthropocene (Crutzen, 2002), a proposed new geological epoch in which the human enterprise has become significant in comparison to the great forces of nature. In essence, the arrival of the Anthropocene implies that the Earth is leaving its current geological epoch, the Holocene.

Lasting now for over 10,000 years, the Holocene epoch is the most recent geological period in Earth history and has proven to be an exceptionally accommodating environment for human development. It was, after all, during the Holocene that we developed agriculture, villages and cities, and the more complex civilizations, including the contemporary civilization we now know.

The arrival of the Anthropocene implies a profound shift in the human–environment relationship at the global level, challenging the assumption that the Earth System will automatically continue to provide an accommodating environment for further human development. Our interaction with the Earth System has passed beyond a simple pressure–impact relationship into one governed by the more chaotic, unpredictable dynamics of complex systems.

We are learning rapidly about these complex system dynamics. An important advance in Earth System science over recent decades has been the elucidation of a number of 'tipping elements' in the Earth System – critical regions or processes where small changes in driving variables can trigger abrupt changes in critical processes, with significant implications for the functioning of the whole Earth System (Plate 1; Schellnhuber, 2002; Lenton et al., 2008; Richardson et al., 2011).

Tipping elements in the Earth System are basically of three types:

1. Those that involve the melting of ice, such as the loss of Arctic sea ice, the potential loss of most or all of the Greenland ice sheet and large emissions of methane from melting permafrost.
2. Changes in circulation, such as a shift in the behaviour of the South Asian monsoon, or a slowing or shutting down of the North Atlantic thermohaline circulation (i.e. the 'Gulf Stream').
3. Flips of major biomes from one state to another, such as the conversion of the Amazon basin from a tropical rainforest to a grassland or a savannah.

Although much uncertainty surrounds the nature of these tipping elements, we know from the palaeo-record that abrupt changes in all of them have occurred in the past and we are beginning to estimate where the tipping point might lie for some (Lenton et al., 2008), where the term *tipping point* refers to a threshold along an important controlling variable (such as temperature) beyond which the tipping element changes into a fundamentally different state.

For example, based on expert judgement, it is highly likely that the threshold for loss of Arctic summertime sea ice has already been crossed, and there is a medium-to-high likelihood that the threshold for loss of the Greenland ice sheet may be crossed by 2100. By contrast, the likelihood of a significant slowing or shutting down of the North Atlantic thermohaline circulation by 2100 is judged to be low (Richardson *et al.*, 2011).

Returning now to the complex systems analysis outlined in Section 1.1 above, what are the ultimate implications of the modifications of the global environment by human activities? That is, where is the Anthropocene going?

It is clear that the Earth as a whole operates as a single complex system, oscillating between two well-defined states in the late Quaternary. Up to now, the assumption has been that, depending on the amount of GHG emissions, the climate will stabilize at a temperature 2, 3, 4 or more degrees C above pre-industrial levels (IPCC, 2007). That is, the aggregate amount of emissions determines the final state of the climate system (Meinshausen *et al.*, 2009).

A complex systems perspective suggests that this may not be so. It is conceivable that there is a planetary-level threshold or tipping point beyond which the Earth System as a whole will move to a much warmer, stable, ice-free state – a state that has existed for long periods in the past. This implies that a number of the individual tipping elements shown in Plate 1, especially those that change the reflectivity (albedo) of the Earth's surface (mainly loss of ice), or those that add significant amounts of GHGs to the atmosphere (methane outburst from permafrost), act in concert once a temperature threshold (or series of closely placed temperature thresholds) is crossed and push the Earth System into a much warmer state. Once such a threshold(s) is crossed, there is nothing that humanity could do to prevent the transition.

If there is such a tipping point, it is very unlikely to lie below a temperature rise of 2°C, and is much more likely to lie above a rise of 4°C (Richardson *et al.*, 2011).

1.5 Challenges of the 21st Century

Achieving sustainability, however that concept is defined, is widely agreed to be the most fundamental challenge facing humanity in the 21st century, and will soon be formalized in the post-2015 sustainable development goals (SDGs) (UN, 2012). While attempting to bring the bulk of the world's population out of poverty will remain at the core of the SDGs, there is a growing recognition that protecting the environment is not an optional extra but essential to support the ongoing human enterprise. This imperative is nowhere more profound than at the global level, where a destabilized and rapidly changing global environment would make sustainability at any other level very difficult or virtually impossible. Yet, this is the risk posed as humanity heads further into the Anthropocene, with no sign of slowing down.

There are two fundamentally different approaches to managing the human–environment relationship in the Anthropocene. One aims actively to manage the Earth System itself, steering it in ways perceived to be beneficial to humanity. *Geo-engineering* is a term often used for this approach, and reflects the dominant belief of the Great Acceleration, namely that humanity can now shape, steer and transform the environment to suit its own purposes.

The most prominent example of geo-engineering is the proposal to counteract GHG-induced warming by inserting aerosols into the stratosphere to scatter some of the incoming solar radiation and thus cool the planet (Royal Society, 2009). This approach, using simple cause–effect logic, ignores the complexity of Earth System dynamics. Injecting aerosols into the stratosphere would change rainfall patterns in ways that we cannot predict, would not counteract the rapid increase in ocean acidity caused by the dissolution of atmospheric CO_2 and would likely lead to other as yet unanticipated changes in Earth System dynamics.

The other approach is based on *planetary stewardship*. This is a fundamentally different philosophy aimed at reducing human pressure on the Earth System and allowing it to function with less interference. Stewardship is a basic human approach to respecting the integrity of the environment on which we depend, and is deeply rooted in our existence, found in all major religions and in indigenous cultures. It approaches humanity's relationship with the environment by managing human values and behaviour, rather than by attempting to control, manipulate, dominate and transform the environment to suit us.

The most prominent global-scale example of the stewardship philosophy is the planetary boundaries approach (Rockström et al., 2009a,b), which recognizes that the Holocene is the only state of the Earth System that we know for sure can support contemporary society. The approach aims to define a safe operating space for humanity, allowing Earth System functioning to continue in a way that we know can support our continued development and well-being. In short, the approach recognizes that the Earth System has intrinsic, hard-wired properties and processes (e.g. tipping elements, Plate 1 and Section 1.4), many of which have thresholds beyond which abrupt and/or irreversible change, detrimental to human well-being, can occur. These features form the basis for boundaries.

Plate 2 summarizes the planetary boundaries approach, showing the nine features and processes of the Earth System that are sufficient to define the state of the System. They are: climate change, ocean acidification, stratospheric ozone depletion, the nitrogen and phosphorus cycles, global freshwater use, change in land use, biodiversity loss, atmospheric aerosol loading and chemical pollution. The inner green circle defines the safe operating space – the set of boundaries that define a Holocene-like state of the Earth System. The red wedges show the current position of the human modification of the process in relation to the boundary.

We have already transgressed three of the boundaries – climate change, biodiversity loss and the nitrogen cycle. The climate change issue is already well known. Much less known is the seriousness and urgency of the biodiversity crisis. At its most fundamental level, the genetic code embodied in biodiversity acts as a giant control panel, complete with its set of in-built redundancies and back-up systems, that guides biosphere functioning, in turn providing the range of ecosystem services on which we depend (Millennium Ecosystem Assessment, 2005). In fact, without the biosphere, the composition of the atmosphere would be vastly different than it is today, and the functioning of the Earth System as a whole would be significantly altered. The planet would be unrecognizable. Although there is, indeed, considerable redundancy and resilience in the Earth's biosphere, it too has its intrinsic limits, and planetary-scale shifts are possible (Barnosky et al., 2011).

Implementation of the planetary boundaries approach has already generated criticism (e.g. Lewis, 2012). Perhaps the most common is based on the long-standing tension between development and environment; that is, setting environmental limits will constrain development and constrain the task of developing countries to lift their populations from poverty (e.g. Kosoy et al., 2012). However, a recent analysis argues that, in fact, there are significant synergies – not conflicts – between respecting the planetary boundaries and improving many aspects of social equity, including improving the material well-being of the poor in developing countries (Steffen and Stafford Smith, 2013). The synergies are especially apparent for the so-called aggregated planetary boundaries, such as those for the phosphorus and nitrogen cycles, land-use change and biodiversity loss. A transfer of resources or access to ecosystem services from those areas of overuse (i.e. the wealthy countries) to those that lack resources or services will simultaneously help keep Earth within the boundaries and improve social equity outcomes. Such analyses hold out hope for the future – that we can build an effective planetary stewardship regime while addressing the great inequities that still plague civilization.

1.6 Conclusion

Confronting the Anthropocene is the most complex, difficult and important challenge that humanity has yet faced in its existence on Earth. Are we headed for a new era of sustainability and enhanced human well-being, or are we headed for environmental degradation, conflict and collapse? The jury is out. As Maurice Strong, convenor of the Earth Summit, wrote: 'Where on Earth are we going?'

References

Barnosky, A.D., Matzke, N., Tomiya, S., Wogan, G.O.U, Swartz, B., Quental, T.B., *et al.* (2011) Has the Earth's sixth mass extinction already arrived? *Nature* 471, 51–57.

Cardinale, B.J. (2011) Impacts of biodiversity loss. *Science* 336, 552–553.

Crutzen, P.J. (2002) Geology of mankind: the anthropocene. *Nature* 415, 23.

EPICA Community Members (2004) Eight glacial cycles from an Antarctic ice core. *Nature* 429, 623–628.

Galloway, J.N. and Cowling, E.B. (2002) Reactive nitrogen and the world: two hundred years of change. *Ambio* 31, 64–71.

Global Footprint Network (2011) Our human development initiative (http://www.footprintnetwork.org/en/index.php/GFN/page/fighting_poverty_our_human_development_initiative/, accessed 22 February 2011).

Hibbard, K., Crutzen, P., Lambin, E., Liverman, D., Mantua, N., McNeill, J., *et al.* (eds) (2006) *Integrated History and Future of People on Earth, Dahlem Workshop Report 96.* MIT Press, Boston Massachusetts, pp. 341–375.

Holdren, J.P. and Ehrlich, P.R. (1974) Human population and the global environment: population growth, rising per capita material consumption and disruptive technologies have made civilization a global ecological force. *American Scientist* 62, 282–292.

IPCC (Intergovernmental Panel on Climate Change) (2007) *Climate Change 2007: The Physical Science Basis. Contribution of Working Group I to the Fourth Assessment Report of the Intergovernmental Panel on Climate Change.* (Solomon, S., Qin, D., Manning, M., Chen, Z., Marquis, M., Averyt, K., Tignor, M.M.B., Miller, H.L. Jr and Chen, Z. (eds)). Cambridge University Press, Cambridge, UK, and New York.

IPCC (2012) *Managing the Risks of Extreme Events and Disasters to Advance Climate Change Adaptation. A Special Report of Working Groups I and II of the Intergovernmental Panel on Climate Change.* (Field, C.B., Barros, V., Stocker, T.F., Qin, D., Dokken, D.J., Ebi, K.L., Mastrandrea, M.D., Mach, K.J., Plattner, G.-K., Allen, S.K., Tignor, M. and Midgley, P.M. (eds)). Cambridge University Press, Cambridge, UK, and New York.

Kolbert, E. (2011) Enter the Anthropocene: age of man. *National Geographic* 219, 60–77.

Kosoy, N., Brown, P.G., Bosselmann, K., Duraiappah, A., Mackey, B., Martinez-Alier, J., *et al.* (2012) Pillars for a flourishing Earth: planetary boundaries, economic growth delusion and green economy. *Current Opinion in Environmental Sustainability* 4, 74–79.

Lenton, T.M., Held, H., Kriegler, E., Hall, J.W., Lucht, W., Rahmstorf, S., *et al.* (2008) Tipping elements in the Earth's climate system. *Proceedings of the National Academy of Sciences USA* 105, 1783–1785.

Lewis, S.L. (2012) We must set planetary boundaries wisely. *Nature,* 485, 417.

McNeill, J.R. (2000) *Something New Under the Sun: An Environmental History of the Twentieth-Century World.* W.W. Norton, New York.

Meinshausen, M., Meinshausen, N., Hare, W., Raper, S.C.B., Frieler, K., Knutti, R., *et al.* (2009) Greenhouse-gas emission targets for limiting global warming to 2°C. *Nature* 458, 1158–1162.

Millennium Ecosystem Assessment (2005) *Ecosystems and Human Well-being. Synthesis.* Island Press, Washington, DC.

Oppenheimer, S. (2004) *Out of Eden. The Peopling of the World.* Constable and Robinson, London.

Petit, J., Jouzel, J., Raynaud, D., Barkov, N., Barnola, J.-M., Basile, I., *et al.* (1999) Climate and atmospheric history of the past 420,000 years from the Vostok ice core, Antarctica. *Nature* 399, 428–436.

Richardson, K., Steffen, W., Liverman, D., Barker, T., Jotzo, F., Kammen, D., *et al.* (2011) *Climate Change: Global Risks, Challenges and Decisions.* Cambridge University Press, Cambridge, UK.

Rockström, J., Steffen, W., Noone, K., Persson, Å., Chapin, F.S., Lambin, E.F., *et al.* (2009a) Planetary boundaries: exploring the safe operating space for humanity. *Ecology and Society* 14(2), 32 (http://www.ecologyandsociety.org/vol14/iss2/art32/, accessed 3 March 2014).

Rockström, J., Steffen, W., Noone, K., Persson, Å., Chapin, F.S., Lambin, E.F., *et al.* (2009b) A safe operating space for humanity. *Nature* 461, 472–475.

Royal Society (2009) *Geoengineering the Climate: Science, Governance and Uncertainty.* The Royal Society, London.

Scheffer, M. (2009) *Critical Transitions in Nature and Society.* Princeton University Press, Princeton, New Jersey.

Schellnhuber, H.J. (2002) Coping with Earth System complexity and irregularity. In: Steffen, W., Jäger, J., Carson, D. and Bradshaw, C. (eds) *Challenges of a Changing Earth: Proceedings of the Global Change Open Science Conference. Amsterdam, The Netherlands, 10–13 July 2001.* Springer-Verlag, Berlin, Heidelberg, New York, pp. 151–156.

Sorrell, S., Speirs, J., Bentley, R., Brandt, A. and Miller, R. (2009) *An Assessment of the Evidence for a Near-term Peak in Global Oil Production.* UK Energy Research Centre, London.

Steffen, W. and Stafford Smith, M. (2013) Planetary boundaries, equity and global sustainability: why wealthy countries could benefit from more equity. *Current Opinion in Environmental Sustainability* 5, 403–408.

Steffen, W., Sanderson, A., Jäger, J., Tyson, P.D., Moore, B. III., Matson, P.A., *et al.* (2004) *Global Change and the Earth System: A Planet Under Pressure.* The IGBP Book Series, Springer-Verlag, New York.

Steffen, W., Crutzen, P.J. and McNeill, J.R. (2007) The Anthropocene: are humans now overwhelming the great forces of nature? *Ambio* 38, 614–621.

UN (United Nations) (2012) United Nations Secretary-General's High-level Panel on Global Sustainability. Resilient People, Resilient Planet: A Future Worth Choosing. United Nations, New York.

Zalasiewicz, J., Crutzen, P. and Steffen, W. (2012) The Anthropocene. In: Gradstein, F.M., Ogg, J.G., Schmitz, M. and Ogg, G.M. (eds) *A Geological Time Scale 2012.* Elsevier, Amsterdam, pp. 1033–1040.

2 Climate Change and Global Health

A.J. (Tony) McMichael

The Australian National University, Canberra, Australia

2.1 Introduction

The phrase *global health*, like *sustainability*, is now widespread in the research and policy arenas. Both terms are often misused. Indeed, both are being exploited, including by many universities and research groups, for institutional gain and status. As Kelley Lee has commented: 'the term "global health" has generated a considerable amount of interest in recent years; so much so that many policy makers, development specialists, health professionals, and, yes, even academics have attempted to jump on the bandwagon claiming global health gains' (Lee *et al.*, 2002).

Meanwhile, the related phrase *globalization of health* conveys a more flexible, process-oriented meaning (Schrecker, 2012). It recognizes that the processes of globalization and the rising international commitment to socio-economic development in lower-income countries have facilitated, first, much more complete and standardized health metrics information about patterns of health, disease and mortality in all countries (Lim *et al.*, 2012), and second, the opportunity to address and coordinate at global level such issues as infectious disease surveillance and control (WHO), rising anti-microbial resistance and the ongoing rise of non-communicable diseases as populations urbanize and join the pervasive Western-style consumer culture (Beaglehole *et al.*, 2011).

Global health is much more than a rebranding of ongoing international health activities. Rather, it extends our conceptual frame and research-and-policy agenda. It emphasizes an important, larger-scale and more ecological dimension of influence on the health risks to whole communities and populations. This dimension distinguishes much of its content from that of the great body of ongoing, generally laudable, international health activity (McMichael, 2013).

Although the dividing line is not clear, international health comprises activities of the following complexion: vaccine trials for those infectious diseases that cause high population burdens of disease and premature deaths; large multi-country trials of polypills, aspirin and such like to reduce the incidence of cardiovascular disease; and the widening evaluation and uptake of pesticide-impregnated bed nets to fend off malaria (especially in sub-Saharan Africa). International health also deals with cross-border regulations relating to infection control, hygiene, safety, the food trade and others. These are bedrock international health activities, essential to the improvement and maintenance of health around the world.

The crucial connotation of the word 'global' is that, today, there is also a new category of large, escalating and unfamiliar risks to population health. Increasingly, populations are being influenced by transboundary forces, as the size and intensity of the human enterprise expands

© CAB International 2014. *Climate Change and Global Health* (ed. C.D. Butler) 11

and as the aggregated human impact on the natural environment's life-supporting systems increases (McMichael and Butler, 2011). In consequence, we are putting unprecedented demands and stresses on much of nature's life-support system, and on the stability and resilience of cultures, societies and communities. This new era of an intensively interconnected 'growth'-oriented world with an increasingly human-dominated biosphere is being referred to as the 'Anthropocene' (Crutzen, 2002).

The resultant emerging challenges to population health include:

- the health and environmental impacts (positive and negative) of liberalization of regional and global trade, capital mobility and the resultant changes in production workforce conditions and in consumer choices and behaviour;
- the spread of 'western' culture and consumerism, the power of modern transnational corporations to influence production choices and methods – as well as national governmental policies;
- the health consequences of urbanization (largely unplanned in relation to actual human biological and psychological needs);
- the many and diverse health risks, escalating and broadening over time, from human-induced global climate change;
- the fundamental risks to population health sustainability (food shortages, water scarcity and a destabilized microbial world) from disruptions of other great biophysical and ecological systems and processes in nature;
- the ongoing worldwide increase in human mobility (migrant and refugee flows).

Some of today's more systemic environmental changes are genuinely 'global' in physical scale, dynamics and action (e.g. stratospheric ozone depletion, human-induced climate change, ocean acidification). Some are greatly amplified and globally generalized versions of previously more localized stressors (e.g. soil exhaustion and erosion, freshwater depletion, biodiversity losses).

They pose a different type of challenge to health systems from those posed by more localized health hazards or health inequalities – such as the remediation of local environmental pollution with toxic chemicals or radiation leaks; the need for strengthening basic primary care and public health services, and for educating girls and women in less literate societies (hygiene, family planning, nutrition, etc.); and evaluating the efficacy of specific interventions (e.g. pesticide-impregnated bed nets against malaria, and DOTS[1] therapy for tuberculosis). Those more locally focused intervention programmes are central to the continuing business of 'international health', with action primarily at national or local level. It is therefore unhelpful for this long-standing international health agenda and the additional agenda of global change and health to be lumped together as 'global health'. That impedes understanding of the qualitative shift in scale, complexity and time frame of the contemporary large-scale environmental change risks to health, including the future sustainability of human population health.

2.2   Responding to the Challenge

These transboundary influences on population health are of a kind and scale that cannot be responded to satisfactorily on a local basis. If our interconnected world becomes more prone to pandemic infectious disease, it will not suffice to rely on local barricades. The impacts of global climate change cannot be avoided by mere local 'adaptive' action. Averting or controlling these complex larger-scale forces requires a much greater supranational understanding and sharing of information, the engagement of regional and world-level agencies, disinterested aid commitments by high-income countries, appropriate international laws and policies and a broad-based integrative research effort (Fidler *et al.*, 2009).

McFarlane and colleagues have written:

> If global health is about the improvement of health worldwide, the reduction of disparities, and protection of societies against global threats that disregard national borders, it is essential that academic institutions reach across geographic, cultural, economic, gender, and linguistic boundaries to develop mutual understanding of the scope of global health and to create collaborative education and research programs.
> (MacFarlane *et al.*, 2008)

Much of this may sound unrealistic in a fragmented and self-interested nation-state-based world. But, in large historical terms, this is a moment of great and now unavoidable transition for the world community. A huge shift is occurring in human ecology and in our relationships with the environment on which it depends absolutely. We can continue to think myopically and sleepwalk into great future problems. Or we can contrive an extraordinary coordinated global response, powered by new understanding and foresight – or by a cascade of crises.

2.3 The Profile of Global Change-related Health Risks

Consider, as examples, what might be required to prevent the risks to global and regional population health from three sources:

- the accelerated generation of new avian-derived influenza strains in the backblocks of South-east and East Asia as commercial food production practices evolve and wild-bird migration patterns respond to changes in climate, winds, food sources and surface water;
- the decline in seafood protein sources due to the ongoing warming, acidification and deoxygenation of the world's coastal seas;
- the increasing numbers of people displaced by crowding, conflict, food shortages and loss of low-lying land.

Obviously, effective prevention will require an understanding, across all of government, the private sector and the general public, of the underlying drivers of these new and growing stressors. How are human actions causing these seismic shifts in demographic profile, settlement patterns, wealth distribution and the viability and functioning of environmental systems at large?

Equally, there is need for the health sector to work increasingly closely, and proactively, with other sectors of government and community to change or transform the ways that humans live, build, move, produce, consume and share. Via such integrated strategies, these emerging systemic global health threats can be lessened or eliminated.

In public health terms, both secondary and tertiary prevention will also be required. The complexity of these large-scale changes, our unfamiliarity with their modes of impact (especially when acting in combination) and current shortcomings in our research and response capacities mean that some changes in the patterns of health risk and health outcome will

inevitably occur because of these global changes. The climate-change-related increase in the tempo and severity of weather disasters around the world over the past decade, often with unprecedented damage to human health and survival, is one obvious example.

The *global* dimension of 'global health' refers primarily, as already discussed, to the scale, complexity and systemic nature of these transboundary influences. Their advent has great implications for professional training, research concepts and methods, and for the types of interventions needed (over a range of timescales) to avert or minimize the risks to human health. This will require coming to terms with some unfamiliar and less exact concepts of 'causality' and of 'public health intervention', a shared understanding of the inherent complexity and related uncertainties the characterize many of these risk sources, pathways and future projections, and the development of new interdisciplinary research methods and collaborative networks.

2.4 Global Changes – Systemic Risks to Health

Most of the worldwide sociodemographic changes are 'new' in their scale, complexion and magnitude of social-political impacts, though usually not in their actual existence. For example, rural populations have been drifting into cities for millennia, but the rapidity of that movement in the past several decades – and the fact that, worldwide, we have just recently become an urban-based (more than 50%) species over the past decade – is new.

The population factor at large is often sidestepped in 'global change' discussions, despite its unprecedented size and absolute growth rate. The UN projects that today's 7 billion will exceed 9 billion by 2050 (United Nations Population Fund, 2013). In absolute and historical terms, these are extraordinary population growth figures that warrant much more attention, concern and policy response.

In partial contrast to the ongoing sociodemographic changes, the various 'global environmental changes' *are* unprecedented, not only in their scale and their system-disrupting nature but also because of their recent début, as critical environmental thresholds have been passed for the first time during our species' existence. Their significance can be understood better if viewed within a long historical perspective.

Since settled agrarian living first emerged around 11,000 years ago, human societies have sought to make life more secure and more comfortable. Over time, that has entailed a slow increase in the extent to which humans have re-engineered local environments, mostly for food production. Over the past two centuries, human societies have succeeded, both in material terms across a broader production front and in terms of life expectancy, way beyond the imaginings of those early, struggling, farmers and herders. However, some of that success has now begun to turn sour as adverse consequences for the natural environment and for human health accrue.

This success has been due largely to the extraordinary subsidy of high-density, portable energy extracted from fossil fuels (see Fig. 2.1). Consequently, the global human 'ecological footprint' has become huge and environmentally damaging (McMichael and Butler, 2011). We are currently placing a demand on Earth's capacity to supply, replenish and absorb that is 40–50% greater than the planet's actual capacity. Hence, the syndrome of environmental depletion and disruption that is now resulting, globally.

Among the major environmental systems that are now showing strain, change and deterioration are, particularly, the climate system, the global nitrogen cycle and the accelerating losses of biodiversity (Rockström *et al.*, 2009). These planetary environmental–ecological systems underpin life support for all species. The sustained health of human populations everywhere depends on them absolutely.

Such environmental disruptions of entire biogeophysical systems, occurring at global scale, have no precedent in human history. Climate change makes that point clearly. Before the mid-20th century, it was not possible for humans, in aggregate, to generate enough greenhouse gas (GHG) emissions and to accelerate the clearance of forests (as carbon 'sinks') to cause a change in the composition of the lower atmosphere sufficient to induce geologically rapid and continuing warming. Now it is.

2.5 Climate Change: Overview of Recent Science

There is now no reasonable doubt that human actions are causing a rise in GHG concentrations in the troposphere, predominantly by mobilizing geologically stored carbon as carbon dioxide (CO_2). Other industrial and agricultural emissions, particularly methane, nitrous oxide and elemental black carbon, are also significant and very potent

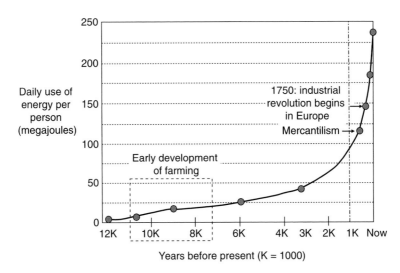

Fig. 2.1. Changes in globally averaged daily energy use per person during the Holocene, entailing a 16-fold increase since the early farming period through to the present. (Source: author. Based on data in Smil, 1994; Boyden, 2004; Ponting, 2007.)

GHGs, but with much shorter atmospheric life-times than the inert gas, CO_2.

Nor is there doubt that this build-up of GHGs is increasing the 'radiative forcing' property of the lower atmosphere, increasing the absorption of outward-bound infrared radiation from Earth's solar-heated surface. That radiation energy then manifests as heat. It is also near certain that most of the unusually rapid global warming since the mid-1970s (0.7°C global-average increase) is attributable to this human-caused increase in GHG concentration. The logical expectation, both from theory and from the 'back-casting' empirical testing of global climate models against the actual observed record of temperature change, is that substantial human-caused warming will occur this century. Indeed, even if all excess GHG emissions were terminated today, additional warming of around 1°C would still occur in the next several decades, because of the additional energy already trapped within the climate system but not yet expressed as warming at the Earth's surface.

In essence, global climate change is a consequence of massive human intervention in the planet's great carbon cycle. The perennial cycling of mobile carbon between soil, vegetation, the atmosphere and the ocean has been occurring within the frame of a fairly steady dynamic equilibrium throughout the post-glacial Holocene epoch, the past 11,000 years. Now, human activity is withdrawing carbon stocks that have long been 'on deposit' (for several hundred million years) and adding them to the already circulating, life-supporting carbon 'currency'. CO_2 concentrations in our atmosphere are now higher than at any time in the past million years, and are probably higher than at any time in the past 20 million years. This resultant glut of circulating carbon has tipped the balance; the system is now out of equilibrium; and a cascade of (mostly) adverse consequences for environmental conditions, ecosystem viability and the health and survival of species is gathering momentum.

2.6 Climate Change and Human Health

Epidemiologists and other health researchers have had little interest, over the past half-century, in studying the risks to health from variations in weather and in climate. Those variations, viewed as natural and unalterable, offered little opportunity for interesting or useful scientific discovery. Since the early 1990s that situation has changed markedly. Today, there is a surge of research interest in understanding how climate change – predicted to increase and extend for many decades or more into this century – is beginning to, and will in future, affect human health and survival.

'Climate change' is not, however, a one-dimensional risk factor of the ilk of asbestos exposure, cigarette smoking or urban air pollution. Indeed, the spectrum of risks to human health from climate change is not just wide but also seemingly unbounded. So, too, are the social and political ramifications.

The various regional manifestations of climate change – whether extremes of warming or drying, the shift of monsoonal rains and seasons, the heightening of particular extreme weather events (e.g. coastal cyclones, inland wildfires, flooding in vulnerable terrain and amplified heatwaves in densely settled cities), shortages of freshwater as river flows decrease and evaporation increases, or changes in the geographic range and rate of infectious disease agent transmission – will bear differentially on communities. Populations living in diverse social, economic and physical conditions will be affected in different ways by changes in their regional climate. Poorer and remote populations will be particularly vulnerable, via exposures to physical hazards (especially during weather disasters), undernutrition, diarrhoeal diseases, various vector-borne infectious diseases and the adverse consequences of displacement and relocation. In the Arctic region, communities will experience changes in diets as native animal populations migrate or decline and access to traditional food sources is impeded (Evengård and McMichael, 2011).

This presents the world with a huge moral and political problem of heightened inequity in population health and safety. That topic will be dealt with in greater detail in other chapters. Suffice it, here, to say that (despite assertions of those with single-track 'economic growth must come first' views and a limited imagination of how we might solve a dual non-deferrable problem) there is no intrinsic incompatibility between these two great and urgent global objectives: that is, developing a path of successful low-carbon, low environmental intensity socio-economic development and taking action to rein in fossil carbon-based energy generation, destruction of natural habitat, overharvesting of many species, overuse of synthetic nitrogen-based fertilizers and so on. The Crises of the Present are not in competition with the looming Crises of the Near Future. We can, and must, find a way to bring these two global agendas into synergistic synchronization with each other (Friel *et al.*, 2008).

The broad scientific consensus assessment is that climate change will have predominantly adverse impacts on human health – and these will become more severe and generalized as climate change gathers force. Meanwhile, there will also be several benefits to human health from climate change – in at least some regions of the world and at least during the earlier stages of the process. Examples noted in the literature include the likely reduction in wintertime death from extremes of cold if winters become milder in temperate countries, gains in local food yields in some mid- to higher-latitude regions, and the contraction of transmission zones for some vector-borne infections if conditions become too hot and dry for vector survival.

Categories of human health impacts

Overall, there are three broad categories of causal pathways by which changes in climate affect human health. There are the direct-acting climatic environmental exposures: to excessive heat, physical hazards, and so on. There are health risks resulting from disruptions and other changes to ecosystems and interspecies relations. And there are risks to both physical and mental health from social and economic disruptions, and the more ominous risks via diverse forms of health damage from tensions and conflict situations driven by climatic changes. These have been well classified by Butler and Harley as *primary*, *secondary* and *tertiary* health impacts (see also Fig. 2.2) (Butler and Harley, 2010).

A different category of risk to health arises as governments, commercial interests and communities implement adaptive measures to lessen the impacts of climate change and its diffuse consequences. Such adaptations may inadvertently be detrimental, in particular to powerless, remote or marginal groups within populations. For example, the diversion of much agriculture into biofuel production is now a significant contributor to some forms of food shortage, and to rising food prices. Two-fifths of the US maize crop is being directed to biofuel production – underpinned by federal tax incentives. If this practice continues in a future world where 'peak oil' actually materializes, then increasingly there will be food shortages and price rises.

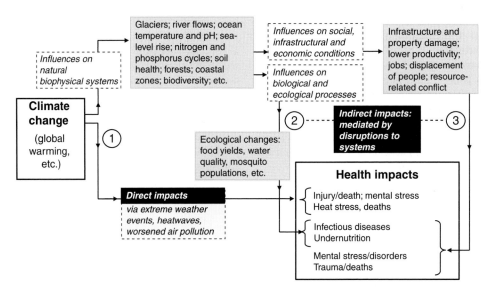

Fig. 2.2. Range of paths by which climate change can influence the health and survival of human populations/communities. The circled numbers refer to primary ①, secondary ② and tertiary ③ effects.

Starting at the least complex end of this three-way classification, there are various relatively direct health impacts of higher temperatures, increased concentrations of certain air pollutants and the immediate and physical risks from exposure to more frequent and more extreme weather events. Data on injuries, deaths and manifest disease events such as gastroenteritis are often available; rates and risks can be estimated. This 'primary' category of direct-acting influences on health has now begun to attract a wider range of epidemiological research. Questions abound. Will warmer summers in temperate countries affect patterns of clothing and outdoor behaviour, and thus contribute to increased risks of skin cancer? Does the increase in coastal groundwater salinity due to rising sea level (Vineis et al., 2011) pose a risk of hypertension in poorer rural populations (e.g. in Bangladesh) dependent on well water for drinking – and might that increase the risk of hypertensive eclampsia of pregnancy?

The secondary health impacts entail more complex processes entailing changes in climatic conditions affecting various geophysical phenomena, biotic processes and ecological relationships – all with consequences for human health. Examples include the impacts on patterns of occurrence of infectious diseases, especially those borne by climate-sensitive 'vectors' such as mosquitoes, and on food yields, and hence nutritional health, child development and the many later-life health sequelae.

The tertiary impacts span a range of risks to community morale, behaviour patterns and mental and emotional health. These can occur as part of the complex aftermath of extreme weather disasters, from downturns in livelihoods and security in rural settings as climatic adversity affects agricultural yields and the supportive services, and from displacement or outmigration of groups or whole communities in response to actual, impending or perceived threats from climate change.

For example, as physical environments change and animal and plant species recede or shift, so the traditional sources of food for various indigenous populations decline. In northern Scandinavia, thinner ice sheets on rivers and lakes and shorter glacial seasons endanger the management of reindeer herds as they are moved across the landscape by Saami herders. In northern Canada, supplies of seal, fish and caribou dwindle as temperatures rise, snow cover declines and ice floes diminish. Consequently, there is greater reliance on imported, processed, less healthy foods (Evengård and McMichael, 2011).

There are similarly fundamental threats to social stability and geopolitical security from climate-related changes in river flows, regional food yields and access to habitable land (e.g. as the sea level rises, impinging on small island states, delta populations and low-lying coastal communities). Shortages in food and fundamental resources will breed tensions, as they have throughout history. Severe cold weather in early 17th century Europe resulted in food shortages, tripling in staple food prices, a doubling in the frequency of famine years and in epidemic outbreaks and a surge in intercountry warfare and in the displacement of people (Zhang et al., 2011). Climate change, acting as a risk multiplier, is a recognized promoter of conflict and warfare (Bowles, D.C., and Butler, C.D., unpublished).

Detecting health (or directly health-related) impacts attributable to climate change

There should, by now, be many emerging opportunities to assess whether and to what extent climate change is now affecting human health outcomes. All such empirical evidence will strengthen the message to communities and policy makers about the potential seriousness of unabated climate change for humans themselves. However, attributing changes in health outcomes to climate change is complex. Most health outcomes have a range of plausible contributory causal factors: environmental, social, behavioural, biomedical or genetic. By comparison, the glaciologists have it easy explaining acceleration in the melting of Arctic sea ice. In general, the signal-to-noise ratio in assessing climatic contributions to human health outcomes is lower than for glaciers and bird nesting (see Fig. 2.3).

2.7 Focus and Purpose of Research and Policy Response

The preferred strategy, always, is primary prevention: elimination of the risk at source. The 1987 Montreal Protocol for the elimination of stratospheric ozone-destroying gases

Climate change impacts on environmental and biological
systems: How easy to assign 'cause' to climate?

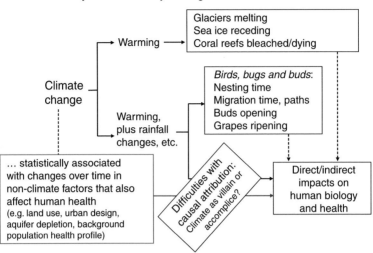

Fig. 2.3. The detection of climate change impacts on qualitatively differing systems: geophysical processes, natural ecological relationships and human well-being and health. The process of attribution is much more complex and difficult for that third, human, impact category – especially since many other aspects of human culture, circumstances and behaviour bear on human health, and often change over time.

is an example. The risks of increased exposure to ultraviolet radiation to biological targets, including the human skin and eye, were recognized as real, serious and preventable by elimination of the underlying cause of that particular environmental health hazard. The same thinking must become paramount in our response to climate change. A change in the world's climatic conditions, associated with now plausible temperature increases this century of 4–5°C (New *et al.*, 2011), would almost certainly cause major surprise step changes as critical thresholds are passed. Feedback processes of the reinforcing ('positive') kind are most likely, and the prospect of runaway climate change therefore lurks in the background (Schellnhuber, 2008).

Adaptive strategies to lessen otherwise now unavoidable risks from human-induced climate change should not distract from the great first-order need to actually avert climate change per se. In pursuing such abatement (mitigation), now urgently needed before the necessary international schedule of annual percentage emissions reduction becomes unachievable, the health research and practitioner constituency has at least two important messages to convey.

The first message is profound. If our collective accrued actions have reached a level whereby they are disrupting and weakening this planet's life-support processes, then that surely is a signal that we have overplayed our hand – and that we must therefore curb our current climate-disrupting economic activities and heighten our attempts to lower fertility rates further (especially in poorer countries, where unwanted pregnancies persist at high rates).

The second message is positive, a 'good news' one. Most mainstream actions taken to reduce GHG emissions and atmospheric concentrations will, at the same time, confer local and often immediate health 'co-benefits' on the community or population undertaking that action. While that may seem to offer most health benefits to richer countries, where private car usage should be constrained, physical activity encouraged, houses made more energy efficient and urban air pollution reduced, many of those benefits would also flow to lower-income countries, where urbanization is now proceeding rapidly (Haines *et al.*, 2009). Further, there would be wider-reaching benefits to crop yields, via the multilateral reduction of black carbon emissions and methane emissions, with some consequent reduction in the amount of ground-level ozone in air (Shindell *et al.*, 2012).

2.8 Conclusion

Humankind has entered an accelerated and expanded phase of globalization. The scale, intensity and connectivity of human activity, and associated demographic, social, economic and environmental 'global changes', are exerting increasing influences on human population well-being, health and survival.

Those influences, whether positive or negative, occur via diverse pathways. For example, they facilitate many aspects of international health care and mainstream public health activities, the spread of urban middle-class consumer culture and its health-related behaviours, the rapidity of global electronic information sharing (including the surveillance of infectious diseases) and the health risks to communities and populations from disruptions to the biosphere's life-support capacity, including the climate system. All those examples transcend national, even regional, boundaries; all reflect aspects

of recent globalization; all impinge on whole groups or communities of people – i.e. they reflect disruptions, distortions or realignments in *human ecology*. They entail new and unfamiliar shifts in the natural and social environments in which we live.

The climate system is fundamental to the biological and psychological health of human populations, via food yields, water flows, constraints on infectious disease agents and physical buffering (reefs, forests, etc.) against natural disasters. Hence, a fuller understanding of the risks to health and survival from climate change should help focus public concern on the long-term and fundamental consequences of climate change, and spur its abatement ('mitigation') by governments.

The advent of these transboundary (*global*) social, demographic and environmental influences on human health – especially those due to systemic environmental disruption and depletion – broadens the role and responsibilities of the health sector. Indeed, of all of government. We have an urgent collective responsibility to lessen and alleviate these risks, including those to future generations, who may yet inherit an impoverished, disrupted and less habitable world.

Note

[1] Directly observed treatment, short course.

References

Beaglehole, R., Bonita, R., Horton, R., Adams, C., Alleyne, G., Asaria, P., *et al.* (2011) Priority actions for the non-communicable disease crisis. *The Lancet* 377, 1438–1447.

Boyden, S.V. (2004) *The Biology of Civilisation.* UNSW Press, Sydney, Australia.

Butler, C.D. and Harley, D. (2010) Primary, secondary and tertiary effects of the eco-climate crisis: the medical response. *Postgraduate Medical Journal* 86, 230–234.

Crutzen, P.J. (2002) Geology of mankind: the anthropocene. *Nature* 415, 23.

Evengård, B. and McMichael, A.J. (2011) Vulnerable populations in the Arctic. *Global Health Action* 4, doi:10.3402/gha.v4i0.11210.

Fidler, D.P., Drager, N. and Lee, K. (2009) Managing the pursuit of health and wealth: the key challenges. *The Lancet* 373, 325–331.

Friel, S., Marmot, M., McMichael, A.J., Kjellstrom, T. and Vågerö, D. (2008) Global health equity and climate stabilisation: a common agenda. *The Lancet* 372, 1677–1683.

Haines, A., McMichael, A.J., Smith, K.R., Roberts, I., Woodcock, J., Markandya, A., *et al.*, and on behalf of the Task Force on Climate Change Mitigation and Public Health (2009) Public health effects of strategies to reduce greenhouse-gas emissions: overview and implications for policy makers. *The Lancet* 374, 2104–2114.

Lee, K., Buse, K. and Fustukian, S. (2002) An introduction to global health policy. In: Lee, K., Buse, K. and Fustukian, S. (eds) *Health Policy in a Globalising World*. Cambridge University Press, Cambridge, UK, pp. 4–5.

Lim, S.S., Vos, T., Flaxman, A.D., Danaei, G., Shibuya, K., Adair-Rohani, H., *et al.* (2012) A comparative risk assessment of burden of disease and injury attributable to 67 risk factors and risk factor clusters in 21 regions, 1990–2010: a systematic analysis for the Global Burden of Disease Study 2010. *The Lancet* 380, 2224–2260.

MacFarlane, S.B., Jacobs, M. and Kaaya, E.E. (2008) In the name of global health: trends in academic institutions. *Journal of Public Health Policy* 29, 383–401.

McMichael, A.J. (2013) Globalization, climate change and health. *New England Journal of Medicine* 368, 1335–1343.

McMichael, A.J. and Butler, C.D. (2011) Promoting global population health while constraining the environmental footprint. *Annual Review of Public Health* 32, 179–197.

New, M., Liverman, D., Schroder, H. and Anderson, K. (2011) Four degrees and beyond: the potential for a global temperature increase of four degrees and its implications. *Philosophical Transactions of the Royal Society A* 369, 6–19.

Ponting, C. (2007) *A New Green History of the World*. Vintage Books, London.

Rockström, J., Steffen, W., Noone, K., Persson, Å., Stuart Chapin, F. III, Lambin, E.F., *et al.* (2009) A safe operating space for humanity. *Nature* 461, 472–475.

Schellnhuber, H.J. (2008) Global warming: stop worrying, start panicking? *Proceedings of the National Academy of Sciences* 105, 14239–14240.

Schrecker, T. (ed.) (2012) *The Ashgate Research Companion to the Globalization of Health*. Ashgate Publishing, Farnham, UK.

Shindell, D., Kuylenstierna, J.C.I., Vignati, E., Dingenen, R.v., Amann, M., Klimont, Z., *et al.* (2012) Simultaneously mitigating near-term climate change and improving human health and food security. *Science* 335, 183–188.

Smil, V. (1994) *Energy in World History*. Westfield Publishing, Boulder, Colorado.

United Nations Population Fund (2013) *UN World Population Prospects, 2012 Revision*. United Nations Department of Economic and Social Affairs Population Division, New York.

Vineis, P., Chan, Q. and Khan, A. (2011) Climate change impacts on water salinity and health. *Journal of Epidemiology and Global Health* 1, 5–10.

WHO (World Health Organization) (2014) Global Outbreak Alert & Response Network (GOARN) (http://www.who.int/csr/outbreaknetwork/en/, accessed 28 February 2014).

Zhang, D.D., Lee, H.F., Wang, C., Li, B., Pei, Q., Zhang, J., *et al.* (2011) The causality analysis of climate change and large-scale human crisis. *Proceedings of the National Academy of Sciences* 108, 17296–17301.

3 Heat-related and Cold-related Mortality and Morbidity

Shakoor Hajat[1] and Lucy Telfar Barnard[2]

[1]*London School of Hygiene and Tropical Medicine, London, UK;* [2]*He Kainga Oranga/Housing and Health Research Programme, Department of Public Health, School of Medicine and Health Sciences, University of Otago, Wellington, New Zealand*

3.1 Introduction

A fundamental way in which climate change is expected to impact on public health relates to changes in population mortality and morbidity rates associated with direct exposure to ambient temperature. Many countries around the world already experience appreciable annual heat- and cold-related health burdens associated with current weather patterns.

Future trends in climate change have focused recent attention on the impacts of hot weather which, due to the concentration of risk in elderly people and in heavily built-up areas, may be amplified in future as a result of ageing populations in many parts of the world and rapid urbanization in many low-income settings. In addition, recent severe heatwaves have illustrated dramatically the dangers of hot weather, in particular the European heatwave of 2003, which was responsible for at least 35,000 excess deaths.

In many countries, however, health burdens during times of cold weather exceed those in periods of hot weather. This is also reflected by greater health services usage in cold weather. Recent global increases in fuel prices may also impact on cold-related health patterns, due to greater difficulties in heating the home adequately during winter months.

Very few health impacts of heat and cold exposure arise directly from hyperthermia or hypothermia, but rather temperature exposure can contribute to increases in mortality and morbidity levels from many different causes, especially cardiovascular and respiratory diseases.

Figure 3.1 shows the typical U-shaped relationship observed between the relative risk (RR) of mortality from all causes and daily temperature in populations in a temperate climate, in this case London, UK, during 1976–2003. There is a well-defined value of temperature above which mortality risk begins to rise, known as the heat threshold (at about 19°C daily mean temperature in the case of London). This relationship is based on aggregated data and so indicates that *some* people in the population of London (the most vulnerable) will be at increased risk of death once daily mean temperatures reach 19°C. Above this threshold, there is a smooth, usually linear, increase in mortality risk with increasing temperature; known as the heat slope. Some extremely hot days (especially those occurring during a heatwave) may be associated with mortality greater than predicted by a linear slope, but those days will generally number few. Similar effects of low temperatures are observed below cold thresholds. Both the heat and cold thresholds and slopes vary considerably across cities and countries, depending on differing climatic, demographic and socio-economic profiles.

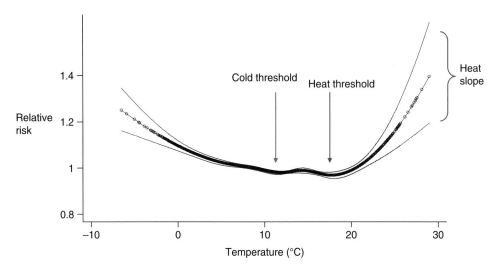

Fig. 3.1. Relationship between relative risk (RR) of all-cause mortality and mean temperature in London, UK, 1976–2003. The thick line is the estimated relationship based on spline functions; thin lines are 95% confidence limits.

Climate change will likely result in future changes to these heat and cold thresholds and slopes in many settings around the world. However, a population's future vulnerability will also be determined by its ability to adapt to changes to local climate conditions.

This chapter reviews epidemiologic and physiologic evidence to characterize current weather-related health impacts on public health and identifies the most susceptible subgroups of the population, as well as other risk factors that heighten vulnerability to temperature-related exposure. The chapter concludes with a discussion of possible changes to heat- and cold-related health burdens that may occur in future as a result of climate change.

3.2   Exposure to Hot and Cold Weather

A healthy human body adjusts to climate. Thus, a 19°C day will feel cool in the middle of a hot summer but warm in winter. Humans also adapt their environment to manage the effects of the climate they live in: populations that experience very cold winters develop warm housing and wear appropriate clothing; while populations that experience hot summers find

ways to moderate the heat or develop cultural behaviours (e.g. siesta) to avoid exertion during the hottest part of the day. Perhaps because of such adaptation, there is no official or World Health Organization (WHO) definition of extreme heat or extreme cold, and researchers have found that heat and cold thresholds – the temperature values above and below which adverse health events increase – vary by location.

However, there are limits to our ability to adjust our environment. These limits may be economic, when we struggle to afford the energy costs necessary for heating or cooling; or practical, as it can be difficult in everyday life to avoid at least some exposure to outdoor temperatures, whether hot or cold.

In addition to economic and practical limits, there are also physical limits: the body may adjust to cooler average temperature over a climate or season, but it is less able to absorb short-term changes in temperature without physical response.

There is also evidence that the winter season has adverse health effects in addition to daily temperature changes, but it is not known whether these health effects are due primarily to stress to the body from longer-term exposure to below-optimal temperatures or to other winter effects, such as dietary and behavioural changes or lower vitamin D levels, resulting from less sunlight exposure (see below).

The physiology and epidemiology of heat and cold exposure are discussed in turn below.

3.3 Heat

Physiology of heat exposure

Whereas human adaptation to cold environments is assisted greatly by behavioural responses (e.g. wearing additional layers of clothing), adaptation to heat is more dependent on the body's ability to act as a natural cooling system. Humans maintain an internal temperature in a narrow range around 37°C, independent from ambient temperature fluctuations (Sessler, 2009). Heat homeostasis is achieved by controlling heat gain or loss through thermoregulatory functions. Heat exchange occurs via:

- conduction (direct contact with hot or cold surfaces);
- convection (air or fluid movement);
- radiation (emission and reception of electromagnetic waves);
- evaporation (cooling due to heat loss from perspiration).

During thermal stress, the main ways in which the body eliminates heat are principally by sweating (Sessler, 2009), but increased cardiac output, and redirection of blood flow to the skin, also promote heat loss by radiation and conduction. Although experimental evidence on human heat physiology is well documented, little information exists on responses in the most vulnerable individuals, such as the elderly and those with chronic health problems. Also, physiological acclimatization may be endangered in people habituated to air-conditioned environments (O'Neill, 2003). This may heighten heat risk in situations such as in parts of the USA in summer 2012, when ferocious thunderstorms led to prolonged power blackouts. Deaths from both accidental and non-accidental causes were shown to have increased during a previous power outage in New York (Anderson and Bell, 2012).

Signs of heat-related illness in an individual often begin with heat exhaustion, which, if left untreated, may progress to heatstroke. Heatstroke is clinically defined if core body temperature is recorded at or above 40.6°C. Progression to death in heatstroke can be very rapid, within hours. Even with prompt medical care, 15% of heatstroke cases are fatal (Kilbourne, 1997). In many situations, however, heat may not be certified as the underlying cause of death, and so heat-related illness is under-reported.

Epidemiology of heat exposure

Epidemiological studies concerning heat are usually conducted on mortality or morbidity from all causes or among broad disease groupings. They characterize one of two different aspects of heat exposure. (i) The effects of exposure during a short specific period of sustained hot weather. The number of health events in such heatwaves are usually compared with the number occurring during the same period in surrounding years, thus obtaining an 'excess' attributable to the heatwave episode. (ii) Alternatively, the health effects of heat exposure occurring throughout the summer months over several years rather than during a specific period are assessed. Such studies consider the daily number of health events in relation to day-to-day fluctuations in the exposure measure, and are commonly assessed using time-series regression methods or the conceptually similar case-crossover design. Figure 3.1 shows such a general relationship between daily mortality and daily mean temperature in London across all months of the year.

Heatwave effects

Table 3.1 summarizes the public health effects of some previously assessed major heatwaves. As well as the underlying demographic, health and socio-economic profile of the population in question, the impacts of any particular heatwave will be dependent on certain characteristics

Table 3.1. Published heatwave effects (selected). (Sources: published review papers, Basu and Samet, 2002; Kovats and Hajat, 2008; Åström *et al.*, 2011.)

Heatwave event	Outcome	Results
1972 + 1973: New York City, USA	Daily deaths	Increased deaths after several days of excessively warm weather and on day following hottest day; ischaemic heart disease and age 65+ years most prominent
1976: London, UK	Daily deaths	15.4% increase in mortality
1987: Athens, Greece	Daily deaths	>2000 excess deaths
1993, February: Adelaide, Australia	All emergency department presentations (e.g. deaths, hospital admissions)	94 patients heat-related illness (78% heat exhaustion), 85% >60 years, 20% from institutional care; severity related to pre-existing conditions; mortality 12%
1995, 12–16 July: Chicago, USA	Deaths from hyperthermia	Males, blacks and persons 75+ years at greatest risk; highest mortality observed 2 days after heat index peaked; excess cardiovascular disease deaths
2003, 1 June–15 August: France	Deaths	14,802 excess deaths (60% increase)
2003, 15 June–15 September: Shanghai, China	Daily total mortality	RR = 1.13 (95% CI 1.06–1.21) 65+ most vulnerable
2006, 15 July–1 August: California, USA	Hospital admissions and emergency department visits	65+ increased rate ratio for heat-related illnesses: emergency department RR = 10.87 (8.49–14.3), hospitalizations RR = 14.23 (9.6–22.1)

of the heatwave itself. For example, risk is heightened in the following heatwave circumstances: greater temperature intensity, longer duration, earlier in the summer season and high night-time temperatures.

Although heatwave effects on mortality outcomes are most studied, risks in morbidity outcomes during heatwaves have also been observed – most commonly for hospital admissions and emergency department visits, but also outcomes such as ambulance transportation and calls to telephone helplines (Cerutti *et al.*, 2006; Leonardi *et al.*, 2006).

General heat effects

These studies most often analyse daily mortality data in relation to the daily exposure measure (e.g. temperature) recorded over several years in a particular location. After control for broad seasonal patterns and trends and potential time-varying confounders such as air pollution concentrations, a heat threshold can often be identified: this is the value of temperature at which heat-related health effects become apparent. Quantification of heat risk is then commonly presented as the percentage change in health events for every 1°C increase in temperature above that heat threshold.

In order to characterize the actual experience better, analysts often construct a composite measure to reflect physiological stress and discomfort better. Typically, this combines temperature and humidity – 'apparent temperature' is one such measure (Steadman, 1984). There may also be some synergistic effects of high temperatures and air pollution on health, although studies assessing this, to date, are inconsistent (Basu, 2009).

Table 3.2 summarizes published heat effects characterized in a variety of settings globally (Hajat and Kosatsky, 2010). Most evidence to date is based on North American and

Table 3.2. Published heat effects (selected). For full reference details see Hajat and Kosatsky, 2010.

Key sources	Setting	Study period	Mortality outcome	Daily temperature exposure	Lag structure of temperature	Heat threshold (°C)	Per cent change mortality per 1°C temp increase (95% CI)
Baccini *et al.*, 2008	Athens	1992–1996	Mortality outcome, all cause (excluding external); maximum and mean apparent temperature; lag 0–3 days			32.7	5.54 (4.3, 6.8)
	Barcelona	1992–2000				22.4	1.56 (1.04, 2.1)
	Budapest	1992–2001				22.8	1.74 (1.5, 2.0)
	Dublin	1990–2000				23.9	−0.02 (−5.4, 5.7)
	Helsinki	1990–2000				23.6	3.72 (1.7, 5.8)
	Ljubljana	1992–1999				21.5	1.34 (0.3, 2.4)
	London	1992–2000				23.9	1.54 (1.01, 2.1)
	Milan	1991–1998				31.8	4.29 (3.4, 5.2)
	Paris	1992–2000	+	+	+	24.1	2.44 (2.1, 2.8)
	Rome	1992–2000				30.3	5.25 (4.57, 5.9)
	Stockholm	1990–2000				21.7	1.17 (0.41, 1.9)
	Turin	1991–1999				27.0	3.32 (2.53, 4.1)
	Valencia	1995–2000				28.2	0.56 (−0.4, 1.5)
	Zurich	1990–1996				21.8	1.37 (0.5, 2.3)
Curriero *et al.*, 2002	Boston	1973–1994	Mortality outcome, all cause (excluding external); maximum and mean apparent temperature; no lag			20.9	2.93
	Chicago					18.4	1.28
	New York					19.1	2.61
	Philadelphia					21.4	2.81
	Baltimore					21.4	2.57
	Washington					21.4	1.37
	Charlotte		+	+	+	32.4	N/A
	Atlanta					24.6	2.98
	Jacksonville					24.9	1.91
	Miami					27.2	2.31
Zanobetti and Schwartz, 2008 (USA)	Birmingham	1999–2000	Mean apparent temperature; no lag; summer months only; heat threshold for meta-regression taken as minimum value within summer months; heat slopes approximated from graphical presentation			N/A	0.36 (−0.5, 1.2)
	Dallas	1999–2002					−0.08 (−0.5, 0.4)
	Detroit	1999–2002	+	+	+		0.60 (0.3, 0.9)

Continued

Table 3.2. Continued.

Key sources	Setting	Study period	Mortality outcome	Daily temperature exposure	Lag structure of temperature	Heat threshold (°C)	Per cent change mortality per 1°C temp increase (95% CI)
	Houston	1999–2002					0.32 (−0.1, 0.7)
	Minneapolis	1999–2002					0.58 (0.2, 1.0)
	Phoenix	1999–2000					0.64 (−0.04, 1.3)
Kim *et al.*, 2006 (South Korea)	Seoul	1994–2003	Mortality outcome, all cause (excluding external); mean temperature; no lag			28.1	9.60 (8.0, 11.2)
	Daegu				+	28.1	4.62 (2.7, 6.5)
	Incheon			+		26.6	5.31 (3.1, 7.6)
	Gwangju					26.6	2.64 (0.34, 5.00)
	Daejeon					28.1	3.91 (−0.6, 8.7)
McMichael *et al.*, 2008	Bucharest	1994–1997	All cause (excluding external); mean temperature; lag 0–1 days			22	3.30 (2.4, 4.3)
	Sofia	1996–1999				16	2.88 (2.1, 3.7)
	New Delhi	1991–1994				29	3.94 (2.8, 5.1)
	Mexico City	1994–1998		+		18	0.77 (0.1, 1.4)
	Chiang Mai	1995–1997			+	28	2.39 (−0.5, 5.4)
	Bangkok	1991–1992				29	5.78 (3.5, 8.1)
	São Paulo	1991–1994				23	3.46 (2.6, 4.3)
	Santiago	1988–1991				16	1.04 (0.3, 1.8)
	Cape Town	1996–1999				17	0.47 (−0.3, 1.2)
Hu *et al.*, 2008	Sydney	1994–2004	All cause (excluding external); maximum temperature; no lag; summer months only; heat threshold for meta-regression taken as minimum value within summer months			N/A	0.9 (0.6, 1.3)
Kan *et al.*, 2003	Shanghai	06/2000–12/2001	All cause (excluding external); mean temperature; lag 0–2 days; slope is 0.73% in abstract but 1.21% in full paper; slope of 0.73% assumed based on figure			26.7	0.73

European populations. In general, higher heat thresholds are observed in populations with higher summertime temperatures, reflecting long-term adaptation.

No evident heat effect was detected in several cities, including Dublin (Ireland), Dallas (USA) and Busan (South Korea). However, an increased mortality of 12.3% (95% CI 5.7–19.4) per 1°C increase in high temperature was reported in Beirut (Lebanon) and even higher in Monterrey (Mexico) (18.8% (13.0–25.0)). Both cities have correspondingly high heat thresholds. In almost half of all locations studied, heat slopes were estimated to be between 1% and 3% per 1°C change in temperature. Even in Helsinki (Finland) and Stockholm (Sweden), cities with relatively cool summers, a clearly defined heat threshold and slope was identified. Heat slopes were appreciably lower in the US-based studies. In part, this may be because these estimates were based largely on temperatures observed throughout the summer months rather than above a heat threshold. However, the papers also reported that slope estimates remained largely unchanged in sensitivity analyses limited to higher temperatures.

Heat relationships from other settings using both mortality and morbidity outcomes have been presented in recent review papers (Basu, 2009; Gosling et al., 2009; Åström et al., 2011).

3.4 Vulnerable Groups and Risk Factors

Identification of population subgroups vulnerable to hot weather is crucial for effective public health interventions. Heat-related mortality and morbidity occurs overwhelmingly in the elderly and the chronically ill, particularly those with cardiovascular, respiratory and renal disease (Kovats and Hajat, 2008). Younger people with diabetes, neurological disorders and psychiatric illness may also face increased risk. Evidence from Europe suggests that women are at greater risk than men, even after controlling for age differences (Hajat et al., 2007). However, during the 1995 Chicago heatwave, elderly men fared worse than similarly aged women (Semenza

et al., 1996), perhaps reflecting increased culturally determined variation in the level of social isolation among elderly men. Young children are also at risk, including because some may be unable to increase their cardiac output adequately, a heat loss mechanism (Jokinen et al., 1990). Social deprivation has been identified as a risk factor for heat death in many US studies, but European evidence is less clear (Kovats and Hajat, 2008). A strong protective factor for heat-related health may be greater access to home air conditioning (AC) (Semenza et al., 1996), more so in US cities than in Europe. People who are poorer in the USA are less likely to have AC; this social gradient is less marked in Europe.

Extrinsic risk factors are also important to identify, as these vary according to location and adaptations to local climate and are, therefore, more easily modifiable. Lack of mobility is one such risk factor (Semenza et al., 1996). In a meta-analysis of six case-control studies, the factors associated with the highest risk of death during heatwaves were: confinement to bed (odds ratio 6.4 (95% CI 4.5–9.2)); pre-existing psychiatric illness (3.6 (1.3–9.8)); not leaving home every day (3.4 (1.6–6.9)); and an inability to care for oneself (3.0 (1.8–4.8)) (Bouchama et al., 2007). Other contextual factors which may heighten vulnerability include living in homes with high thermal mass and poor ventilation, and living on the upper floors of high-rise buildings (Semenza et al., 1996; Vandentorren et al., 2004). Elderly people in hospital and residential homes are also at increased risk, due to their frailty, and hence need attention from carers (Kovats and Hajat, 2008).

Such concentrations of vulnerable individuals offer the opportunity for more targeted interventions – one study of the 2003 heatwave in France reported that excess mortality in the most frail individuals in nursing homes was comparatively low, due to greater medical care (Holstein et al., 2005). Little heat-related mortality is observed in care homes in the southern USA, except when air-conditioning systems fail (Sullivan-Bolyai et al., 1979). People in cities are more vulnerable than rural populations (Hajat et al., 2007); this is due, in part, to the urban heat island effect, whereby city temperatures are raised due to increased heat storage capacity and lower night-time radiation (Oke, 1973).

3.5 Public Health Responses to Hot Weather

Many cities have introduced public health protection measures to anticipate and minimize the dangers of hot weather – most notably heat-health warning systems, which trigger community alerts and emergency actions when hot weather is forecast (Kovats and Hajat, 2008). However, there is little published evidence, to date, on the effectiveness of different intervention options (Hajat and Kosatsky, 2010).

3.6 Cold

Physiology of cold exposure

Less than 1% of cold deaths are caused by a drop in core body temperature (hypothermia) (Collins, 1987). The bulk of cold deaths are due to the stress to the body of maintaining core body temperature in suboptimal indoor or outdoor temperatures, and subsequent cardiac and cerebrovascular responses (Conlon et al., 2011). Prolonged or chronic lower-level cold stress causes an immune and inflammatory response (Pozos and Danzl, 2001). Lung inflammation from cold air, possibly compounded by increased virus survival at cold temperatures and/or more time indoors with other people, increases the risk of contracting respiratory infection. In cold months, seasonal dietary changes may reduce vitamin C levels; less sun exposure may lower vitamin D levels, also contributing to lower immunity.

When challenged by cold, blood flow in the skin and extremities is reduced (vasoconstriction) to reduce heat loss to the air and protect organs such as the heart, kidneys and liver (Pozos and Danzl, 2001). Such vasoconstriction increases blood pressure, which can cause circulatory illness or even death in those with a compromised circulatory system. If cooling continues, muscle contractions,

shivering and rapid breathing follow, in order to generate heat.

Reduced lung function can also occur. Cold air in the lungs may cause the bronchioles to constrict. Asthma and/or susceptibility to respiratory infections follows, increased by the ability of some viruses to survive for longer on cold surfaces, though personal risk can be reduced by more hand washing.

Sufficient exposure to very cold temperatures slows the breathing rate and is then followed by unconsciousness. Several circulatory effects can occur, including cardiac arrhythmias and ischaemic stroke (Pozos and Danzl, 2001; Conlon et al., 2011). The final cold response is death. The time taken to progress from the first stages of cold response is generally inversely proportional to the degree of cold experienced: in general, the colder the temperature, the faster the onset of severe symptoms (Conlon et al., 2011).

The degree of cold experienced may be relative to previous body temperature. In Japan, the phenomena of winter bath deaths, where predominantly elderly people suffer sudden death immediately before or after bathing, may indicate an adverse physiological response to rapid lowering of body temperature when undressing or moving from a hot bath into a cool bathroom (Chiba et al., 2005).

Epidemiology of cold exposure

There are two main epidemiologic ways of measuring the effects of cold weather on public health.

Excess winter mortality (EWM) or morbidity

This approach compares winter with either summer or 'non-winter', in order to determine both the number of 'excess' health events, and a rate ratio or index of the difference. The most common measure (Curwen and Devis, 1988) compares events over the 4 winter months, with events totalled over the other 8 months. An index for excess winter mortality (EWMI) for a southern hemisphere country is calculated as follows:

$$\text{EWMI} = \frac{\sum_{deaths}(Jun,\ Jul,\ Aug,\ Sep) - \sum_{deaths}(Feb,\ Mar,\ Apr,\ May,\ Oct,\ Nov,\ Dec,\ Jan)/2}{\sum_{deaths}(Feb,\ Mar,\ Apr,\ May,\ Oct,\ Nov,\ Dec,\ Jan)/2}$$

The resulting index may be expressed as a percentage (e.g. 'mortality is 18% higher in winter than the rest of the year') or, by adding 1, as a rate ratio ('the winter: non-winter mortality rate ratio is 1.18').

Variations on this method include dividing the number of events by the number of days rather than the number of months; comparing different winter and non-winter or winter and summer periods, whether by defining winter by the period of highest mortality rather than set calendar dates; comparing the 4 'winter' months with 4 'summer' months or, more often, comparing the 3 standard winter months with the 3 standard summer months.

The advantages of using such an index to measure the effects of cold are:

1. Ease of calculation. Daily mortality or morbidity data are generally available, and the formula does not need advanced statistical methods.
2. It is easily understood by the lay public and policy makers.
3. It can measure the full effect of winter, without losing any delayed effects.
4. It enables easy comparison of winter burdens between different countries.

Its disadvantages are:

1. It is crude. Neither seasons nor temperature changes begin or end according to set dates in a fixed calendar. The effect of 'winter' may be sensitive to the denominator months including a period of increased health events, e.g. a heatwave.
2. Variations in annual excess can be hard to interpret; for example, are the differences due to improved health care, a milder winter, a cooler (or very hot) summer or an unusual influenza season?

Excess winter mortality, as measured by the standard winter:non-winter index, ranges from around 8% in parts of Scandinavia and Russia to about 20% in temperate Europe, Australia and New Zealand (see Table 3.3).

Table 3.3. Selected excess winter mortality based on deaths from all causes. (Estimates from Woodhouse, 1993, and Healy, 2003.)

Location	Dates	Excess winter mortality (%)
Australia	1976–1984	20
Canada	1976–1983	7
Chile	1976–1983	18
Egypt	1976–1980	−12
England and Wales	1976–1984	21
Europe	1988–1997	16
Finland	1988–1997	10
Germany	1988–1997	11
Hong Kong	1976–1983	18
Iceland	1976–1984	1
Japan	1976–1984	17
Tunisia	1976–1982	19
USA	1978–1982	9
USSR	1976–1984	8

Assessment of explicit weather factors

An alternative to using the winter:non-winter index is to use statistical time-series regression methods (or the conceptually similar case-crossover design) to model the specific relationship between daily temperature and daily deaths. These are the same methods as those used to assess general heat effects, as described above.

As shown in Fig. 3.1, the general relationship between temperature and mortality is usually shown to be U-shaped from such studies. This is often referred to as the 'temperature–mortality curve'. As with heat, the effects of cold are commonly expressed as an X% change in deaths for every 1 degree decrease below a cold threshold of Y degrees. The value or range of values of temperature at which mortality is at its minimum is known as the minimum mortality temperature.

The relationship shown in Fig. 3.1 considers only temperature measurements recorded on the same date as the date of death – referred to as lag 0. Unlike heat effects, which occur mostly within 1 or 2 days following exposure (lags 1–2 days), cold effects can be delayed by a few days or weeks, and so the cold slope shown in Fig. 3.1 would be much steeper if lagged measures of temperature had been displayed. So, in assessment of cold impacts, studies of this type usually consider the effects on mortality on many subsequent days following the day of initial exposure.

Notable published papers where the minimum mortality temperature and cold slopes have been estimated for a number of settings globally are, for example, Curriero *et al.* (2002), Analitis *et al.* (2008) and McMichael *et al.* (2008).

A sample of cold slopes and minimum mortality temperatures is shown in Table 3.4. Observed minimum mortality temperatures fall in the range from 16.5°C in the Netherlands to 32.4°C in Miami, Florida, USA. Observed cold slopes range from 0.91% in the Netherlands to 7.12% in Tampa, Florida. As these studies allow for slow-changing seasonal patterns in the distribution of deaths, they only permit for assessment of acute effects (days or weeks) of temperature exposure. It is, however, possible that cold weather also contributes to chronic

effects on human health; such an effect is not captured by these data.

Other health outcomes

For both excess winter indices and cold slope estimates, the end point most commonly analysed is mortality, but death is only the most severe outcome. Similar effects have been noted for hospitalizations, particularly for respiratory and circulatory diagnoses. Figure 3.2 shows the typical annual distribution of daily emergency hospital admissions and daily deaths from respiratory and cardiovascular diseases in London, UK (Kovats *et al.*, 2004).

3.7 Vulnerable Groups and Risk Factors

Age

Children and the elderly are less able to thermoregulate their bodies, and in other ways are also

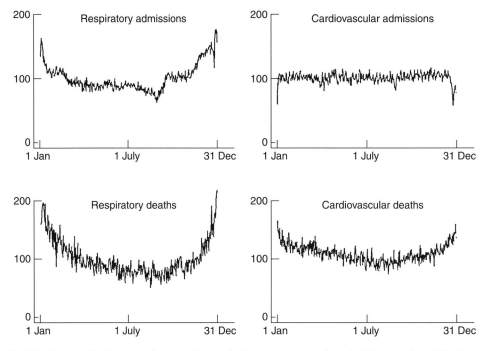

Fig. 3.2. Seasonality (in the northern hemisphere) of emergency hospital admissions and mortality for respiratory and cardiovascular disease (% of average) (Kovats *et al.*, 2004). (Reprinted with permission of *BMJ*.)

Table 3.4. Selected low temperature effects on mortality.

Reference	Location	Population	Dates	Outcome	Measure	Cold effect
Brown et al., 2010	UK	Total	1993–2007	Death	Daily deaths versus Central England temperature Lags included	1.2% increase in daily deaths for each 1°C decrease in mean Central England temperature <5°C
Gouveia et al., 2003	São Paulo, Brazil	65 years+	1991–1994	All non-violent deaths	Daily deaths versus São Paulo temperature Lags included	5.5% increase in daily deaths per degree drop in temperature <20°C
Donaldson et al., 2002	Two Russian districts	50–59 years; 65–74 years	1990–1994	All-cause deaths	Daily deaths versus temperature	1.15% (95% CI 0.97–1.32%) increase in daily deaths per 1°C drop in temperature from 0°C to −29.6°C
Curriero et al., 2002	11 US cities	Total	1973–1994	All-cause deaths	Daily deaths versus temperature	'Minimum mortality temperature' ranged from 18.4°C to 32.4°C. Cold slope ranged from −2.25 to −7.12

more susceptible to many effects of cold (Davie et al., 2007). Children lose heat faster, due to their higher body surface:weight ratio. Both groups are also more susceptible to cold-associated infections, because of either immature or weakening immunity. Because the effects of cold are associated so strongly with older age groups, many studies measure only EWM or cold temperature-related mortality in those aged over 65. In developed countries, mortality rates are generally low among children, particularly for causes associated with cold. Children's greater susceptibility to cold is instead evidenced by hospital admission data (Crighton et al., 2003); however, seasonality of illness in children is complex, probably most highly influenced by school term dates and accompanying infectious disease transmission; and, in the very young, by seasonal differences in birth rate.

Sex

Some studies have found that the effect of cold is stronger for women than for men in all three of the main measures of cold outcome: all-cause mortality, circulatory mortality and respiratory mortality (Davie et al., 2007).

Ethnicity

We know of no known biological reason to expect differences in cold response by ethnicity, but cultural differences in diet and modes of dress, and biological differences in vitamin D levels and, perhaps, layers of insulator subcutaneous fat, could combine to create differences in the cold response. However, few studies have measured differences in cold temperature effects by ethnicity, including one in New Zealand (Davie *et al.*, 2007).

Deprivation, fuel poverty and housing

Fuel poverty has been widely proposed as an explanation for some cold-related deaths, but a number of studies have found no difference in EWM or cold risk by socio-economic status (Davie *et al.*, 2007; Hajat *et al.*, 2007). While health outcomes in general may be improved by increasing home insulation, statistical associations between housing quality and cold-related morbidity or mortality are small to negligible (Barnard *et al.*, 2008). One possible explanation for this apparent contradiction is that better housing quality may improve health over the full climatic range, meaning differences in health levels between hot and cold periods remain roughly similar. On the other hand, in many countries, the life expectancy of the poor is lower. More of these poor are likely to die in winter, but this does not mean their EWM will be different to those from the same location who are more affluent.

3.8   Public Health Responses to Cold Weather

As awareness of the effects of cold has increased, a handful of public health cold death mitigation projects have been undertaken. In the UK, in addition to the introduction of a winter fuel payment to the elderly, some areas have offered free installation of improved home heating systems or insulation. But, cash for winter fuel purchases may be diverted for other purposes. In the UK, improved home heating has had limited uptake there. In contrast, home insulation programmes have had

greater acceptance in New Zealand, and have been found to decrease hospital admissions and prescriptions for medications associated with cold-related conditions, though with no impact on measured lung function (Howden-Chapman *et al.*, 2008).

Some researchers, most prominently Keatinge and Donaldson, have argued for greater emphasis on the risks of outdoor cold exposure. These workers found, over a series of UK studies, that EWM had not reduced with increases in central heating; that elderly people with access to unrestricted home heating did not have a lower EWM than those who were charged separately for heating; and that surveys in Europe and Siberia had provided statistical evidence that dressing warmly, with hats and warm and waterproof coats, as well as warm homes, was associated with low winter mortality (Keatinge and Donaldson, 2004). They therefore advocate increased public health advice promoting protection from outdoor cold exposure.

3.9   Important Commonalities and Differences Between Heat and Cold Impacts

Although similar epidemiologic methods can be used to quantify both heat and cold effects on health, important distinctions also exist. As stated previously, heat effects have been demonstrated to be mostly immediate (occurring within a day or two of exposure), whereas cold effects are typically delayed by up to a few weeks following initial exposure. Lagged measures of temperature are, therefore, very important to consider in assessments of cold effects. Cold effects on cardiovascular diseases have been shown to be more immediate than cold effects on respiratory causes (Analitis *et al.*, 2008).

Although some heatwave studies have demonstrated that the effects of hot days can be enhanced when occurring in succession, including continually high night-time temperatures, there is little evidence of additional impacts of cold days when they occur as part of a wave (Barnett *et al.*, 2012).

Studies have shown that a substantial fraction of heat-related deaths occur in already frail individuals whose exposure to the hot weather has simply hastened their deaths by a matter of a few days or weeks. Between 20% and 50% of deaths during heatwaves have been attributed to such short-term mortality displacement in selected locations (Kalkstein, 1993). The role of displacement is likely to be smaller in situations in which heat-related deaths are not restricted to chronic diseases in elderly individuals – for example, during very extreme heatwaves, when a high proportion of deaths might arise from heatstroke in otherwise healthy individuals (Toulemon and Barbieri, 2008), or in low-income countries, where heat-related deaths from infectious diseases might be common (Hajat *et al.*, 2005).

There is little suggestion that deaths from cold-related exposure are explained by short-term mortality displacement (Braga *et al.*, 2002). An analysis of years of life lost in Brisbane, Australia, estimated 1519 years of life lost due to current high temperatures and 5053 years of life lost due to low temperatures (Huang *et al.*, 2012). There is also now evidence that a high summer mortality burden may be more likely to occur in years when mortality levels in the previous winter are low (Stafoggia *et al.*, 2009). This may be explained by vulnerable individuals, during a mild winter, remaining in the pool of people at risk as the year transitions from the winter to the summer season.

3.10   Conclusion: Climate Change and Future Temperature-related Deaths

In many parts of the world, temperature-related burdens will alter as a result of climate change. More heat-related deaths will occur in summer months from living in a warmer world. However, it is likely that, correspondingly, milder winters in future will also lead to a decrease in cold-related health burdens in some parts of the world. In countries such as the UK, where current cold-related health impacts exceed current heat-related impacts, climate change is expected to cause a net reduction in temperature-related health burdens. This would result from any increase in heat-related impacts being offset by a greater reduction in cold risk. However, this is only likely to be the case in the near term (coming decades) as, if temperatures continue to rise, heat-related impacts will eventually become more dominant.

As well as rises in mean temperature, climate change is expected to result in greater variance of temperatures, and so extremes of cold temperatures will still occur in addition to a greater frequency of heatwaves.

Estimates of temperature–health relationships associated with current weather patterns can be used to provide projections of possible future health burdens caused by climate change (see Table 3.5). In such assessments, the current general temperature–health risk profile in any particular setting is applied to local climate change projections in order to estimate future burdens, with the most realistic estimates being those which also consider changed population vulnerability. For example, population ageing may increase future vulnerability, while health protection and other adaptation measures may contribute to reducing future risk.

Other studies are presented in a recent systematic review (Huang *et al.*, 2011).

AC is an important adaptation, but is unaffordable for many poor people and is itself highly energy intensive, thereby contributing to further greenhouse gas emissions, at least where electricity is supplied from coal and other fossil fuels. Consequently, the most cost-effective public health use of AC to reduce heat burdens may be focused best on communal areas such as publicly accessible cooling centres.

It is important to stress that the potential changes in temperature-related health impacts discussed above only consider the most direct way in which climate change is expected to impact on future public health burdens. Other, more indirect processes are also probable, most of which are likely to be detrimental to global human health, as discussed in other chapters of this book.

Table 3.5. Published estimates of future temperature-related deaths (selected).

Source	City	Years	Emission scenarios (number)	Future adaptation modelled?	Other modelling elements considered	Change in heat deaths	Cold deaths assessed?
Kalkstein and Greene, 1997	44 US cities	2020s 2050s	3 climate models	Yes (20–25%)	Mortality displacement	363–753 460–999	Slight decrease
Guest et al., 1999	5 Australian cities	2030	Yes (2)			–10%	Offset heat deaths
Donaldson et al., 2002	UK	2050s	Yes (4)			250%	
Dessai, 2003	Lisbon, Portugal	2020s 2050s	2 climate models	Yes (15%) Yes (40%)	Different approaches to model current risk	5.8–15.1 7.3–35.6/100,000	
Hayhoe et al., 2004	Los Angeles, California	2100	Yes (2)			200–700%	
Doyon et al., 2008	3 cities Quebec	2020 2080	Yes (2)			2% 10%	Do not offset heat deaths
Huang et al., 2012	Brisbane, Australia	2050			Years of life lost estimated	1014 (1°C ↑) 2450 (2°C ↑) 6782 (4°C ↑)	1112 –2069 –3540

References

Analitis, A., Katsouyanni, K., Biggeri, A., Baccini, M., Forsberg, B., Bisanti, L., *et al.* (2008) Effects of cold weather on mortality: results from 15 European cities within the PHEWE project. *American Journal of Epidemiology* 168, 1397–1408.

Anderson, G.B. and Bell, M.L. (2012) Lights out: impact of the August 2003 power outage on mortality in New York, NY. *Epidemiology* 23, 189–193.

Åström, D.O., Forsberg, B. and Rocklöv, J. (2011) Heat wave impact on morbidity and mortality in the elderly population: a review of recent studies. *Maturitas* 69, 99–105.

Baccini, M., Biggeri, A., Accetta, G., Kosatsky, T., Katsouyanni, K., Analitis, A., *et al.* (2008) Heat effects on mortality in 15 European cities. *Epidemiology* 19, 711–719.

Barnard, L.F., Baker, M.G., Hales, S. and Howden-Chapman, P.L. (2008) Excess winter morbidity and mortality: do housing and socio-economic status have an effect? *Reviews on Environmental Health* 23, 203–221.

Barnett, A.G., Hajat, S., Gasparrini, A. and Rocklov, J. (2012) Cold and heat waves in the United States. *Environmental Research* 112, 218–224.

Basu, R. (2009) High ambient temperature and mortality: a review of epidemiologic studies from 2001 to 2008. *Environmental Health* 8, 40, doi:10.1186/1476-069X-8-40.

Basu, R. and Samet, J.M. (2002) Relation between elevated ambient temperature and mortality: a review of the epidemiologic evidence. *Epidemiologic Reviews* 24, 190–202.

Bouchama, A., Dehbi, M., Mohamed, G., Matthies, F., Shoukri, M. and Menne, B. (2007) Prognostic factors in heat wave related deaths: a meta-analysis. *Archives of Internal Medicine* 167, 2170–2176.

Braga, A.L., Zanobetti, A. and Schwartz, J. (2002) The effect of weather on respiratory and cardiovascular deaths in 12 U.S. cities. *Environmental Health Perspectives* 110, 859–863.

Brown, G., Fearn, V. and Wells, C. (2010) Exploratory analysis of seasonal mortality in England and Wales. *Health Statistics Quarterly* 48, 58–80.

Cerutti, B., Tereanu, C., Domenighetti, G., Cantoni, E., Gaia, M., Bolgiani, I., *et al.* (2006) Temperature related mortality and ambulance service interventions during the heat waves of 2003 in Ticino (Switzerland). *Soz Präventivmed* 51, 185–193.

Chiba, T., Yamauchi, M., Nishida, N., Kaneko, T., Yoshizaki, K. and Yoshioka, N. (2005) Risk factors of sudden death in the Japanese hot bath in the senior population. *Forensic Science International* 149, 151–158.

Collins, K.J. (1987) Effects of cold on old people. *British Journal of Hospital Medicine* 38, 506–508, 510–512, 514.

Conlon, K.C., Rajkovich, N.B., White-Newsome, J.L., Larsen, L. and O'Neill, M.S. (2011) Preventing cold-related morbidity and mortality in a changing climate. *Maturitas* 69, 197–202.

Crighton, E.J., Moineddin, R., Upshur, R.E. and Mamdani, M. (2003) The seasonality of total hospitalizations in Ontario by age and gender: a time series analysis. *Canadian Journal of Public Health* 94, 453–457.

Curriero, F.C., Heiner, K.S., Samet, J.M., Zeger, S.L., Strug, L. and Patz, J.A. (2002) Temperature and mortality in 11 cities of the eastern United States. *American Journal of Epidemiology* 155, 80–87.

Curwen, M. and Devis, T. (1988) Winter mortality, temperature and influenza: has the relationship changed in recent years? *Population Trends* 54, 17–20.

Davie, G.S., Baker, M.G., Hales, S. and Carlin, J.B. (2007) Trends and determinants of excess winter mortality in New Zealand: 1980 to 2000. *BMC Public Health* 7, 263, doi:10.1186/1471-2458-7-263.

Dessai, S. (2003) Heat stress and mortality in Lisbon Part II. An assessment of the potential impacts of climate change. *International Journal of Biometeorology* 48, 37–44.

Donaldson, G., Kovats, R.S., Keatinge, W.R. and McMichael, A.J. (2002) Heat- and cold-related mortality and morbidity and climate change. In: Maynard, R. and Green, E. (eds) *Health Effects of Climate Change in the UK*. Department of Health, London, pp. 70–80.

Doyon, B., Belanger, D. and Gosselin, P. (2008) The potential impact of climate change on annual and seasonal mortality for three cities in Quebec, Canada. *International Journal of Health Geography* 7, 23, doi:10.1186/1476-072X-7-23.

Gosling, S.N., Lowe, J.A., McGregor, G.R., Pelling, M. and Malamud, B.D. (2009) Associations between elevated atmospheric temperature and human mortality: a critical review of the literature. *Climatic Change* 92, 299–341.

Gouveia, N., Hajat, S. and Armstrong, B. (2003) Socioeconomic differentials in the temperature–mortality relationship in São Paulo, Brazil. *International Journal of Epidemiology* 32, 390–397.

Guest, C.S., Willson, K., Woodward, A.J., Hennessy, K., Kalkstein, L.S., Skinner, C., *et al.* (1999) Climate and mortality in Australia: retrospective study, 1979–1990, and predicted impacts in five major cities in 2030. *Climate Research* 13, 1–15.

Hajat, S. and Kosatsky, T. (2010) Heat-related mortality: a review and exploration of heterogeneity. *Journal of Epidemiology and Community Health* 64, 753–760.

Hajat, S., Armstrong, B.G., Gouveia, N. and Wilkinson, P. (2005) Mortality displacement of heat-related deaths: a comparison of Delhi, Sao Paulo, and London. *Epidemiology* 16, 613–620.

Hajat, S., Kovats, R.S. and Lachowycz, K. (2007) Heat-related and cold-related deaths in England and Wales: who is at risk? *Occupational and Environmental Medicine* 64, 93–100.

Hayhoe, K., Cayan, D., Field, C.B., Frumhoff, P.C., Maurer, E.P., Miller, N.L., *et al.* (2004) Emissions pathways, climate change, and impacts on California. *Proceedings of the National Academy of Sciences* 101, 12422–12427.

Healy, J.D. (2003) Excess winter mortality in Europe: a cross country analysis identifying key risk factors. *Journal of Epidemiology and Community Health* 57, 784–789.

Holstein, J., Canoui-Poitrine, F., Neumann, A., Lepage, E. and Spira, A. (2005) Were less disabled patients the most affected by 2003 heat wave in nursing homes in Paris, France? *Journal of Public Health* 27, 359–365.

Howden-Chapman, P., Pierse, N., Nicholls, S., Gillespie-Bennett, J., Viggers, H., Cunningham, M., *et al.* (2008) Effects of improved home heating on asthma in community dwelling children: randomised controlled trial. *BMJ* 337, a1411.

Hu, W., Mengersen, K., McMichael, A.J. and Tong, S. (2008) Temperature, air pollution and total mortality during summers in Sydney, 1994–2004. *International Journal of Biometeorology* 52, 689–696.

Huang, C., Barnett, A.G., Wang, X., Vaneckova, P., FitzGerald, G. and Tong, S. (2011) Projecting future heat-related mortality under climate change scenarios: a systematic review. *Environmental Health Perspectives* 119, 1681–1690.

Huang, C., Barnett, A.G., Wang, X. and Tong, S. (2012) The impact of temperature on years of life lost in Brisbane, Australia. *Nature Climate Change* 2, 265–270.

Jokinen, E., Valimaki, I., Antila, K., Seppanen, A. and Tuominen, J. (1990) Children in sauna: cardiovascular adjustment. *Pediatrics* 86, 282–288.

Kalkstein, L.S. (1993) Health and climate change. Direct impacts in cities. *The Lancet* 342, 1397–1399.

Kalkstein, L.S. and Greene, J.S. (1997) An evaluation of climate/mortality relationships in large U.S. cities and the possible impacts of a climate change. *Environmental Health Perspectives* 105, 84–93.

Kan, H., Jia, J. and Chen, B.H. (2003) Temperature and daily mortality in Shanghai: a time-series study. *Biomedical and Environmental Sciences* 16, 133–139.

Keatinge, W.R. and Donaldson, G.C. (2004) Winter mortality in elderly people in Britain: action on outdoor cold stress is needed to reduce winter mortality. *BMJ* 329, 976; author reply 977.

Kilbourne, E.M. (1997) Heat waves and hot environments. In: Noji, E. (ed.) *The Public Health Consequences of Disasters.* Oxford University Press, New York, pp. 245–269.

Kim, H., Ha, J.S. and Park, J. (2006) High temperature, heat index, and mortality in 6 major cities in South Korea. *Archives of Environmental and Occupational Health* 61, 265–270.

Kovats, R.S. and Hajat, S. (2008) Heat stress and public health: a critical review. *Annual Review of Public Health* 29, 41–55.

Kovats, R.S., Hajat, S. and Wilkinson, P. (2004) Contrasting patterns of mortality and hospital admissions during hot weather and heat waves in Greater London, UK. *Occupational and Environmental Medicine* 61, 893–898.

Leonardi, G.S., Hajat, S., Kovats, R.S., Smith, G.E., Cooper, D. and Gerard, E. (2006) Syndromic surveillance use to detect the early effects of heat-waves: an analysis of NHS direct data in England. *Soz Präventivmed* 51, 194–201.

McMichael, A.J., Wilkinson, P., Kovats, R.S., Pattenden, S., Hajat, S., Armstrong, B., *et al.* (2008) International study of temperature, heat and urban mortality: the 'ISOTHURM' project. *International Journal of Epidemiology* 37, 1121–1131.

Oke, T.R. (1973) City size and the urban heat island. *Atmospheric Environment* 7, 769–779.

O'Neill, M.S. (2003) Air-conditioning and heat-related health effects. *Applied Environmental Science and Public Health* 1, 9–12.

Pozos, R.S. and Danzl, D. (2001) Human physiological responses to cold stress and hypothermia. *Medical Aspects of Harsh Environments* 1, 351–382.

Semenza, J.C., Rubin, C.H., Falter, K.H., Selanikio, J.D., Flanders, W.D., Howe, H.L., *et al.* (1996) Heat-related deaths during the July 1995 heat wave in Chicago. *New England Journal of Medicine* 335, 84–90.

Sessler, D.I. (2009) Thermoregulatory defense mechanisms. *Critical Care Medicine* 37, S203–S210.

Stafoggia, M., Forastiere, F., Michelozzi, P. and Perucci, C.A. (2009) Summer temperature-related mortality: effect modification by previous winter mortality. *Epidemiology* 20, 575–583.

Steadman, R.G. (1984) A universal scale of apparent temperature. *Journal of Climate and Applied Meteorology* 23, 1674–1687.

Sullivan-Bolyai, J.Z., Lumish, R.M., Smith, E.W., Howell, J.T., Bregman, D.J., Lund, M., *et al.* (1979) Hyperpyrexia due to air-conditioning failure in a nursing home. *Public Health Reports* 94, 466–470.

Toulemon, L. and Barbieri, M. (2008) The mortality impact of the August 2003 heat wave in France: investigating the 'harvesting' effect and other long-term consequences. *Population Studies* 62, 39–53.

Vandentorren, S., Suzan, F., Medina, S., Pascal, M., Maulpoix, A., Cohen, J.C., *et al.* (2004) Mortality in 13 French cities during the August 2003 heat wave. *American Journal of Public Health* 94, 1518–1520.

Woodhouse, P.R. (1993) Why do more old people die in winter? *Journal of the Hong Kong Geriatric Society* 3, 23–29.

Zanobetti, A. and Schwartz, J. (2008) Temperature and mortality in nine US cities. *Epidemiology* 19, 563–570.

4 Occupational Heat Effects: A Global Health and Economic Threat Due to Climate Change

Tord Kjellstrom,[1,2] **Rebekah Lucas,**[2] **Bruno Lemke**[3] **and Subhashis Sahu**[4]

[1]*National Centre for Epidemiology and Population Health, Australian National University, Canberra, Australia, and Institute of International Health, University College London, UK;* [2]*Umeå Centre for Global Health Research, Umeå University, Umeå, Sweden;* [3]*School of Health, Nelson-Marlborough Institute of Technology, Nelson, New Zealand;* [4]*Ergonomics and Occupational Physiology Laboratory, Department of Physiology, University of Kalyani, West Bengal, India*

4.1 Climate Change Will Increase Workplace Heat Exposures for Many Millions of People

Global climate change is causing hot seasons in hot places to become even hotter. Since 1980, many populated places (particularly cities) with a hot climate (temperatures regularly above 35°C) have recorded a 1–2°C increase in average temperatures as a result of climate change and the urban heat island effect (Intergovernmental Panel on Climate Change (IPCC), 2007). Global average temperature is predicted to increase a further 1–3°C during this century (IPCC, 2007). Recent comparisons of different models based on the 'Representative Concentration Pathways' (RCPs) indicate that the global average temperature increase may be significantly greater, with 3°C as an average (Rogelj *et al.*, 2012). Furthermore, in urban areas with rapid development of buildings, roads and other major physical structures, as well as clearance of forest and green spaces, this temperature increase is likely to proceed faster and to higher levels due to the 'urban heat island effect' (Oke, 1973; Kjellstrom, 2009a).

In tropical areas at low altitude, where the majority of the global population live and work, current heat exposure levels are already so high that the health and well-being of working people are at risk (Parsons, 2003) and work capacity is reduced (Kjellstrom *et al.*, 2009b). Some populations exposed to very hot weather have evolved partially effective cultural mechanisms in response, such as siestas and reduced expectation of excessive work during the hottest months; however, these long-standing adaptations may be breaking down as globalization advances. The monthly pattern of ambient heat exposure in Kolkata, India (Plate 3) gives an example of the very high afternoon maximum temperatures (above 35°C) recorded during the hottest months.

Air humidity (expressed as the dew point in Plate 3) is another climate variable of great importance for occupational heat stress. High relative humidity impairs sweat evaporation, thus inhibiting heat loss, which is especially important for people working physically. When the dew point is at the same level as the minimum temperature (as during many months each year in Kolkata), relative humidity is at 100%, which means that the potential for evaporative heat loss is zero (Parsons, 2003).

Workplace heat conditions of relevance to health and work capacity can be expressed as a 'heat stress index'. This index takes into account four key climate variables: air temperature, air humidity, air movement over the skin (wind speed) and heat radiation (i.e. from the sun or hot machinery) (Parsons, 2003). Numerous heat stress indexes have been published in the past 60 years, but the most widely used for workplace heat stress assessments is the WBGT (wet bulb globe temperature) (Parsons, 2003), which is the basis for an international standard (ISO, 1989) and many national standards or guidelines (e.g. ACGIH, 2009). The prescribed way of measuring WBGT involves three thermometers (ISO, 1989), but a viable shortcut is to assume that air movement over the skin is at 1 m/s (approximately the arm and leg movement speed for typical physical labour) and that the work is carried out inside or in the full shade (assuming no solar radiation). With these assumptions, it is possible to calculate indoor, or in-shade, WBGT from temperature and humidity data (Lemke and Kjellstrom, 2012). Adjustments of the results can be made to estimate WBGT outdoors in the sun and, with actual sun heat radiation data available, these estimates can be very close to measured values.

Time trends of WBGT for different locations around the world have been published during recent years, and for most places the values are increasing, creating additional heat stress in workplaces and heightening the risk of heat-related illnesses unless efficient local cooling methods are applied (Kjellstrom, 2009a,b; Hyatt *et al.*, 2010; Kjellstrom *et al.*, 2013). Figure 4.1 shows one example of the 30-year trend from 1983 to 2012. WBGT >29°C is extremely hot and limits physical work capacity and mental performance for working people, whether in heavy or even moderate labour (Parsons, 2003; Hancock *et al.*, 2007). In Dallas, Texas, USA, the number of days per year with a WBGT maximum greater than 29°C has doubled since 1980 (i.e. <20 days in 1980 versus >40 days in 2012). Climate modelling of future WBGT levels indicates increasing levels and likely additional impacts on work capacity or productivity (Kjellstrom *et al.*, 2009b, 2011a,b, 2013). The effects of excessive heat exposures in workplaces are 'direct', as defined by Butler *et al.* (Butler *et al.*, 2005), later labelled 'primary' (Butler and Harley, 2010), but the impacts of heat on family and community economy will also lead to secondary (indirect) or even tertiary (systemic) effects on human health and well-being.

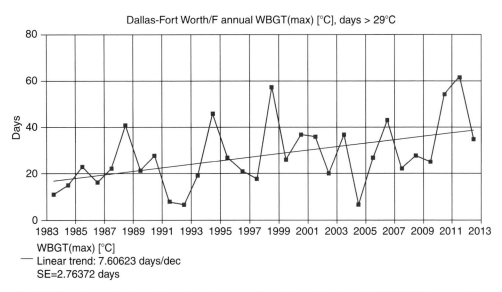

Fig. 4.1. Time trend of extremely hot days at one weather station. (Data from NOAA/GSOD; presentation by Hothaps-Soft, www.climatechip.org; Otto, M., *et al.*, unpublished.)

4.2 Human Physiology and Performance Links to Ambient Heat Exposure

The core body temperature of all humans needs to be close to 37°C to avoid heat strain or serious heatstroke and death[1] (Bridger, 2003). The human body has physiological heat control mechanisms that can maintain the core body temperature even when the external air temperature is greater than 37°C. However, these mechanisms rely heavily on sweat evaporation, which (as mentioned) is less efficient at high relative humidity levels. Clothing also influences heat loss. Light clothing is best in most hot environments (Parsons, 2003), but many jobs require special protective clothing, which adds to the heat stress (Bernard, 1999).

Physical movement and activity (particularly muscular work) increases the body's core temperature, and the extra heat is more difficult to dissipate in hot environments (Bridger, 2003). If cooling methods in a workplace are insufficient and workers already wear light clothing, the only way for a working person to reduce heat stress is to take breaks or slow down their work to reduce the metabolic rate (Parsons, 2003). This reduces 'work capacity', hourly work output and work productivity, with associated economic losses for the individual, the enterprise and the community (Kjellstrom, 2009a; Kjellstrom *et al.*, 2011b; Dunne *et al.*, 2013). The quantitative reduction of work capacity and productivity in a specific work situation, rice farming, in India has been documented recently (see Fig. 4.2) (Sahu *et al.*, 2013). A change of WBGT from 26°C to 32°C reduces hourly productivity by 30%. Even routine household activities can cause heat stress, necessitating a reduction of such activities on hot days. Commuting to and from work is also a daily source of heat exposure for many people (Bridger, 2003). Children, the elderly and pregnant women are likely to be more sensitive to overheating than adults, and people carrying out heavy labour are particularly vulnerable (Parsons, 2003).

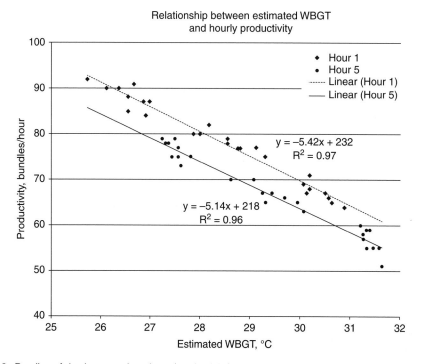

Fig. 4.2. Bundles of rice harvested per hour (productivity) at different environmental heat levels (WBGT) in West Bengal, India. Each point is a group average of 10–18 workers; Hour 1 is the first work hour of the day and Hour 5 is the 5th work hour (Sahu *et al.*, 2013).

An important issue for work capacity as well as clinical heat effects is the profuse sweating that some work involves, and the associated risk of dehydration from insufficient water replacement (Bridger, 2003). Workers carrying out heavy labour may lose 1 l of sweat per hour, and in certain workplaces drinking water may not be readily available to replace this loss of body water (Parsons, 2003). This is a common problem among, for example, sugarcane cutters in Central America (Crowe et al., 2013), who, in many plantations, have to bring their own drinking water to work and do not carry sufficient quantities (10 l of water is heavy, 10 kg) to maintain their fluid balance (Bridger, 2003). A number of reports on this 'Mesoamerican nephropathy' are now available (http://www.regionalnephropathy.org/wp-content/uploads/2013/04/Technical-Report-for-Website-Final.pdf).

These impacts on people exposed to heat are the subject of the 'Hothaps' research and prevention programme (Hothaps = high occupational temperature health and productivity suppression) (Kjellstrom, 2009b). A large number of publications have emerged from this programme (clickable links can be found at the website www.climatechip.org).

Reports of serious heat-related illnesses and deaths first emerged in the 19th century (as reviewed by Hollowell, 2010), and during the 20th century, military studies highlighted the problems for sailors and soldiers of the British colonial forces in tropical parts of the world. Reports on workplace heat problems also occurred then, but Wyndham and colleagues published the first systematic analysis in the 1960s (Wyndham, 1969).

Very serious clinical heat effects, including heatstroke and death, are still reported among people carrying out work or heavy physical activity (e.g. soldiers in training). A study of US agricultural workers using reported occupational health statistics (MMWR, 2008) showed that life-threatening work conditions involved heavy labour carried out in excessive heat (particularly during the afternoon), with limited possibilities to take rest or drink water. Indeed, on average, approximately 40 fatal cases of occupational heatstroke per year have been reported during the 15-year period 1990–2004.

Non-fatal effects of heat exposure are also reported in Table 4.1. Climate change can be expected to increase these morbidity risks among working people with limited ability, or economic disincentives, to take preventive action.

4.3 Clinical Effects of Heat Exposure

Human heat exposure and internal heat stress also leads to clinical effects, including serious heatstroke and even death. The symptoms and signs include heat exhaustion (a basis for the work capacity loss described above), heat rash, heat fainting (heat syncope), heatstroke and heat-induced death (Bouchama and Knochel, 2002). Most epidemiological research to date on the health effects of 'heatwaves' focuses on: mortality (e.g. Kovats and Hajat, 2008); hospital admissions (e.g. Green et al., 2010); outpatient visits to hospitals (Lin et al., 2012); or ambulance call-outs (e.g. Dolney and Sheridan, 2006). A significant problem in comparing the available studies is the lack of an agreed approach to quantifying heat exposure for epidemiological studies (Ye et al., 2012).

The various clinical effects of occupational heat exposure are summarized in Table 4.1.

4.4 Economic and Well-Being Impacts

These effects on occupational health will have economic consequences, not only for the worker and the employer, as typical work targets cannot be met. They are likely to be of sufficient scale to harm the wider economy also. The various clinical effects listed above are therefore important in calculating future climate change impacts. The only available economic analysis that assessed heat impacts on work capacity and productivity loss (DARA, 2012) concluded that reduced work capacity would be the largest economic threat from climate change by 2030. The global cost in 2030 was estimated at US$2.4 trillion, which is 56% of the total calculated economic cost attributed to climate change. Further analysis of costs in different countries and locations, as well as analysis of the approaches for prevention, are of increasing importance as climate change progresses.

Table 4.1. Health effects and negative impacts of excessive heat exposure at work.

Effect	Evidence, where described	References (examples; further references can be added)
Death from heatstroke at work	South African mine workers; US agricultural workers (and media reports from China, India and other countries)	Wyndham, 1969; MMWR, 2008
Specific serious heatstroke symptoms; heat exhaustion	Many hot workplaces around the world	Parsons, 2003
Clinical damage of organs	Heart overload and kidney damage; US military; El Salvador sugar workers	Schrier *et al.*, 1967; García-Trabanino *et al.*, 2005
Injuries due to accidents	Increased accidents in heat; Europe	Ramsey *et al.*, 1983
Mood/behaviour/mental health	Heat exhaustion; South African mine workers; Australian farmers	Wyndham, 1969; Berry *et al.*, 2010
Work capacity and economic loss	Low work capacity; less productivity in low- and middle-income countries; heat impact on GDP	Nag and Nag, 1992; Dell *et al.*, 2009; Kjellstrom *et al.*, 2009b

4.5 Preventive Actions

Climate change is likely to force heat-exposed workers and householders to slow down work and daily activities unless effective cooling methods can be applied, such as: seeking shade; provision and use of fans; air conditioning; or wearing specially designed cooling clothes (Parsons, 2003). However, in many work situations, air conditioning is not feasible (for instance, outdoors), and protective clothing may increase heat stress. In addition, there are important costs for air conditioning and, in most places, related electricity consumption adds to greenhouse gases (Global Energy Assessment, 2012). Notably, fans are not always effective at cooling when air temperature exceeds 34°C (the normal skin temperature; Parsons, 2003), as they blow hot air. To date, the impacts of the 'Hothaps effect' (Kjellstrom, 2009b; Kjellstrom

et al., 2009a) and climate change on people at work and in households, and the consequent implications for daily life and the economy, has received very little attention globally.

The threats to health and well-being from the 'Hothaps effect' emphasize the need for: (i) mitigation of climate change; (ii) public and occupational health and safety programmes that protect individuals currently at risk; and (iii) preventive policies and interventions (adaptation) through the design of urban areas and workplaces, as well as residential housing and commercial places, to reduce existing and future heat exposure in the locations where people live and work. Further evidence of local heat exposure and its effects is likely to be of importance in the development of policies and actions on heat prevention in conjunction with climate change. The following chapter describes methods and analysis techniques that can be used for local studies.

Note

[1] Or hypothermia.

References

ACGIH (2009) *TLVs for Chemical Substances and Physical Agents and Biological Exposure Indicators.* American Conference of Governmental Industrial Hygienists, Cincinatti, Ohio.

Bernard, T.E. (1999) Heat stress and protective clothing: an emerging approach from the United States. *Annals of Occupational Hygiene* 43, 321–327.

Berry, H.L., Bowen, K. and Kjellstrom, T. (2010) Climate change and mental health: a causal pathways framework. *International Journal of Public Health* 55, 123–132.

Bouchama, A. and Knochel, J.P. (2002) Heat stroke. *New England Journal of Medicine* 346, 1978–1988.

Bridger, R.S. (2003) *Introduction to Ergonomics.* Taylor and Francis, London.

Butler, C.D. and Harley, D. (2010) Primary, secondary and tertiary effects of the eco-climate crisis: the medical response. *Postgraduate Medical Journal* 86, 230–234.

Butler, C.D., Corvalán, C.F. and Koren, H.S. (2005) Human health, well-being and global ecological scenarios. *Ecosystems* 8, 153–162.

Crowe, J., Roman, B., Wesseling, C., Robles, A., Kjellstrom, T., Morales, D., *et al.* (2013) Heat exposure in sugar cane harvesters in Costa Rica. *American Journal of Industrial Medicine* 56, 1157–1164.

DARA (2012) *Climate Vulnerability Monitor – A Guide to the Cold Calculus of a Hot Planet.* DARA International Foundation, Madrid (website: www.daraint.org).

Dell, M., Jones, B.F. and Olken, B.A. (2009) Temperature and income: reconciling new cross-sectional and panel estimates. *American Economic Review* 99, 198–204.

Dolney, T.J. and Sheridan, S.C. (2006) The relationship between extreme heat and ambulance response calls for the city of Toronto, Ontario, Canada. *Environmental Research* 101, 94–103.

Dunne, J.P., Stouffer, R.J. and John, J.G. (2013) Reductions in labour capacity from heat stress under climate warming. *Nature Climate Change* doi:10.1038/nclimate1827.

García-Trabanino, R., Dominguez, J., Jansa, J. and Oliver, A. (2005) Proteinuria and chronic kidney disease on the coast of El Salvador (Spanish). *Nefrología* 25, 31–38.

Global Energy Assessment (2012) *Global Energy Assessment – Towards a Sustainable Future.* Cambridge University Press, International Institute for Applied Systems Analysis, Laxenberg, Austria.

Green, R.S., Basu, R., Malig, B., Broadwin, R., Kim, J.J. and Ostro, B. (2010) The effect of temperature on hospital admissions in nine California counties. *International Journal of Public Health* 55, 113–121.

Hancock, P.A., Ross, J.M. and Szalma, J.L. (2007) A meta-analysis of performance response under thermal stressors. *Human Factors: The Journal of the Human Factors and Ergonomics Society* 49, 851–877.

Hollowell, D.R. (2010) Perceptions of, and reactions to, environmental heat: a brief note on issues of concern in relation to occupational health. *Global Health Action* 3, doi:10.3402/gha.v3i0.5632.

Hyatt, O., Lemke, B. and Kjellstrom, T. (2010) Regional maps of occupational heat exposure: past, present and potential future. *Global Health Action* 3, doi:10.3402/gha.v3i0.5715.

IPCC (Intergovernmental Panel on Climate Change) (2007) *Fourth Assessment Report. Climate Change.* Cambridge University Press, Cambridge, UK (available free on the web: www.ipcc.ch).

ISO (1989) *Hot Environments – Estimation of the Heat Stress on Working Man, Based on the WBGT-Index (Wet Bulb Globe Temperature).* ISO Standard 7243. International Standards Organization, Geneva, Switzerland.

Kjellstrom, T. (2009a) Climate change exposures, chronic diseases and mental health in urban populations – a threat to health security, particularly for the poor and disadvantaged. Technical Report. World Health Organization, Kobe, Japan.

Kjellstrom, T. (2009b) Climate change, direct heat exposure, health and well-being in low and middle income countries. *Global Health Action* 2, 1–4.

Kjellstrom, T., Gabrysch, S., Lemke, B. and Dear, K. (2009a) The 'Hothaps' programme for assessing climate change impacts on occupational health and productivity: an invitation to carry out field studies. *Global Health Action* 2 (http://www.ncbi.nlm.nih.gov/pmc/articles/PMC2799255/, accessed 3 March 2014).

Kjellstrom, T., Holmer, I. and Lemke, B. (2009b) Workplace heat stress, health and productivity – an increasing challenge for low and middle-income countries during climate change. *Global Health Action* 2, doi:10.3402/gha.v2i0.2047.

Kjellstrom, T., Lemke, B. and Otto, M. (2011a) Climate change, occupational heat stress and impacts on health and productivity in Africa. *African Newsletter on Occupational Health and Safety* 21, 44–47.

Kjellstrom, T., Lemke, B. and Otto, M. (2011b) Increased workplace heat exposure due to climate change: a potential threat to occupational health, worker productivity and local economic development in Asia and the Pacific region. *Asian-Pacific Newsletter on Occupational Health and Safety* 18(1), 6–11.

Kjellstrom, T., Lemke, B. and Otto, M. (2013) Mapping occupational heat exposure and effects in South-East Asia. *Industrial Health* 51, 56–67.

Kovats, R.S. and Hajat, S. (2008) Heat stress and public health: a critical review. *Annual Review of Public Health* 29, 41–55.

Lemke, B. and Kjellstrom, T. (2012) Calculating workplace WBGT from meteorological data. *Industrial Health* 50, 267–278.

Lin, Y.-K., Chang, C.-K., Li, M.-H., Wu, Y.-C. and Wang, Y.-C. (2012) High-temperature indices associated with mortality and outpatient visits: characterizing the association with elevated temperature. *Science of The Total Environment* 427–428, 41–49.

MMWR (2008) Heat-related deaths among crop workers – United States, 1992–2006. *MMWR Weekly. Center for Disease Control and Prevention* 57, 649–653.

Nag, A. and Nag, P.K. (1992) Heat stress of women doing manipulative work. *American Industrial Hygiene Association Journal* 53, 751–756.

Oke, T.R. (1973) City size and the urban heat island. *Atmospheric Environment* 7, 769–779.

Parsons, K. (2003) *Human Thermal Environment. The Effects of Hot, Moderate and Cold Temperatures on Human Health, Comfort and Performance.* CRC Press, New York.

Ramsey, J.D., Burford, C.L., Beshir, M.Y. and Jensen, R.L. (1983) Effects of workplace thermal conditions on safe working behavior. *Journal of Safety Research* 14, 105–114.

Rogelj, J., Meinshausen, M. and Knutti, R. (2012) Global warming under old and new scenarios using IPCC climate sensitivity range estimates. *Nature Climate Change* 2, 248–253.

Sahu, S., Sett, M. and Kjellstrom, T. (2013) Heat exposure, cardiovascular stress and work productivity in rice harvesters in India: implications for a climate change future. *Industrial Health* 51, 424–431.

Schrier, R.W., Henderson, H.S., Tisher, C.C. and Tannen, R.L. (1967) Nephropathy associated with heat stress and exercise. *Annals of Internal Medicine* 67, 356–376.

Wyndham, C.H. (1969) Adaptation to heat and cold. *Environmental Research* 2, 442–469.

Ye, X., Wolff, R., Yu, W., Vaneckova, P., Pan, X. and Tong, S. (2012) Ambient temperature and morbidity: a review of epidemiological evidence. *Environmental Health Perspectives* 120, 19–28.

5 Measuring and Estimating Occupational Heat Exposure and Effects in Relation to Climate Change: 'Hothaps' Tools for Impact Assessments and Prevention Approaches

Tord Kjellstrom,[1,2] Rebekah Lucas,[2] Bruno Lemke,[3]
Matthias Otto[3] and Vidhya Venugopal[4]

[1]*National Centre for Epidemiology and Population Health, Australian National University, Canberra, Australia, and Institute of International Health, University College London, UK;* [2]*Umeå Centre for Global Health Research, Umeå University, Umeå, Sweden;* [3]*School of Health, Nelson-Marlborough Institute of Technology, Nelson, New Zealand;* [4]*Department of Environmental Health Engineering, Sri Ramachandra University, Chennai, India*

5.1 The Effects of Heat Exposure are Essential for Climate Change Impact Analysis

The links between human physiology, climate conditions and effects on health and work capacity are discussed in detail in physiology and ergonomics texts (e.g. Bridger, 2003; Parsons, 2003) and the preceding chapter (Kjellstrom *et al.*, 2014). High heat exposure risks heat exhaustion and heatstroke and is perceived subjectively as unpleasant or dangerous. People involved in heavy physical activity are particularly affected (Kjellstrom, 2009). An enterprise can compensate for this heat effect by carrying out heat-sensitive work during the cooler night hours of the hot season or by scheduling such work in the cooler season, but as climate change intensifies, the duration of 'cool periods' will diminish. Also influencing heat stress is the humidity level, which often rises at night, reducing the cooling impact. Further, some work must

be performed during daylight, also reducing available 'cool hours'. For instance, many agricultural workers need to work outdoors. If their pay is based on output, rather than time, they may work beyond safe heat exposure limits, risking death from heatstroke, as has occurred in the USA (MMWR, 2008).

The mechanisms underpinning the harmful health and productivity effects of heat have been described in recent reports, many of which are available for free on the Internet (see references in the *Global Health Action* journal 2009–2011) and will not be repeated here. There are also indirect health risks that working people face from increasing heat exposure, such as the chance of mosquito bites at dawn and dusk, forced by avoidance of work during the middle of the day (Bennett and McMichael, 2010).

This chapter describes five components of heat exposure and effect studies in workplace settings: a descriptive pilot study; heat monitoring studies; exploratory interview surveys;

quantitative studies of heat exposure–response relationships; and local health and productivity impact assessments. They can be carried out separately or in combination. The results of the local studies can be used to improve occupational health protection actions and can contribute to the global assessments of climate change impacts. The study approaches were developed in the Hothaps programme (High Occupational Temperature Health and Productivity Suppression) (Kjellstrom *et al.*, 2009).

5.2   Descriptive Pilot Studies Using Primarily Available Data

Descriptive studies document the general heat exposure situation for people in the study location (district, province, country). These studies can estimate likely heat exposure and assess how people cope. They may also identify occupational and other groups particularly affected by heat. Initial heat exposure measurements in selected workplaces can also be included. A pilot study report is very useful before proceeding to the other components of a Hothaps programme and related fundraising. The pilot study report could include a proposed timetable and budget for any of the other programme components a research group is considering.

A variety of data can be included in a pilot study to highlight not only climatic conditions but also population numbers and distributions, factors that influence sensitivity to climate variables and actions intended to reduce impacts on health and well-being. Table 5.1 lists potential items to include.

Some of the items in Table 5.1 can be obtained from national and international databases, and others will require identification by a review of the available literature on the relevant topics. Climate data on a daily, and even hourly, basis for many thousand weather stations around the world can be found for free on certain databases. More detailed local data may require sourcing from local or national authorities, and in some countries these data are not provided for free. A major source of free data is the US National Oceanographic and Atmospheric Administration (NOAA); we use

their daily data (GSOD = global summary of the day) from 1980 to 2012 in the new software called Hothaps-Soft.

National population data can be sourced from selected United Nations websites and population health status data from the World Health Organization website. The International Labour Organization website has data on workforce distributions for some countries, and websites devoted to air-conditioning (AC) issues have data on the use and need for such equipment in different locations. An interesting website for a variety of national health and other data is Gapminder.org.

Box 5.1

Further information on tools for pilot study development and free data on climate conditions and future modelling can be found at: www.climatechip.org.

5.3   Local Heat Monitoring and Occupational Exposure Assessment

This study component describes heat exposure in selected workplaces, representative of common types of work in the places where the field studies are carried out. The basic contributors to human heat exposure are air temperature and humidity, wind speed (or air movement over the skin) and heat radiation (in outdoor work, usually from solar radiation). Work settings indoors or in full shade often involve no significant heat radiation exposure, and wind speed can often be assumed at 1 m/s (Lemke and Kjellstrom, 2012).

Workplaces are chosen to represent the most common conditions in the location and/or the most exposed to ambient heat. As this study component is descriptive, different approaches can be taken to workplace representativeness and the duration of measurements. In most places, no local systematic studies have ever been carried out, so any new data may provide important evidence. Comparisons of weather station data and heat exposure levels in local workplaces (indoors and outdoors) measured at the same times would make it possible to create a calculation model for workplace heat exposures from past and future weather station data.

Table 5.1. Suggested pilot study content.

Report content	Description
Geographic conditions	*Proposed study location:* Region, country, geographic and other features (altitude, inland, coastal, economy, industries, history, etc.)
Climate conditions	*Climate situation at the study location:* Typical seasonal and daily variations in temperature and humidity, cloud cover, hot season duration, wind speed, rainfall patterns and other factors of relevance to heat exposure
Population	*Demographic information:* Population size, age and sex distribution; other basic health statistical data if available (e.g. infant mortality rate, adult mortality rates (age 15–64 years), life expectancy, occupational injury rates and any easily accessible variables)
Occupational features	*Occupational distribution in different age and sex groups* (e.g. for agriculture, construction, mining, manufacturing or other industry and services): Occupations requiring work in hot outdoor or indoor environments during the hot season; current practices used to cope with heat (e.g. night work, early starts, siesta, seasonal work)
Current heat experiences at work	*Based on findings from key informants* (from worker groups, managers, officials responsible for occupational health) Measures taken to avoid overheating in workplaces Information can be complemented by the experiences of heat exposure among the researchers themselves Preliminary climate variable measurements in selected workplaces may also be carried out to ascertain the heat exposure levels in workplaces (see component 2)
Current heat experiences commuting to/from work	*Heat exposures and discomfort reported by key informants*
Current heat experiences in households	*Reported heat exposures, heat-related disturbances to daily living* (i.e. sleep or household chores) and/or measures to avoid overheating in households
Literature on local situation	*Brief literature review* (in local language or any other language, including grey literature): Dealing with the climate situation in the location Health aspects of climate factors in location, including seasonal (monthly) variations in relevant health data A reference list in English should be included
Feasibility of local field studies	*Feasibility assessment* for future Hothaps field studies in this location
Tentative budget estimate and timetable	*Tentative budgets* for feasible local Hothaps field study, including costs for personnel, equipment, travel, analytical services and other costs, as well as a preliminary timetable for key stages of each study component

In order to combine the climate variables into a heat-related index, different formulas have been used. The most commonly used heat stress index is wet bulb globe temperature (WBGT) (Parsons, 2003), as it can be interpreted in terms of occupational health risks and decreased work capacity. Three thermometers measuring natural wet bulb temperature (Tnwb), globe temperature (Tg) and common air temperature (Ta) are required for formal WGBT measurements. Tnwb is measured inside a wetted small cloth sock, which simulates the heat exchange effect of evaporation from sweat-covered skin. Tg is measured inside a black globe, which simulates the uptake of heat on the skin from heat radiation (e.g. solar radiation), and Ta can be measured using a normal thermometer in the shade. Ideally, heat exposure should be measured according to the international standard methods for WBGT (ISO, 1989).

Environmental measurement equipment should be in places typical of the heat exposure that working people experience outdoors or indoors in this location. The ideal recording frequency may be hourly or half-hourly intervals, as this makes it easy to relate the results to parts of each 24-h period. Measurement duration varies depending on the equipment used, the type of workplace assessment and the research question. It is possible to record hourly values for a month at a time and calculate monthly averages (standard deviations or 90% variation limits) for every hour of the day. Hothaps assessments for specific workplaces during shorter periods can be very informative (e.g. a week of half-hourly measurements, 336 values, using data-logger equipment). As environmental heat exposure is always highest during daylight hours, daytime measurements may be adequate if there are risks that equipment will be damaged or stolen during the night.

Special WBGT monitoring equipment is available, costing approximately US$2000, but estimates of WBGT and other heat stress indexes can be calculated from measurements using less expensive equipment. For instance, data loggers can be used to measure temperature and humidity continuously. A few different models are available, and the usual cost is approximately US$100. With a data logger, hourly measurements are stored automatically and can be transferred easily to Excel or other software for graphing or statistical analysis. It is necessary to keep records of when and where measurements were made and who carried them out, as shown in Table 5.2.

Ideally, several data loggers should be used to measure heat exposures simultaneously in several sites in a location, both indoors and outdoors at the same time. The outdoor measurements can then be used to 'calibrate' the hourly recordings with nearby weather station recordings.

With just temperature and humidity measurements, WBGT can be calculated with mathematical formulae based on assumptions about the heat exposure situation (the formulae for indoors or in the shade are from Bernard and Pourmoghani, 1999, and for outdoors from Liljegren *et al.*, 2008, and Lemke and Kjellstrom, 2012; see also website: www.climatechip.org).

5.4 Explorative Interview Surveys of Heat Impacts and Preventive Approaches

In this component, questionnaires or surveys are used to collect information about the perceived heat exposure situation and impacts on health and work capacity in different workplaces. Open-ended questions allow workers to express their own perception or story of the heat situation at work and/or household and what can or should be done to prevent heat exhaustion. The surveys may focus on safety officers and health and safety officials, and/or employers and workers. The results indicate how people experience the impacts of heat on health and

Table 5.2. Detailed recording of heat exposure measurements.

Location for measurements (country, region, city, workplace, etc.):					
Organization or person carrying out measurements:					
Measurement series number:	1	2	3	4	5
Start, date/time					
End, date/time					
Frequency of recording					
Placement site at workplace					
Type of work at the site					
Person testing					
Downloaded file name					

productivity, and what measures are taken to reduce the impacts of heat. In some situations, formal qualitative studies may be performed. Appropriate questionnaires and surveys provide information that can be used for hypothesis generation and planning of subsequent quantitative studies, including assessments of intervention effectiveness.

Surveys and questionnaires collect data to answer the following questions about occupational heat impacts in the study location (these questions can also be used in the pilot study in interviews with key informants, as mentioned earlier; see Table 5.1):

- What is the climate profile in the location (seasonal and typical daily variations)?
- Which occupations in this location are most affected by heat and humidity?
- How are people in these occupations affected by heat at work?
- How does heat exposure change their capacity to carry out work?
- What are the working people doing, or could they do, to avoid heat exposures and heat effects at work? (Includes descriptions of current preventive interventions and ideas for new interventions; should include practices for access to water for rehydration.)
- Are AC or other cooling systems used in these workplaces? What proportion of time is AC, or other cooling, available and used? (In some places, electricity cuts limit the cooling periods.)
- What government regulations or other standards for maximum occupational heat exposure exist, are these enforced, and if so, how?
- How do people commute to and from work and does heat exposure during travel affect them?

- Is AC available in commuting vehicles?
- How does heat exposure affect people's capacity to carry out daily household activities during the hot season?
- How do people minimize these heat effects?
- Are AC or other cooling systems used in the workers' households? What proportion of time is AC, or other cooling, available and used?
- Is high temperature at night a problem during the hot season? Does heat affect sleep?
- Any other information about heat impacts during the hot season and how it compares with the less hot season.

An explorative survey ideally should include a variety of occupation groups, with at least ten respondents from each group in each location. A study can also use 'focus groups' representing the occupation groups of interest. If formal qualitative analysis approaches are used, complete interviews need to be transcribed and analysed according to published methods (not described here).

Standard questionnaires for workplace and household heat exposures have been developed (see example on the website, www.climatechip.org). These have several parts, as described in Table 5.3, and the way questions are formulated may vary depending on the study groups, which can be:

1. Representatives of enterprises, e.g. Health and Safety Officers.
2. Trade union or other collective worker representatives.
3. Officials from a government department or similar organization.
4. Individual working people (employed or self-employed).

Interviews can be conducted person-to-person, over the telephone or via mail or e-mail. Answers from individuals may be reported as 'case studies'. If a focus group is used, the responses

Table 5.3. Content of questionnaires about work, household activities and heat.

Part 1. General information about person interviewed and the organization she or he represents
Part 2. Questions concerning the type of work
Part 3. Questions about heat exposure at work (and potential linked chemical exposures)
Part 4. Questions concerning impacts of heat on health
Part 5. Questions concerning impacts on work activities and productivity
Part 6. Heat prevention approaches
Part 7. Experience of climate change to date, heat exposure outside of work

should be recorded on behalf of the whole group, including any dissension, so that the whole range of views is represented. Each person interviewed should be informed of the purpose of the study and asked for informed consent to allow use of their answers in analysis and reportage.

5.5 Quantitative Studies of Health Impacts, Work Capacity and Prevention

This study component measures the actual heat exposures, health impacts and work outputs during periods of different levels of heat exposure in selected workplaces. Ideally, the results will describe hourly exposures and effects. Occupation types to include are those that are common in the study location and are vulnerable to heat exposure as identified in the pilot study or the qualitative study. The quantitative study provides exposure–effect and exposure–response relationships (Bonita *et al.*, 2007) for effects of heat exposure and can be used to quantify the effectiveness of different methods to reduce heat exposure and impact.

A longitudinal cohort ('case-crossover' or 'panel study') design includes defined groups of workers with a high heat exposure at work. These workers are assessed at least twice, once during the hottest time of year and again during cooler periods (or seasons). The health status and work output (per working hour) for each worker is measured. Work output should be quantifiable in terms of production output per working hour, number of working hours per day, identifying the 24-h periods when work is carried out. Health status measurements can include medical reports, questionnaires and/or physiological measurements.

It is assumed that each group studied should include at least 20 workers whose work activities are similar during both hot and less hot work periods or seasons and whose work output can be quantified. Smaller group numbers might be sufficient if the heat effects are substantial. Statistical power calculation methods to define study population size requirement can be found in textbooks and on the Internet. Data to be collected are suggested in Table 5.4 (Checkoway, 2004).

Physiological measurements create important biomedical explanations for any demonstrated links between workplace heat exposures and effects (Table 5.5). These include body temperature, heart rate, physical activity level (can be measured with accelerometers), hydration indicators and psychological measures (see Hancock *et al.*, 2007).

Data according to the description above are collected during consecutive days in each of two study periods including, ideally, at least 10 hot working days and 10 cooler days (the difference between work period temperatures should be 10 degrees or more, if possible) (periods with intermediate heat exposure can be included to create a more continuous exposure range). Detailed methods will depend on local conditions, but the methods need to be explained clearly in the report.

Box 5.2

Further information on tools for quantitative exposure and effect studies can be found on the website: www.climatechip.org.

5.6 Occupational Health and Economic Impact Assessment for Local Climate Change

In this Hothaps component, the relationship between heat exposure and occupational health, work capacity and well-being is estimated from the new field study data on exposure–effect and exposure–response relationships, and international guidelines for maximum safe exposures. With the documented exposure–response relationships and local demographic and labour force data, it will be possible to estimate the heat stress impacts not only on individuals but also on populations, locally, nationally or globally (Dunne *et al.*, 2013). Statistical distributions of exposure and incidence of effects can be established and used in the impact assessments. When combined with modelling of likely future climate conditions, the occupational health and productivity impacts of climate change can be estimated. Reductions of productivity can also be translated into potential economic losses (DARA, 2012). The results from different locations will also feed into global health, productivity and economic

Table 5.4. Minimum and ideal additional data collected in a quantitative Hothaps study.

Type of data and variables	Minimum data	Ideal additional data
Heat exposure	1. Air temperature in work area during hottest hour and coolest hour 2. Absolute humidity (e.g. dew point) in work area at the same time as air temperature measurements 3. Description of climate conditions (solar radiation, radiated heat exposure from processes or equipment, breeze)	In work area for each hour: air temperature, absolute humidity, natural wet bulb temperature, globe temperature, wind speed. WBGT measurement
Work type	Short description of work, how it is carried out and how output can be quantified	More detailed description of work, how it is carried out and how output can be quantified
Daily work output and hours	Time when work started and finished, duration of work, output in the same period (calculate output per hour)	Timing and length of breaks, exact measurement of what the worker does at different times Accelerometer measurements of physical activity may be tested
Preventive measures	Brief description of methods used to reduce heat exposure at work, rehydration procedures (liquid intake during work day); any other methods for cooling	Detailed description of methods used, timing, volume and provision of rehydration, etc.; liquid intake during shift
Symptoms of heat exposure	The worker's own history of how the work environment feels during heat exposure; any symptoms from heat strain	More detailed information about symptoms or clinical signs, and when they happened; past incidence data based on stored records
Physiological reaction to heat	Nil	Pulse rate at different times during shift; core body temperature (tympanic, oral or rectal) at end of shift; heat shock protein level in blood; indicators of dehydration (e.g. urinary specific gravity or creatinine, net body weight loss taking liquid intake into account)
Other physical activity during the day	Description of daily commuting to/from work (walk, bicycle, bus, car, etc.)	More details about heat exposure during travel to work, daily physical labour at home (e.g. family farm work); exercise activities

impact assessments that will inform the future evaluations of global climate change impacts.

This component uses standard health impact assessment quantitative methods (see www.who.int/hia/en) to calculate future impacts from estimates of future heat exposure in vulnerable occupations, numbers of exposed workers and exposure–response relationships for specific heat-related health risks and productivity impacts (see also Haines et al., 2006).

Future occupational heat exposures in vulnerable groups are estimated from IPCC climate change projections (or other internationally recognized projections) at this location. The location may be defined geographically at different levels – village, city, province and country. Many reports use grid cell (0.5 × 0.5 degree) climate estimates from different scenarios, pathways and models for future 30-year averages.

Table 5.5. Advantages and limitation of physiological measurements in the field.

	Advantages	Limitations
Workload		
Oxygen consumption	Gold standard of cardiovascular fitness and capacity Highly reliable and valid Mobile units are available	Mobile units are expensive Uncomfortable to wear for prolonged periods (i.e. during a work day) A maximal exercise test is required to ascertaining accurately an individual's relative work rate intensity
Heart rate	Highly reliable and valid Convenient for field testing and comfortable to wear (chest strap and watch worn) Linear relationship between oxygen consumption and heart rate	Relatively expensive if data transfer equipment included
Internal body core temperature		
Gastrointestinal temperature	Reliable Valid Convenient Mobile	Slight lag during rapid changes in body core temperature Limited transmission range (approximately 1 m) Exclusion criteria (e.g. individuals with any gastrointestinal illness or previous surgery should not swallow the pill) Cultural inhibitions/taboos (e.g. swallowing an unnatural object) Expensive
Tympanic temperature	Non-invasive Mobile Convenient Relatively cheap	Poor reliability and validity (depends on ambient temperature) Inaccurate during physical activity Errors can result from inaccurate placement, technician error or ambient interference
Fluid balance		
Urinary measures	Cheap Non-invasive Easy to measure	Only sensitive to large acute changes in fluid balance (i.e. 3% or 5% body mass deficit) Lags plasma osmolality measures
Blood samples	Accurately measures acute changes in hydration state	Invasive Relatively expensive Specialized training
Changes in body mass	Cheap Reliable	Measurement can be difficult Food and fluid ingestion may need recording

The future distribution of various occupations also needs consideration. For example, how will the numbers of workers change in different exposure situations either outdoor or indoor? Is it predicted that there will be AC or other efficient cooling systems? What is the likely level of work intensity for heat-exposed workers? Analysis of the differing impacts on

people, proportional to their degree of labour or protection from AC, enables quantification of the disparate health effects on working people due to climate change.

This type of analysis also allows assessment of combined effects of different workplace exposures (i.e. heat and general air pollution or heat and evaporating chemicals).

Box 5.3

Accessible databases are becoming available that show the modelled geographical distributions of future climate change, including of temperature, humidity and other variables for specific locations. Future changes in the occupational distributions need to be integrated with these models by local research groups. National labour force data and projections can be very helpful for this. The calculations can be done on age–sex groups, if there are sufficient available data. The gridded climate data may also need adjustment for altitude, the 'heat island effect' and other factors that influence local heat exposures. The reports from different locations should ideally include data that can be combined with higher geographic level analysis (national, regional, global).

References

Bennett, C.M. and McMichael, A.J. (2010) Non-heat related impacts of climate change on working populations. *Global Health Action* 3, doi:10.3402/gha.v3i0.5640.

Bernard, T.E. and Pourmoghani, M. (1999) Prediction of workplace Wet Bulb Global Temperature. *Applied Occupational and Environmental Hygiene* 14, 126–134.

Bonita, R., Beaglehole, R. and Kjellstrom, T. (2007) *Basic Epidemiology*, 2nd edn. World Health Organization, Geneva, Switzerland.

Bridger, R.S. (2003) *Introduction to Ergonomics*, 2nd edn. Taylor and Francis, London.

Checkoway, H. (2004) Case-crossover designs in occupational health. *Occupational and Environmental Medicine* 61, 953–954.

DARA (2012) *Climate Vulnerability Monitor – A Guide to the Cold Calculus of a Hot Planet*. DARA International Foundation, Madrid (http://daraint.org/, accessed 28 February 2014).

Dunne, J.P., Stouffer, R.J. and John, J.G. (2013) Reductions in labour capacity from heat stress under climate warming. *Nature Climate Change*, doi:10.1038/nclimate1827.

Haines, A., Kovats, R.S., Campbell-Lendrum, D. and Corvalán, C.F. (2006) Climate change and human health: impacts, vulnerability and public health. *Public Health* 120, 585–596.

Hancock, P.A., Ross, J.M. and Szalma, J.L. (2007) A meta-analysis of performance response under thermal stressors. *Human Factors: The Journal of the Human Factors and Ergonomics Society* 49, 851–877.

ISO (International Standards Organization) (1989) *Hot Environments – Estimation of the Heat Stress on Working Man, Based on the WBGT-Index (Wet Bulb Globe Temperature)*. ISO Standard 7243. International Standards Organization, Geneva, Switzerland.

Kjellstrom, T. (2009) Climate change, direct heat exposure, health and well-being in low and middle income countries. *Global Health Action* 2, 1–4.

Kjellstrom, T., Holmer, I. and Lemke, B. (2009) Workplace heat stress, health and productivity – an increasing challenge for low and middle-income countries during climate change. *Global Health Action* 2, doi:10.3402/gha.v2i0.2047.

Kjellstrom, T., Lucas, R., Lemke, B. and Sahu, S. (2014) Occupational heat effects: a global health and economic threat due to climate change. In: Butler, C.D. (ed.) *Climate Change and Global Health*. CAB International, Wallingford, UK, pp. 38–44.

Lemke, B. and Kjellstrom, T. (2012) Calculating workplace WBGT from meteorological data. *Industrial Health* 50, 267–278.

Liljegren, J., Carhart, R., Lawday, P., Tschopp, S. and Sharp, R. (2008) Modeling Wet Bulb Globe Temperature using standard meteorological measurements. *Journal of Occupational and Environmental Hygiene* 5, 645–655.

MMWR (Morbidity and Mortality Weekly Report) (2008) Heat-related deaths among crop workers – United States, 1992–2006. *MMWR Weekly. Center for Disease Control and Prevention* 57, 649–653.

Parsons, K. (2003) *Human Thermal Environment. The Effects of Hot, Moderate and Cold Temperatures on Human Health, Comfort and Performance*. CRC Press, New York.

6 Climate Extremes, Disasters and Health

Yasushi Honda,[1] Taken Oki,[2] Shinjiro Kanae[3] and Colin D. Butler[4]

[1]*Faculty of Health and Sport Sciences, The University of Tsukuba, Japan;*
[2]*Institute of Industrial Science, The University of Tokyo, Japan;*
[3]*Department of Civil Engineering, Tokyo Institute of Technology, Japan;*
[4]*Faculty of Health, The University of Canberra, Australia, and National
Centre for Epidemiology and Population Health, The Australian National
University, Australia, and Benevolent Organisation for Development,
Health & Insight (BODHI)*

6.1 Introduction

In 2012, the Intergovernmental Panel on Climate Change (IPCC) published the Special Report on Extreme Events (SREX) (IPCC, 2012). This extensive report, almost 600 pages long, assesses the scientific literature on the relationship between society and extreme events, related to weather and climate ('climate extremes').[1] The SREX stresses the interacting, iterative role between human and environmental factors, sometimes called 'ecosocial' determinants. The assessment concludes by discussing ways to manage the risks to human well-being posed by disasters.

This chapter draws extensively from this report, to which the first three authors contributed. The supporting references for many statements in this chapter can be found in that report. We start with some definitions.

Definitions

The SREX defines several key terms (see Box 6.1).

6.2 Climate Extremes, Impacts and Trends

By definition, extreme events are rare. This makes changes in their frequency or intensity difficult to detect, especially in the more distant past, and in regions or times when data quality and quantity are lacking. Identification varies by the kind of extreme. For example, global trends in temperature are easier to identify and more reliable than are droughts at a regional scale. Some trends are more geographically uniform.

Despite these difficulties, the SREX concluded that observational evidence gathered since 1950 allowed several firm conclusions. Note that assigning 'low confidence' in observed changes in a specific extreme on regional or global scales neither implies nor excludes the possibility of changes in this extreme. These conclusions are presented as part of the discussion for each kind of climate extreme, for temperature, floods, cyclones, droughts and tornadoes.

6.3 Disasters

The impact of disasters, including climatic extremes on human health and infrastructure,

---

Box 6.1

Climate change: An alteration in weather over a prolonged period, identifiable by statistical tests, such as changes in the mean and/or the variability of its properties. This change must persist for at least decades. Change to the climate can be caused by natural or human (anthropogenic) processes. Natural causes include external forcings[2] such as variations in sunlight or the Earth's orbit. Human causes include large-scale alterations to the composition of the atmosphere, ocean or vegetation, or a combination of these.

Climate extreme: The occurrence of a weather or climate measure above (or below) a predetermined threshold (e.g. 97.5%) near the upper (or lower) ends of the observed range of values of the variable. The SREX classifies extreme weather events as a form of extreme climatic event.

Disaster: Severe changes in the functioning of a community or a society because of hazardous physical events, requiring immediate emergency responses. Disasters cause widespread adverse human, material, economic or environmental effects, often in combination.

Disaster risk: The chance of disaster, influenced by social vulnerability, exposure (the likelihood of experiencing a disaster) and the severity of the extreme event.

Disaster risk management: Processes that lessen disaster risk and promote improved response and recovery with benefits for human security and other aspects of well-being.

Adaptation: The attempt to adjust human and/or natural systems to actual or expected climate and its effects, in order to minimize harm. Sometimes, beneficial opportunities may be exploited.

Impact: The interaction between exposure and vulnerability of natural or human-made assets (including livelihoods) to risk from extreme events. Vulnerability is a combination of susceptibility and recovery capacity, related to various forms of wealth, including human and social capital, financial resources and infrastructure quality.

Resilience: The ability of a system and its parts to anticipate, accommodate and recover from a hazardous event in a timely and efficient manner without catastrophic loss. This is also related to the forms of wealth described above.

Transformation: A change in fundamental system attributes, such as values, regulation, technology or biology.

Box 6.2

There are two main disaster databases. The EM-DAT database is maintained by the University of Louvaine, Belgium. Another, the DesInventar database, registers and compiles small and medium-scale events. Many disasters in these databases record extremely severe interruption in local social or economic functioning. However, such 'extreme social events' are *not* necessarily driven by the climate or physical extreme events defined by the SREX.

can vary widely, even within geographically and temporally proximal locations. For example, category 4 cyclones, Sidr (Bangladesh 2007) and Nargis (Myanmar 2008), affected adjacent countries at almost the same time, but with very different mortality. Sidr caused a relatively modest 4200 deaths, but Nargis killed 138,000 (CRED, 2009; Yokoi and Takayabu, 2010). A major reason for this difference is the adaptation effort (see Fig. 6.1). Bangladesh experiences

more frequent cyclones than Mynamar and has progressively improved its cyclone shelters and early warning system.

This variation is also illustrated in Tables 6.1–6.3. In absolute terms, China and India each experience huge disasters, mainly because they have enormous, densely populated regions.

Losses and economic valuation

Economic losses from weather- and climate-related disasters have increased (see Fig. 6.2), but so too have population, infrastructure and the trend to settle places vulnerable to disasters (Barthel and Neumayer, 2012). The SREX concluded, with high confidence, that increased *exposure* of people and economic assets to risk is the major cause of long-term increases in economic losses from climate-related disasters. However, a role for climate change in this trend has not been excluded (high agreement, medium confidence).

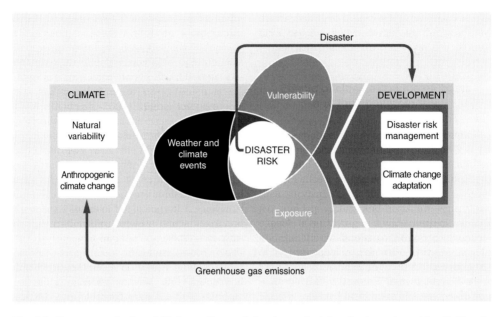

Fig. 6.1. Exposure and vulnerability to weather and climate events determine impacts and the likelihood of disasters. This is modified by the exposure and vulnerability of human society and natural ecosystems. (© IPCC, 2012.)

Table 6.1. Top 20 most important hydrological, climatological and meteorological disasters, 1900–2012, ranked by number killed. (Source: Guha-Sapir *et al.*, 2011.)

Rank	Country	Type	Date	Killed	Affected	Cost (US$ m)
1	China PDR	Flood	1931	3.7 m		1400
2	China PDR	Drought	1928	3 m		
3	China PDR	Flood	1959–1961	2 m		
4	Bangladesh	Drought	1943	1.9 m		
5	India	Drought	1942	1.5 m		
6	India	Drought	1965–1967	1.5 m	100 m	100
7	India	Drought	1900	1.25 m		
8	Soviet Union	Drought[a]	1921	1.2 m	5 m	
9	China PDR	Drought	1920	500 k	20 m	
10	China PDR	Flood	1939	500 k		
11	Ethiopia	Drought	1983–1984	300 k	7.75 m	
12	Bangladesh	Cyclone	1970	300 k	3.65 m	86.4
13	Sudan	Drought	1983–1985	150 k	8.4 m	
14	China PDR	Flood	1935	142 k	10.0 m	
15	Bangladesh	Cyclone	1991	139 k	15.4 m	1780
16	Myanmar	Cyclone	2008	138 k	2.42 m	4000
17	Ethiopia	Drought	1973–1978	100 k	3 m	76
18	China PDR	Flood	1911	100 k		
19	China PDR	Cyclone	1922	100 k		
20	Mozambique	Drought	1981–1985	100 k	4.75 m	

[a]Many Ukrainians would argue this disaster was largely human-made.
k = kilo (thousand); m = million.

Table 6.2. Top 20 most important hydrological, climatological and meteorological disasters, 1900–2012, ranked by number affected. (Source: Guha-Sapir *et al.*, 2011.)

Rank	Country	Type	Date	Killed	Affected (m)	Cost (US$)
1	India	Drought	1987	300	300	
2	India	Drought	2002		300	911 m
3	China PDR	Flood	1998	3656	239	30 bn
4	China PDR	Flood	1991	1729	210	7.5 bn
5	India	Drought	1972–1973		200	100 m
6	China PDR	Flood	1996	2775	155	12.6 bn
7	China PDR	Flood	2003	430	150	7.9 bn
8	China PDR	Flood	2010	1691	134	18 bn
9	India	Flood	1993	827	128	7 bn
10	China PDR	Flood	1995	1437	114	6.7 bn
11	China PDR	Flood	2007	535	105	4.4 bn
12	China PDR	Flood	1999	725	101	8.1 bn
13	China PDR	Flood	1989	2000	100	2.8 bn
14	India	Drought	1965–1967	1.5 m	100	100 m
15	India	Drought	1982–1983		100	
16	China PDR	Local storm	2002		100	
17	China PDR	Drought	1994		82	13.8 bn
18	China PDR	Flood	2002	793	80	3.1 bn
19	China PDR	Flood	1994	1001	79	5.46 bn
20	China PDR	Extreme winter	2008	129	77	21.1 bn

Note: m = million; bn = billion.

Table 6.3. Top 20 most important hydrological, climatological and meteorological disasters, 1900–2012, ranked by damage cost. (Source: Guha-Sapir *et al.*, 2011.)

Rank	Country	Type	Date	Killed	Affected (m)	Cost (US$ bn)
1	USA	Hurricane Katrina	2005	1833	500 k	125
2	USA	Hurricane Sandy	2012	54		50
3	Thailand	Flood	2011–2012	813	9.5 m	40
4	China PDR	Flood	1998	3656	239 m	30
5	USA	Hurricane Ike	2008	82	200 k	30
6	USA	Hurricane Andrew	1992	44	250 k	26.5
7	China PDR	Extreme winter	2008	129	77 m	21.1
8	USA	Drought	2012			20
9	China PDR	Flood	2010	1691	134 m	18
10	USA	Hurricane Ivan	2004	52		18
11	USA	Hurricane Charley	2004	10	30 k	16
12	USA	Hurricane Rita	2005	10	300 k	16
13	Korean DPR	Flood	1995	68	5.7 m	15
14	USA	Hurricane Wilma	2005	4	30 k	14.3
15	USA	Local storm	2011	176	1.15 k	14
16	China PDR	Drought	1994		82 m	13.8
17	China PDR	Flood	1996	2775	155 m	12.6
18	USA	Flood	1993	48	31 k	12
19	Germany	Flood	2002	27	330 k	11.6
20	USA	Hurricane Frances	2004	47	5 m	11

Note: k = kilo (thousand); m = million; bn = billion.

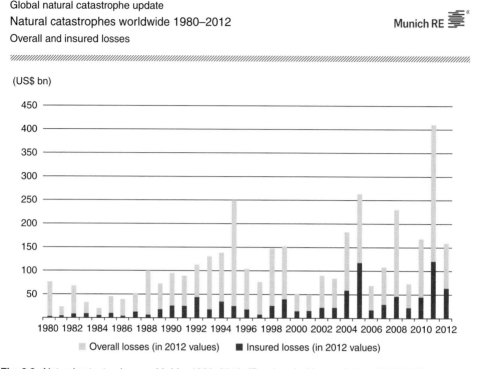

Global natural catastrophe update
Natural catastrophes worldwide 1980–2012
Overall and insured losses

Munich RE

Fig. 6.2. Natural catastrophes worldwide, 1980–2012. (Reprinted with permission, © 2013 Münchener Rückversicherungs-Gesellschaft, Geo Risks Research, NatCatSERVICE.)

These increases in loss illustrate large spatial and interannual variability (high confidence, based on high agreement, medium evidence), according to the SREX. The reported losses in recent decades reflect monetized direct asset damages, and are distributed unequally. Estimates of annual losses since 1980 range from a few to above US$200 billion (in 2010 US dollars), with the highest value for 2005 (the year of Hurricane Katrina). Hurricane ('Superstorm') Sandy, which struck New York City and adjacent parts of the north-east US coast in late 2012, occurred after the SREX was completed. Its damages are currently estimated to be as high as US$50 billion (Knowlton *et al.*, 2013).

These loss estimates are necessarily conservative, because many impacts are difficult to monetize. Major examples of this include lost human lives, lost cultural heritage and damaged ecosystem services. The non-accounting of such 'negative externalities' is central to the critique of conventional economics by ecological economists (Kubiszewski *et al.*, 2013).

Impacts on the informal 'black' economy, as well as indirect economic effects, can be very important in some areas and sectors, but are also generally excluded from reported estimates of loss. An additional problem with the conventional measure of gross domestic product (GDP) is that repairs increase the GDP, even where such expenditure simply restores the original conditions. At times, however, disasters are so large that they depress GDP, such as the 2011 Thai floods. These curtailed work at many export industries for weeks and even months, at an estimated cost of US$40 billion, one of the most expensive disasters ever measured (see Table 6.3).

Disaster costs and deaths comparing developed and developing countries

The measured monetary cost of disasters (from both extreme climatic and geophysical events) in US$ terms, unadjusted for purchasing power,[3] is much higher in developed countries, as are insured losses (also in US$) (see Fig. 6.2). This is not only because the cost of infrastructure is higher in developed countries (due to higher wage and tax structures) but also because comparatively few people in developing countries can afford to pay insurance: in many cases, no formal insurance is possible for the very poor.

In contrast, with high confidence, fatality rates and economic losses expressed as a proportion of GDP are higher in developing countries. Between 1970 and 2008, over 95% of deaths from natural disasters occurred in developing countries (IPCC, 2012).

6.4 Types of Disasters

Many forms of disaster can harm human health, including meteorological, hydrological, climatological, geophysical and biological (see Table 6.4). Their sequence, pattern and specific circumstances are also relevant. For example, the impact of an earthquake, a geophysical disaster, is worse if it occurs soon after a storm, especially if in a developing country. Some epidemics are more likely to follow disasters, which may damage crops and critical infrastructure, such as to water, sewerage, roads and bridges. The devastation of Hurricane Mitch in Nicaragua and nearby countries, which occurred in 1999 at the end of the wet season, when soils were waterlogged, is one such example (Epstein, 1999).

The paths that link disasters to epidemics can be circuitous. The cholera epidemic that started in 2010 in Haiti, later shown to be of South Asian origin (Chin et al., 2011), was introduced inadvertently by United Nations peacekeepers brought in after the severe earthquake earlier that year (Farmer et al., 2011).

Correspondingly, non-extreme climatic events can, at times, have disproportionately severe human impacts. For example, ordinary rains in several upstream areas may still cause a downstream flood.

The SREX describes three kinds of 'compound events': (i) two or more extreme events occurring simultaneously or successively; (ii) combinations of extreme events with underlying conditions that amplify the impact of the events; or (iii) combinations of events that are not themselves considered extreme. Identification of compound events is restricted because they necessitate multiple sets of specialized experts. Even when multiple groups notice the interaction, communication and collaboration are often hindered.

Earthquakes, climate change and energy

At present, earthquakes are not believed to be related causally to climate change, although at least some earthquakes have human causes, including practices associated with gas fracking (Jones, 2013), dams (Kerr and Stone, 2009) and CO_2 storage (Zoback and Gorelick, 2012).

6.5 Temperature Extremes

The SREX concluded that it was very likely there had been an overall decrease in the number of cold days and nights, and an overall increase in the number of warm days and nights, at the

Table 6.4. Definition of disaster subgroups. (Source: Guha-Sapir et al., 2011.)

Subgroup	Definition	Example
Geophysical	Events originating from solid earth	Earthquake, volcano, mass movement (dry)
Meteorological	Events caused by short-lived/small- to meso-scale atmospheric processes (minutes to days)	Storm
Hydrological	Events caused by deviations in the normal water cycle and/or overflow of bodies of water caused by wind	Flood, landslide
Climatological	Events caused by long-lived/meso- to macro-scale processes (intraseasonal to multidecadal climate variability)	Extreme temperature, drought, wildfire
Biological	Disaster caused by exposure of living organism to pathogens and toxins	Epidemics, insect infestation, animal stampede

global scale; that is, for most land areas with sufficient data. It is likely that these changes have also occurred at the continental scale in North America, Europe and Australia. There is medium confidence in a warming trend in daily temperature extremes in much of Asia. Confidence in observed trends in daily temperature extremes in Africa and South America generally varies from low to medium, depending on the region. In many (but not all) regions over the globe with sufficient data, there is medium confidence that the length or number of warm spells or heatwaves has increased. The health effects of heatwaves are described in Chapter 3, this volume.

6.6   Floods

The main cause of death from floods is from drowning (French *et al.*, 1983). However, a wide range of health impacts can occur, including trauma, injuries and contact with chemical substances or power lines (Ahern *et al.*, 2005). Infectious diseases include diarrhoea, malaria and leptospirosis. Adverse mental health effects can occur. Floods can also cause undernutrition, due to crop damage, interacting with insufficient famine relief.

Flood trends

In theory, at least three types of flood may be worsened by climate change (riverine, coastal and glacial outlet). There have been statistically significant trends in the number of heavy precipitation events in some regions (Allan, 2011). Some workers have proposed that more intense 'atmospheric rivers' will increase flooding (Lavers *et al.*, 2013). However, the SREX is more cautious, noting only low confidence that floods have increased since 1950 and medium confidence that heavy rain would increase in some regions, leading to localized flooding. The sea level is continuing to rise, and seems likely to contribute to increased coastal flooding. The SREX concluded that it was likely that there had been an increase in extreme coastal high water related to increases in mean sea level.

The SREX gave 'high' confidence to a future increase in glacial lake outburst floods (GLOFs), but not to riverine or coastal flood events (see Box 6.3).

> **Box 6.3**   Glacial Lake Outburst Flood Adaptation in Bhutan
>
> A social survey concerning GLOFs in central Bhutan in 2010 was undertaken to obtain information about community vulnerability and the implementation of adaptation techniques (Takenaka *et al.*, 2012) This found that traditional communities were safe from the direct impacts of floods, but that temporary residents were vulnerable. Groups especially at risk were construction workers camped on riverbanks. Adaptation measures proposed included the establishment of off-limit zones based on accurate hazard maps. Early warning systems utilizing cell phones or radios are also effective. Schools can also play a major role in these systems: they are well equipped with communication tools in case of GLOFs, and teachers can provide regular disaster education.

It found limited to medium confidence available to assess climate-driven observed changes in the magnitude and frequency of floods at regional scales, because of limited instrumental records at gauge stations and because of the confounding effects of changes in land use and engineering.

The question of floods associated with tropical cyclones is also uncertain. There is no consensus about the past trend in tropical cyclones, due to an enhanced ability to detect them (Peduzzi *et al.*, 2012).

6.7   Cyclones, Tornadoes and Hail

Cyclones and other types of atmospheric depression involve heavy rain and strong wind. In addition to the environmental damage caused by floods, these extremes can bring devastation through strong winds, storm surges and injuries from falls and flying debris. Severe cyclones can also cause serious mental trauma. For example, a survey conducted 3 months after Typhoon Morakot struck Taiwan in 2009 found that 26% of 271 adolescents who had experienced evacuation, and in some cases the deaths of relatives, had post-traumatic stress disorder (PTSD) (Yang *et al.*, 2011). Some rescuers and volunteers also suffered from PTSD (Neria *et al.*, 2008).

The SREX concluded with only low confidence that there had been an observed long-term (i.e. 40 years or more) increase in tropical cyclone activity (i.e. intensity, frequency, duration), after accounting for past changes in observing capabilities. However, it did conclude that there was likely to have been a poleward shift in the main northern and southern hemisphere extra-tropical storm tracks.

A recent analysis has concluded that increases in population living in exposed areas, in conjunction with increases in the intensity of cyclones (even if less frequent), will lead to higher risks from cyclones (Peduzzi *et al.*, 2012). However, these risks may be reduced by development and targeted mitigation and adaptation.

The SREX also had low confidence in observed trends in small spatial-scale phenomena such as tornadoes, hail and thunderstorms because of data inhomogeneities and inadequate monitoring systems. While increased US tornadoes have been observed, this is thought to be due to increased exposure, including more people living in remote areas. Climate models are unable to resolve such small-scale phenomena directly, reducing confidence in future projections. However, a warmer world may increase some atmospheric conditions conducive to tornadoes such as atmospheric instability, but other conditions such as vertical shear may decrease, due to a reduced pole-to-equator temperature gradient. Thus, enhanced tornado formation in a warmer world cannot be excluded.

6.8 Droughts

Unlike heatwaves or cyclones, the health effects of droughts are rarely direct. People can generally migrate to obtain water, and in some cases obtain assistance from governmental or non-governmental organizations. Health impacts are mostly secondary or tertiary, such as diarrhoea from contaminated drinking water or undernutrition through decreased crop yield. Conflict can be associated with prolonged drought, as can raised food prices.

In times of drought, children and women in areas with marginal water supplies expend more time and labour gathering this precious resource, and children may thus be deprived of educational opportunities. This is a serious effect, considering that education is important for the development of children, and consequently the country. Sometimes, dry conditions due to droughts can lead to wildfires, claiming many lives, including in developed countries.

The SREX had medium confidence that some regions had experienced more intense and longer droughts, including southern Europe and West Africa, but in other regions droughts had become less frequent, less intense or shorter; for example, in central North America.

6.9 Future Trends of Extreme Events and Adaptation

There has long been debate concerning the attribution of specific effects to anthropogenic climate change, even events as extreme as the 2003 European or 2010 Russian heatwaves. However, more recently, others have argued that because climate change is unequivocal, the null hypothesis should be reversed (Trenberth, 2011; Hansen *et al.*, 2012). That is, the task now is to prove there is *no* anthropogenic component to a particular observed change in climate.

Usually, climate change is considered using the mean value of the weather/climate variables. However, as depicted in Fig. 6.3, the frequency of extreme events can rise disproportionately if the mean, variability or symmetry of the distribution is altered. Because of this, models based on the climate mechanism are preferable to alternatives that extrapolate the statistical relation between climate variables and extreme events.

Exposure to disaster risk is also changing. For example, the average physical exposure to floods by region reported in SREX shows the absolute increase in exposure to be highest in Asia, but relatively highest in Africa, due mainly to projected population increase. Because some hazards are also increasing, the total impact will be even higher.

6.10 Conflicts in Adaptation

International conflicts in adaptation and mitigation, for example, in the areas of water management, can occur. Even after decreased river flow due to climate change, for example, an upstream country may capture the same amount of water, consequently reducing downstream water supply.

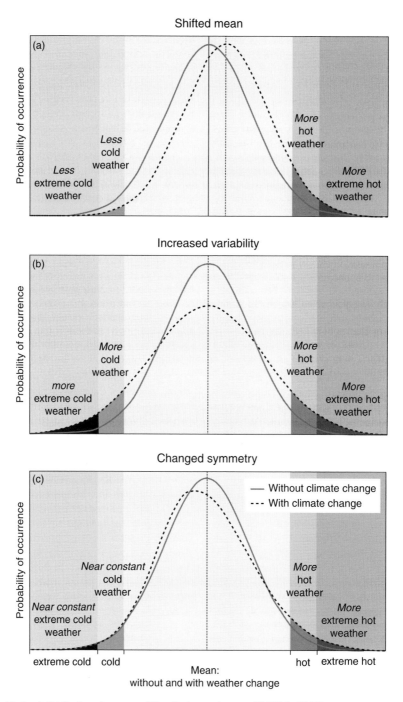

Fig. 6.3. Mode of distribution change and its effect on extremes (© IPCC, 2012).

6.11   Conclusion

Some commentators have suggested that inter-governmental reports such as the IPCC and the Millennium Ecosystem Assessment are more conservative than the general scientific litera-ture (Butler, 2005), which itself may be biased towards caution and optimism, on issues such as climate change (Hansen, 2007), future food security and future economic prosperity.

This is plausible, for several reasons. First, humans are biased towards optimism (Sharot, 2011), and this may influence science too. Reports such as those by the IPCC involve many authors from many nations and cultures. Selection bias may steer the composition of IPCC panels towards more mainstream views. Participants with outlying views are unlikely to become prominent, and their views may not be expressed, heard or mentioned in the final report, which is likely to be closer to the average position of the writing team. Another reason for conservatism is that IPCC reports are open to review, and by representatives of governments, prior to release. Governments of countries who perceive themselves at special risk due to the effects of climate change (e.g. low-lying island states) may be perceived as having an incentive to exaggerate the danger of climate change. However, such countries are comparatively small in population and limited in economic and political power.

In the recent past, and even in some cases today, the most populous and powerful govern-ments have sought to minimize the risk of cli-mate change. If there is no major risk, it follows that there is little benefit from aggressive mitiga-tion, that is from the lowering of carbon emis-sions (seen by most nations as essential for economic growth and the reduction of poverty), in order to reduce their acknowledged responsi-bility to mitigate. This is done because almost all major developed and developing countries hold the view that to act to lower their greenhouse emissions substantially will harm their economy and hinder development (Vihma, 2011).

Despite these reasons for conservatism, many IPCC reports have been characterized, by critics ideologically opposed to the reality and risks of climate change, as alarmist. But the strength of the IPCC reports is based on their conservatism and consensus. They still contain ample evidence to demand preventative action.

The extreme events described in this chapter have a potential for huge economical loss. Even when direct health impacts appear small, mas-sive economic damage may eventually cause secondary and tertiary impacts on health, such as undernutrition, mental health disorder, suicide and even war. Comprehensive understanding of the climate change impact combined with international collaboration are sorely needed to prepare better solutions for the health, economic, environmental and social problems caused by extreme events induced by climate change.

Notes

[1] The report also discusses extreme physical events, unrelated to weather or climate, such as earthquakes and tsunamis.

[2] A term referring to the degree of energy affecting the Earth system; these can be external, such as from the sun, or internal, such as heat-trapping 'greenhouse' gases or heat-shielding substances such as particulate matter from dusts or aerosols.

[3] The purchasing power in different countries of tradable currencies (such as the US dollar) varies not only by exchange rate but also by the wage structure, living standards and the quality of public goods. These differences are partially reflected in statistical data, sometimes called 'international dollars' or purchasing power parity (PPP) currency. In general, PPP costs are 2–4 times higher in developing countries than in tradable currency.

References

Ahern, M., Kovats, R.S., Wilkinson, P., Few, R. and Matthies, F. (2005) Global health impacts of floods: epidemiologic evidence. *Epidemiologic Reviews* 27, 36–46.
Allan, R.P. (2011) Human influence on rainfall. *Nature* 470, 344–345.

Barthel, F. and Neumayer, E. (2012) A trend analysis of normalized insured damage from natural disasters. *Climatic Change* 113, 215–237.

Butler, C.D. (2005) Peering into the fog: ecologic change, human affairs and the future. *EcoHealth* 2, 17–21.

Chin, C.-S., Sorenson, J., Harris, J.B., Robins, W.P., Charles, R.C., Jean-Charles, R.R., *et al.* (2011) The origin of the Haitian cholera outbreak strain. *The New England Journal of Medicine* 364, 33–42.

CRED (Centre for Research on the Epidemiology of Disasters) (2009) EM-DAT, Emergency Events Database. Centre for Research on the Epidemiology of Disasters (CRED), Université Catholique de Louvain, Brussels (www.emdat.be/).

Epstein, P. (1999) Climate and health. *Science* 285, 347–348.

Farmer, P., Almazor, C.P., Bahnsen, E.T., Barry, D., Bazile, J., Bloom, B.R., *et al.* (2011) Meeting cholera's challenge to Haiti and the world: a joint statement on cholera prevention and care. *PLoS Neglected Tropical Diseases* 5, e1145.

French, J., Ing, R., von Allmen, S. and Wood, R. (1983) Mortality from flash floods: a review of national weather service reports, 1969–81. *Public Health Reports* 98, 584–588.

Guha-Sapir, D., Vos, F., Below, R. and Ponserre, S. (2011) *Annual Disaster Statistical Review 2010: The Numbers and Trends.* Centre for Research on the Epidemiology of Disasters, Université Catholique de Louvain, Brussels, Belgium, 42 pp (www.cred.be/sites/default/files/ADSR_2010.pdf, accessed 3 March 2014).

Hansen, J. (2007) Scientific reticence and sea level rise. *Environmental Research Letters* 2, 024002, doi:10.1088/1748-9326/2/2/024002.

Hansen, J., Sato, M. and Ruedy, R. (2012) Perception of climate change. *Proceedings of the National Academy of Sciences* 109, 14726–14727.

IPCC (2012) *Intergovernmental Panel on Climate Change Special Report on Managing the Risks of Extreme Events and Disasters to Advance Climate Change Adaptation* (Field, C.B., Barros, V., Stocker, T.F., Qin, D., Dokken, D.J., Ebi, K.L., *et al.* (eds)). Cambridge University Press, Cambridge, UK, and New York.

Jones, N. (2013) Wastewater injection cracks open quake concerns. *Nature Geoscience* 6, 329.

Kerr, R.A. and Stone, R. (2009) A human trigger for the great quake of Sichuan? *Science* 323, 322.

Knowlton, K., Rotkin-Ellman, M. and Sheffield, P. (2013) Post-Sandy preparedness policies lag as sea levels rise. *Environmental Health Perspectives* 121, a208–a209.

Kubiszewski, I., Costanza, R., Franco, C., Lawn, P., Talberth, J., Jackson, T., *et al.* (2013) Beyond GDP: measuring and achieving global genuine progress. *Ecological Economics* 93, 57–68.

Lavers, D.A., Allan, R.P., Villarini, G., Lloyd-Hughes, B., Brayshaw, D.J. and Wade, A.J. (2013) Future changes in atmospheric rivers and their implications for winter flooding in Britain. *Environmental Research Letters* 8, 034010, doi:10.1088/1748-9326/8/3/034010.

Neria, Y., Nandi, A. and Galea, S. (2008) Post-traumatic stress disorder following disasters: a systematic review. *Psychological Medicine* 38, 467–480.

Peduzzi, P., Chatenoux, B., Dao, H., Bono, A.D., Herold, C., Kossin, J., *et al.* (2012) Global trends in tropical cyclone risk. *Nature Climate Change* 2, 289–294.

Sharot, T. (2011) *The Optimism Bias.* Pantheon Books, New York.

Takenaka, S., Satoh, T. and Lhamo, S. (2012) A social survey for GLOF disaster mitigation in Bhutan. *Global Environmental Research* 16, 77–82.

Trenberth, K.E. (2011) Attribution of climate variations and trends to human influences and natural variability. *WIREs Climate Change* 2, 925–930.

Vihma, A. (2011) India and the global climate governance: between principles and pragmatism. *The Journal of Environment and Development* 20, 69–94.

Yang, P., Yen, C.-F., Tang, T.-C., Chen, C.-S., Yang, R.-C., Huang, M.-S., *et al.* (2011) Posttraumatic stress disorder in adolescents after Typhoon Morakot-associated mudslides. *Journal of Anxiety Disorders* 25, 362–368.

Yokoi, S. and Takayabu, Y.N. (2010) Environmental and external factors in the genesis of tropical Cyclone Nargis in April 2008 over the Bay of Bengal. *Journal of the Meteorological Society of Japan* 88, 425–435.

Zoback, M.D. and Gorelick, S.M. (2012) Earthquake triggering and large-scale geologic storage of carbon dioxide. *Proceedings of the National Academy of Sciences* doi:10.1073/pnas.1202473109.

7 Global Warming and Malaria in Tropical Highlands – An Estimation of Ethiopia's 'Unmitigated' Annual Malaria Burden in the 21st Century

Menno Jan Bouma[1] and Mercedes Pascual[2]
[1]*London School of Hygiene and Tropical Medicine, London, UK;* [2]*Department of Ecology and Evolutionary Biology, University of Michigan, Ann Arbor, Michigan, USA, and Howard Hughes Medical Institute, Chevy Chase, Maryland, USA*

7.1 Introduction

Malaria arguably is the disease with the most striking and lasting impact on human history. With the advent of science and economic development, it has made a remarkable transition between the 19th and the 21st century. Its near global distribution has shrunk dramatically from high latitudes, including countries such as Sweden and Canada to the warmer tropics and subtropics. In such areas, it is predominantly a disease of poverty, but in earlier days kings died ignominious malaria deaths, and attending cardinals at Vatican conclaves could find their numbers decimated. Now, parts of the tropical world are considered trapped in poverty partly because of malaria; resolving the malaria problem is seen as a key to prosperity.

The latitudinal and altitudinal restrictions of malaria in the 19th century pointed to temperature as an important determinant, in addition to water and humidity, a combination encountered in proximity of marshes, emanating the bad odours the name of the disease refers to.[1] In the first half of the 20th century, the effects of temperature on the parasite and different stages of the vector were quantified, the parasite being sensitive during its cold-blooded sexual stage in the female mosquito, with a threshold temperature that varies between malaria species. Above critical temperatures, development of the vector's aquatic stages is shortened, reducing the generation time. Higher temperatures also quicken the digestion of the blood meal and maturation of its developing eggs, thus increasing vectorial biting frequency. Given these well-established, mainly laboratory determined climate sensitivities, malaria has long been identified as the infectious disease most vulnerable to climate change (WHO, 1990), and has been given a high profile as an 'indicator disease' to judge whether climate change poses a real world threat to health and the burden of other putatively temperature-sensitive diseases. The equivocal results since published make it seem that this 'canary in the mine' (Randolph, 2009) has not yet sickened, and the 2007 Intergovernmental Panel on Climate Change (IPCC) report was more cautious than its predecessors regarding malaria's prospective climate change burden (Confalonieri *et al.*, 2007). However, the supposed 'missing' warming trend in the African highlands that pointed at other causes (such as drug resistance) for the observed increase of malaria (Hay *et al.*, 2002) proved, in fact, to exist (Pascual *et al.*, 2006; Stern *et al.*, 2011). Additionally, vectors were shown to be exquisitely sensitive to small temperature rises. With warming now beyond dispute, the *global decline* of malaria in the 20th century is now used by sceptics, the argument in reverse, to signify that malaria's climate determinants have faded (Gething *et al.*, 2010).

This chapter clarifies some of these seemingly conflicting observations and focuses on densely populated highlands, long considered malaria safe havens in the tropics. In Africa, Ethiopia has the largest highland population, and with exceptionally reliable data, and with a single main vector, is an ideal location to investigate malaria's climate sensitivity and its vulnerability to warming.

Rainfall and related parameters such as humidity are of great significance for malaria epidemiology; however, their prediction in climate change models is far less robust than for temperature. But in the highlands, which usually receive more rainfall, temperature is more important for limiting or permitting transmission. We will show that the multifactorial complexity of malaria that complicates its relationship with climate at lower altitudes and on a global (macro) scale is much simpler (strongly temperature dependent) at the regional highland scale in Ethiopia.

However, in this chapter, we will not discuss further the possibility that any increased regional malaria burden might be balanced by declines elsewhere, particularly at the warmer limits of its distribution (Lafferty, 2009). That argument does not inform the central scientific question of whether climate change is altering malaria epidemiology. It further fails to account for population distribution, because even if the 'winners and losers' are cancelled out, it is still important to understand and hence ameliorate predicted increases (Pascual and Bouma, 2009). Instead, we will address the spatial issues of warming at the cool (in the highlands, that is the highly populated) tail of the malaria distribution.

7.2   Climate and Global Malaria Distribution; From the Historical 'Latitudinal' to its Present 'Control Fringes'

Both the malaria vector, around 60 species of the *Anopheline* genus with a wide geographical distribution, and the parasite occupying the vector for the duration of the sexual part (sporogony) of its life cycle, are temperature dependent. Permissive temperatures are in many regions remarkably similar for vector and parasite, and this may reflect their long co-evolution. Within the temperature boundaries of malaria's ecological niche, the temperature-dependent processes are usually non-linear. That is, small increases of temperatures can have disproportionately large biological consequences for parasite and vector. For example, the sensitivity of these responses around the lower critical value has been studied extensively in the first half of the 20th century, in part to account for malaria's global distribution and its natural decline in higher latitudes and higher altitudes in endemic regions. Development time decreases when temperatures exceed the lower threshold, for both parasite and vector, including through more rapid development of the vector's aquatic stages, and reproduction (egg development, and therefore the interval of egg laying and biting). In recent years, perhaps in search of a more favourable climate change consequence, the upper limit of malaria's niche has also received attention. The development times and critical temperatures determine malaria's possible geographical distribution and transmission efficiency, thus playing an important part in malaria's burden, or, where transmission is suppressed, potential burden.

Other features of malaria dynamics, such as vectorial feeding preferences, are also important. The transmission capacity of vectors at higher latitudes, particularly in the northern hemisphere, is much reduced because of their preference to feed on domestic animals (zoophilia).[2] The 20th century has seen a striking shrinking of malaria distribution at higher latitudes. The inherent instability of malaria, and the economic development reflected in improved housing, animal husbandry practices and other public health defences, especially the increased ability to control vectors, are held responsible for this decline. Figure 7.1 shows malaria's potential wide geographical distribution, broadly extending north between the 60°F (15.6°C) and 70°F (21.1°C) isotherm of the warmest month. This extension was not reached in the southern hemisphere, due to the absence of indoor hibernating vectors, permitting malaria transmission far longer than outdoor temperatures would suggest.

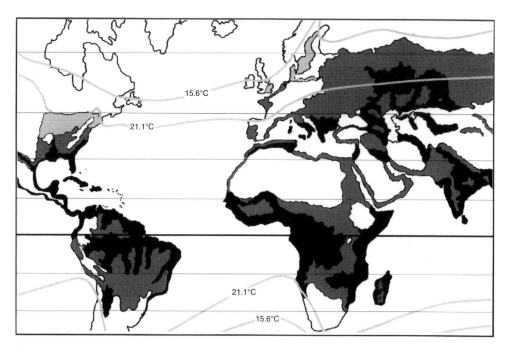

Fig. 7.1. Malaria distribution in its very early stages of reduction (adapted from Boyd, 1930). Note the absence of malaria (white) in the tropical highlands (Ethiopia, the Andes and Himalayas) and desert regions (Sahara, Kalahari, Arabian and Gobi), and areas where malaria had already receded (light grey). Black areas indicate intense transmission, and dark grey areas display areas where transmission is less intense. Also shown are the 60°F (15.6°C) and 70°F (21.1°C) summer isotherms.

Of the four truly human malaria species, *Plasmodium vivax* and *Plasmodium falciparum* are the most widely distributed. *P. vivax* has a wider distribution in colder regions, as the parasite can develop at lower temperatures and has the capacity to remain viable but hidden in infected patients' liver cells during periods when its chance to infect a mosquito is slim. The natural poleward decline of malaria was noticed long before proof of its vector-borne transmission. Drake (1850) was probably the first epidemiologist to quantify this striking temperature dependence, using data from the characteristics of US army posts, typified by uniform living conditions, standards of treatment, diagnosis and reporting (see Fig. 7.2). These data showed the shrinking range of malaria with increasing latitude. Drake observed that this relation was most striking in posts along the Mississippi, which experienced very comparable conditions, other than ambient temperature. The data from all US military posts at lower latitudes introduce excessive variability, reflecting a multitude of local determinants and thus obscuring the relationship. Strikingly, Drake correctly assumed a biological agent rather than a 'miasma' respecting the 60°F (15.6°C) isotherm.

The decline of malaria, particularly in the first half of the 20th century in the USA and Europe, has been associated with socio-economic development and changed animal husbandry practices. In more developed societies, humans are increasingly separated from domestic animals and their zoophilic *Anophelines*. After World War II, intervention with dichlorodiphenyltrichloroethane (DDT) further aided the establishment of latitudinal 'control fringes' shaped by public health defences rather than by climate. These fringes, however, remain sensitive to societal stress and the breakdown of public services, conditions that could also result from the tertiary consequences of climate change.

In recent decades, local malaria transmission has recurred in parts of the former Soviet Union states of Tajikistan, Azerbaijan, Armenia and, more recently, in Greece. Around the current

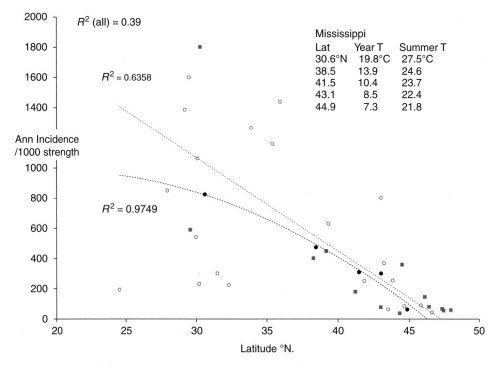

Fig. 7.2. Malaria incidence in mid-19th century US army posts (1870–1874). This figure shows all military stations, posts on the Mississippi river (black circles) with regression (black curved) line and combined posts on the Mississippi and Missouri (grey squares), linear (grey) regression line.

'control fringes', favourable transmission conditions such as rainfall anomalies may facilitate outbreaks of malaria, but temperature does not appear to play a significant role (see Fig. 7.2) in the spatial distribution of endemic malaria in regions below 40°N in the USA, which correspond with even higher latitude in Europe.

7.3   Warming and Malaria Risk in the Higher Latitudes – An Alarm for Developed Countries?

We have shown above that the inhibitory effect of temperature on malaria at higher latitudes requires unusual uniform conditions (military posts on the same Mississippi river) to become noticeable. For at least 50 years, the parasite has been unable to reproduce in vast high-latitude territories. The suitability of conditions for malaria transmission in such regions is unclear, including whether and what degree of future warming might tip the odds in favour of the parasite.

A major concern of malaria climate change research occurred in the 1990s. This resulted from publications that spatially applied projected future climate values to climate-sensitive malaria parameters used in the equation to derive its basic reproduction rate (R_0), a mathematical expression of transmission efficiency. R_0 is the number of cases resulting from an index case in a completely susceptible population. To attempt assessing the impact of climate change on malaria, R_0 was simplified by removing several factors from the equation. Removed were the vector density (though climate sensitive, this is hard to project) and the human malaria recovery rate (dependent on treatment efficacy). The resultant index is referred to as the 'epidemic' or 'transmission potential'. The relative changes (rather than the still very small absolute ones) of this parameter as compared to climate in 1990 showed striking increases, particularly in northern hemisphere developed countries (Martens *et al.*, 1994, 1999; van Lieshout *et al.*, 2004). These projections were widely publicized, and raised, particularly through the lay press, a state of alarm. For example, in the Netherlands, an impression developed that mass malaria chemoprophylaxis was around the corner. The caveats to the conclusions provided by these researchers (e.g. the importance of public health defences) were insufficient to avoid strong peer criticism, generating an acrimonious debate that still lingers.

On a global (low-resolution) scale, the potential expansion of malaria to vast high-latitude areas (that are, in fact, least vulnerable) is prominent, but the high-risk, densely populated tropical highlands are scarcely noticeable. Taking into account the rapid removal of infected patients (carriers) in more developed countries through early diagnosis and treatment (Bouma, 2003), and the 'divorce' between cattle and humans previously mentioned, the R_0 at high latitudes would be three or more orders of magnitude lower than the already marginal values observed early in the 20th century. The temperature-related *absolute* increases of the transmission potential in the higher latitudes are, indeed, trivial (Gething *et al.*, 2010), and climate change cannot therefore realistically threaten such developed regions without a dramatic drop in living standards. Interestingly, Kiszewski *et al.* (2004) used the absolute figures (spatially) from the same simplification of the R_0 as Martens *et al.* (1995, 1999) to denote the 'stability' of malaria, that is the perceived amount of effort it would take to control the disease.

Statistical climate-based malaria modelling on a global scale (Rogers and Randolph, 2000), an approach that does not explicitly incorporate biological processes, suggested on balance little or no change in malarial territories due to climate change predicted this century. But, while this study supported the view that developed high-latitude countries had little to fear from warming, it also showed that tropical highland areas would become more malarious. However, on the global scale, these regions appear as mere specks, highlighting, in our view, the unsuitability of the global scale for studying malaria's climate vulnerability. Such a scale does not support any general insensitivity of malaria to climate or climate change. Use of this scale has generated some damaging misconceptions. The declining malaria trend at higher latitudes, despite rising temperatures, has been used to argue

the irrelevance of climate (change) for malaria globally (Gething *et al.*, 2010). But, while correct for higher latitudes (as we, too, have argued), this does not generalize to tropical malaria (Bouma *et al.*, 2011), particularly in highland areas.

7.4 Malaria Climate Suitability and Biogeographically Based Climate Change Impact Assessment in Africa

Numerous time-series analyses have supported the hypothesis that rising temperatures have contributed to the rise of malaria observed in East African highlands from the 1980s (Loevinsohn, 1994; Tulu, 1996; Bouma, 2003; Zhou *et al.*, 2004; Pascual *et al.*, 2006). But these studies can only estimate the climate change associated burden for a single area; they may not be broadly representative. Unfortunately, many of these time-series studies are not detrended, meaning that other alternative explanations such as increased drug resistance or lowered public health defences could be partially or completely explanatory. To distinguish the potential impact of climate change from other causal explanations requires a geographical spatial analysis that incorporates the distribution of malaria burden and population.

One such tool for this is biogeography. This discipline originated in the early 19th century, when the altitudinal stratification of vegetation led von Humboldt, on his South American journey, to conceive 'isotherms' and organism-specific 'niches' defined by climate conditions (von Humboldt and Bonpland, 1805). This explorer also mapped the distribution of the 'quinine tree'; climate determinants later used to establish Cinchona plantations in British India and the Dutch East Indies.

Recent expansion of computational power has enabled a rapid evolution of geographical information systems that can aid niche definition of species. Projected changes in climate can be applied spatially to assess future distributions. A change of a single degree Celsius corresponds to approximately 150 km in latitude, a warming signal observable for plants, bird and insect species.

To date, malaria climate change assessments have relied mainly on defining suitable conditions for parasite and vector, and estimating the resultant burden by comparing populations inhabiting newly suitable areas with inhabitants in regions that have become unsuitable. But this approach of counting the population at risk (PAR) does not consider the absolute malaria risk, and thus the actual disease burden. More importantly, ignoring worsening malaria in traditionally low malarious areas will underestimate significantly the burden from climate change.

The estimate of global (changes in) malaria suitability, the 'transmission potential' described above (Martens *et al.*, 1995, 1999) is determined mainly by the (non-linear) temperature suitability of the parasite and adequate rainfall for the vector. For Africa, another suitability index was developed for *P. falciparum* and its most important vector (complex), *Anopheles gambiae* (Craig *et al.*, 1999). This 'MARA' (mapping malaria risk in Africa) model is based on empirically observed thresholds and estimated optima with artificially smoothed edges using fuzzy logic, rather than determined by biological measured non-linear functions. These thresholds are: (i) 18–40°C mean diurnal air temperature for the parasite (optimum suitability between 22°C and 32°C); (ii) optimum rainfall of at least 80 mm/month; and (iii) a minimum annual temperature threshold of 4°C for aquatic vector reproduction (at least 6°C is optimum). For every month of the year, the lowest suitability value (most restricting condition) of the three parameters was selected, and the highest value of 5 consecutive suitable months (3 for North Africa) was used to calculate the annual 'transmission probability of stable malaria', expressed as a figure between 0 and 1. This Falciparum Climate Suitability (FCS) index produced a good match for Africa's malaria distribution, a reasonable match on country level, but did not reflect quantitative transmission intensity and prevalence satisfactorily, and hence local disease burden.

Using the FCS as 'state of the art' methodology, Hay *et al.* (2006) applied the predicted changes in climate by 2015 and 2030 (HADCM3, with chosen baselines centred around 1975 and 2005) to estimate changes in PAR in Africa. Only small increases were detected on the continental level, particularly in comparison to the effects of growing populations. Their worst-case scenario with increases of 13.1% and 14.6% (for 2015 and 2030, respectively) were found when the

HADCM3 projections (which are available only at the low resolution of 0.5 × 0.5 degrees) were artificially 'up-sampled' to match the higher resolution of the 1975 climatology. The authors recognized the low spatial resolution of current predictions as a major impediment and the up-sampling procedure they used as a major source of bias. However, using only the lower resolutions that do not suffer from this up-sampling bias, some interesting patterns emerge for data from individual highland countries. These are provided in their annex, but unfortunately are not discussed further in their paper. In the annex, four classes of risk are proposed for the spectrum of FCS: taking into account the population, the authors provided the percentage of the population in each risk class for each country. We have graphed these data for Ethiopia in Fig. 7.3.

The first columns in the figure show the sensitivity of the FCS to the resolution (pixel size) used, with the coarse resolution considerably underestimating the population in malaria-free areas. This proportion in the fine resolution is around 35%, corresponding with the expert estimates of the 1980s. It is obvious that the few (approximately 10) pixels which represent Ethiopia at low resolution do insufficient justice to its high spatial variability of altitude, climate, population density and malaria risk. The last three bars show, in coarse resolution, the striking changes in endemicity between 2005 and 2030 in Ethiopia. In another publication, an approximate doubling of all-cause and malaria-specific mortality between class 2 (epidemic prone) and classes 3 and 4 (seasonal and stable endemic malaria) is reported (Snow *et al.*, 2003). This would suggest that climate change between 2005 and 2030 would result in the doubling of the (all-cause) mortality for over 50% of Ethiopia's population. The coarse resolution may overestimate the higher-risk classes of malaria but it also underestimates Ethiopia's population in malaria-free areas, and further warming may seem to have little impact to worsen the malaria situation if PAR is used as the yardstick. These country data for Ethiopia are in striking contrast with the rather benign implications of climate change presented by Hay *et al.* (2006) for malaria on a continental level.

The only study that ventures beyond changes in the PAR, using the number of months with suitable conditions (person-months of exposure)

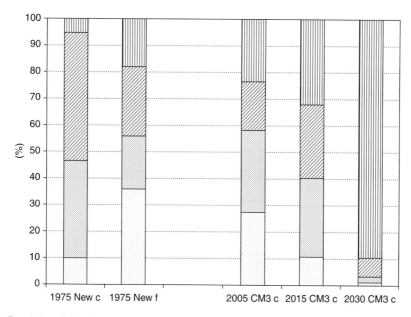

Fig. 7.3. Population distribution in Ethiopia. This is based on categories of suitability (FCS) in malaria-free areas (dotted, FCS = 0); epidemic-prone areas (grey, FCS <0.25); seasonal transmission (diagonal stripes, FCS >0.25 and <0.75); and stable endemic (vertical stripes, FCS >0.75). Climatology of 1975 (New *et al.*, 1999) in fine (f) and 'down-sampled' coarse (c) resolution. Climatology projections HADCM3 for 2001–2010 (2005), 2015 and 2030 in coarse (c) resolution. (Based on country data from Hay *et al.*, 2006.)

as a proxy for transmission intensity (Tanser *et al.*, 2003), indeed shows dramatic effects of climate change (HADCM3, with baseline centred around 1950), particularly for Africa's highlands and areas with pre-existing malaria transmission. Ethiopia emerges as the country worst affected, with increases in person-months of exposure (in 2070–2099) between 149% and 349% for the 'A2a' scenario with a 3-degree increase in temperature. Strikingly, and perhaps unsurprising in view of Drake's 19th century US findings (Fig. 7.2), very little impact was found in changes of the latitudinal distribution of malaria (Tanser *et al.*, 2003).

7.5 Estimation of Ethiopia's Climate Change Burden Using the Malaria Lapse Rate

Ethiopia has, in the higher parts of the country, very high population densities, contrasting sharply with its sparsely settled arid and semi-arid lowlands. The population density pattern closely follows that of the altitude (Fig. 7.4a and b) and

rainfall (not shown). In the 1960s, when Ethiopia was one of the few continental African nations to attempt malaria eradication, malaria was restricted mainly to areas below 1500 m and close to rivers and water bodies in the dry lowlands. Between 1000 and 2000 m, an inverse relation between malaria and population density can be observed. Ethiopia's 'human carrying capacity' appears to increase with reducing levels of malaria transmission, reflecting the impact of the altitudinal changes of the disease and its effect on mortality and fertility (Bouma *et al.*, 2014).

Historically, the *latitudinal* malaria gradient has been weak, and we have described how it further eroded in the 20th century due to public health advances. In contrast, the *altitudinal* gradient of malaria in the tropics has remained strong and resilient. The transmitting efficiency of local vectors has played an important role in maintaining this distinction. 'Lapse rates' on mountains is a geographical term that refers to changes of physical parameters with altitude and is commonly used for the change of temperature with altitude. We will hereafter refer to the malaria gradient as the 'malaria lapse rate'.

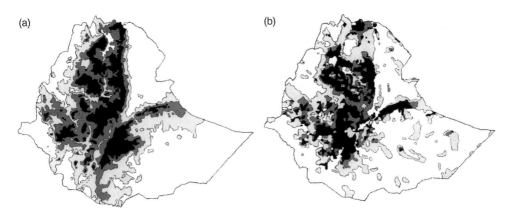

Fig. 7.4. Ethiopian (a) Altitude: 1000–1250 m (light grey); 1250–1750 m (dark grey); over 1750 m (black). (b) Population density (population per km²): 10–50 (light grey); 50–100 (dark grey); over 100 (black). The distribution of rainfall (not shown) follows a similar pattern. (After Cox *et al.*, 1999.)

In the decades following World War II, extensive malaria mass surveys were undertaken, mostly as feasibility studies for the ambitious malaria eradication efforts in the 1950s and 1960s, in an era of intense global health activity. These surveys, including in many mountainous African countries, provide prevalence data, usually in children, in order to classify endemicity. Figure 7.5 shows the Ethiopian malaria and temperature lapse rates. Even though the malaria prevalence surveys were conducted in different years, altitude alone explained 73% of the variability, and very few cases occurred over 2200 m. The dry, sparsely inhabited areas below 1000 m in Ethiopia are excluded from analysis as aridity is restrictive and the malaria lapse rate curve becomes less smooth.

Despite more vigorous intervention in more recent periods, we have demonstrated the persistence of malaria lapse rates using recent survey data for the highlands of Ethiopia, Kenya and Tanzania (Bouma *et al.*, 2011). The strongest gradient was found in the best surveyed regions. These findings challenge, at least for the highlands, the conclusion by some that temperature (climate) is no longer an important epidemiological determinant. The unusually high efficiency of the African vectors (*An. gambiae* and *Anopheles funestus*) play an important role in maintaining this gradient, and very low vector densities appear sufficient to maintain transmission. The observed malaria lapse rates are empirical, i.e. they vary between countries, depending on characteristics such as how the local temperature and rainfall themselves vary with altitude, latitude and the environmental sensitivities of the local vector and malaria parasite.

Montane biotopes are rich in species diversity, because of the exceptional range of ecological niches resulting from steep temperature (and to a lesser extent rainfall) gradients over short distances. Insects, including malaria vectors, are apt to bridge distances between discontinuous but suitable niches on the wing or carried by the wind. This feature can allow rapid colonization of new territory if altered environmental conditions permit. The close temperature dependence of vector and parasite means that differences in ambient temperatures can be expected to shift the malaria lapse rate to higher altitudes in warmer years and when temperatures gradually warm due to climate change.

Rapid spatial changes in vector breeding and related epidemics have previously been observed in African highlands during exceptionally warm years (Garnham, 1945), and spatial shifts in altitudinal malaria distribution can also be discerned in Africa and South American highlands (Siraj *et al.*, 2014). Importantly, the malaria lapse rate is a quantitative suitability curve relating altitude (temperature) directly to prevalence. In contrast with the indices described earlier, this suitability index, in combination with the human altitudinal distribution, can generate a true temperature-dependent disease burden. We have used the malaria and

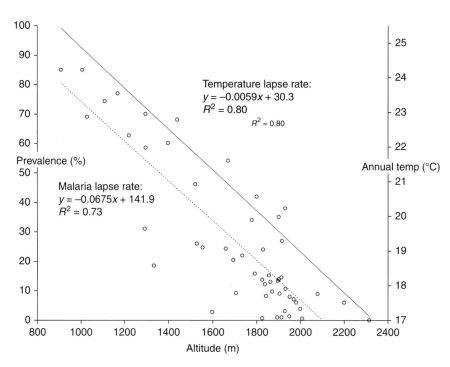

Fig. 7.5. The malaria lapse rate relating prevalence (circles) with altitude in Ethiopia and the regression coefficients for malaria (stippled) and temperature (grey). These data (reviewed by Melville *et al.*, 1945) were collected in an era with very limited human intervention (neither effective treatment nor vector control), and hence represent the 'natural state', an estimation of the potential malaria burden.

the local temperature lapse rates in Ethiopia to calculate the impact of 1, 2 and 3°C of global warming. Rising temperatures mean that stationary human populations will be exposed to higher levels of malaria transmission, and the malaria burden can be calculated as a function of temperature change.

In the absence of water vapour, the decrease of temperature with altitude is strictly linear. A value of 0.5°C/100 m is a common rule of thumb. In view of the temporal rainfall distribution in Ethiopia, we calculate that the lapse rate for Ethiopia during malaria's transmission months is between 0.52°C and 0.59°C/100 m. For every degree rise in environmental temperatures, this equates to an upward shift of the malaria niche (suitability distribution) of 169–192 m. A 2°C rise would shift the temperature range 338–384 m. For the calculations below, we used 182, 364 and 546 m for 1, 2 and 3°C rises in temperature.

Figure 7.6 shows the population size for every 50 m and the corresponding prevalence

for that altitude, between 1000 and 3000 m, where 80% of Ethiopia's population resides. If we assume, conservatively, that prevalence equates to incidence (i.e. a malaria carrier suffers from one attack per year in a short transmission season), the disease burden is the product of the population and the corresponding prevalence for every 'bin'. The sum of all bins reflects the total annual disease burden. Temperature-dependent shifts in the malaria lapse rate (1–3°C) produce different total annual burdens (see Table 7.1).

As the population in Ethiopia below 1000 m is low, an increase of 2°C will almost double Ethiopia's malaria burden. Ethiopia's population is also expected to double in the next 50 years, thus creating additional stresses, assuming there are no altitudinal population shifts. Suitable areas for habitation in excess of 3000 m are very restricted, limiting the scope for population shifts to higher latitudes as an adaptation to warming. Since the 1980s, temperatures in the region have already increased about 1°C, adding

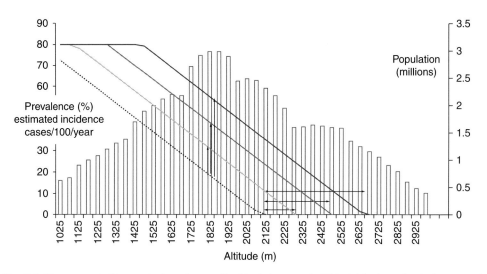

Fig. 7.6. The pre-control lapse rate (prevalence with altitude between in Ethiopia) is shown by the interrupted line and the shifts with 1°C (light grey), 2°C (darker grey) and 3°C (darkest grey) (based on a local temperature lapse rate of 182 m/°C). Columns represent the estimated population (2012) in 50 m 'bins'. Double arrows show increases in malarious territory (500 m for 3°C of warming), and the single arrows show increases in intensity of malaria transmission between 1800 and 1850 m, with 1°C, 2°C and 3°C of global warming.

Table 7.1. Malaria cases (additional) in Ethiopia (millions) with 1°C, 2°C and 3°C increase (based on 2012 populations).

Malaria cases Ethiopia (millions)	Annual cases	Additional cases/year (95% CI regression coefficient malaria lapse rate)	Change of cases with baseline (%)	Total population in malaria-free areas (%)
Baseline	12.29			35.3
1°C warming		5.61 (4.85–6.08)	46	22.6
2°C warming		11.19 (9.74–12.18)	91	18.3
3°C warming		16.23 (14.35–17.49)	132	10.7

an estimated 5.61 million global warming attributable cases. The rise in temperature may indeed account for the striking increase in reported malaria cases in Ethiopia since the 1990s.

7.6 Closing Remarks

This chapter has focused on spatial patterns of malaria, especially those of altitudinal shifts, with climate change, emphasizing the need to consider the effect of warmer temperatures on the disease from a regional context rather than a global one. In the latter, different regions at the edge of the disease distribution such as latitudinal and altitudinal fringes are not sufficiently distinguished, despite clear geographic, demographic and biological differences. Previous suitability indexes have been restricted to climate change associated changes in the population at (any) risk. Without specifying the absolute risk, such impact assessments have limited value. For the Ethiopian highlands of Africa, the expected increase in disease burden with warmer temperatures was estimated based on simple but robust lapse rate patterns.

The current effect of temperature on malaria in the African highlands is partly obscured by unprecedented control efforts, guided and supported by foreign aid, with mandatory insecticide spraying in highland fringes. This impact makes it particularly important that effective intervention, including the timely substitution of medication and insecticides when resistance develops, is integral and sustained. If not, malaria's full highland burden due to global change will inevitably surface.

Acknowledgements

The authors wish to acknowledge Dr Jonathon Cox, for generating the altitudinal distribution of the population in Ethiopia, and Professor Colin D. Butler, for various editorial suggestions. We also thank an anonymous reviewer.

Notes

[1] An early European name for malaria was paludism, derived from Latin for swamp.
[2] Human latitudinal expansion in the northern hemisphere was, in many cases, accompanied by and facilitated by cattle herding, and we believe that this facilitated the expanded range of *Anopheline* vectors which were zoophilic.

References

Bouma, M.J. (2003) Methodological problems and amendments to demonstrate effects of temperature on the epidemiology of malaria: a new perspective on the highland epidemics in Madagascar, 1972–1989. *Transactions of the Royal Society of Tropical Medicine and Hygiene* 97, 133–139.

Bouma, M.J., Baeza, A., Ter Veen, A. and Pascual, M. (2011) Global malaria maps and climate change: a focus on East African highlands. *Trends in Parasitology* 27, 421–422.

Boyd, M. (1930) *An Introduction to Malariology.* Harvard University Press, Cambridge, Massachusetts.

Confalonieri, U., Menne, B., Akhtar, R., Ebi, K.L., Hauengue, M., Kovats, R.S., *et al.* (2007) Human health. In: Parry, M.L., Canziani, O.F., Palutikof, J.P., Linden, P.J.vd., Hanson, C.E. (eds) *Climate Change 2007: Impacts, Adaptation and Vulnerability.* Contribution of Working Group II to the Fourth Assessment Report of the Intergovernmental Panel on Climate Change. Cambridge University Press, Cambridge, UK, pp. 391–431.

Cox, J., Craig, M., Le Suer, D. and Sharp, B. (1999) Malaria mapping risk in the highlands of Africa. Mapping Malaria Risk in Africa (MARA)/Epidemiology of Highland Malaria in Africa Project (HIMAL) Technical Report. UNDP/World Bank/WHO Special Programme for Research and Training in Tropical Diseases, Geneva, Swizerland, and International Development Research Centre, Ottawa, Canada.

Craig, M.H., Snow, R.W. and Le Sueur, D. (1999) A climate-based distribution model of malaria transmission in sub-Saharan Africa. *Parasitology Today* 15, 105–111.

Drake, D. (1850) *A Systematic Treatise Historical, Etiological and Practical on the Principal Diseases of the Interior Valley of North America: 1850.* Kessinger Publishing, Cincinnati, Ohio.

Garnham, P.C.C. (1945) Malaria epidemics at exceptionally high altitudes in Kenya. *British Medical Journal* 2, 45–46.

Gething, P.W., Smith, D.L., Patil, A.P., Tatem, A.J., Snow, R.W. and Hay, S.I. (2010) Climate change and the global malaria recession. *Nature* 465, 342–346.

Hay, S.I., Cox, J., Rogers, D.J., Randolph, S.E., Stern, D.I., Shanks, G.D., *et al.* (2002) Regional warming and malaria resurgence [reply]. *Nature* 415, 628.

Hay, S.I., Tatem, A.J., Guerra, C.A. and Snow, R.W. (2006) *Population at Malaria Risk in Africa: 2005, 2015 and 2030.* Foresight Publications Department for Business Innovation & Skills (www.foresight.gov.uk).

Kiszewski, A., Mellinger, A., Spielman, A.,Malaney, P., Sachs, S.E. and Sachs, J. (2004) A global index representing the stability of malaria transmission. *The American Journal of Tropical Medicine and Hygiene* 70, 486–498.

Lafferty, K.D. (2009) The ecology of climate change and infectious diseases. *Ecology* 90, 888–900.

Loevinsohn, M. (1994) Climatic warming and increased malaria incidence in Rwanda. *The Lancet* 343, 714–718.

Martens, P., Kovats, R.S., Nijhofa, S., de Vries, P., Livermore, M.T.J., Bradley, D.J., *et al.* (1999) Climate change and future populations at risk of malaria. *Global Environmental Change* 9, S89–S107.

Martens, W., Rotmans, J. and Niessen, L. (1994) *Climate Change and Malaria Risk, An Integrated Modeling Approach.* GLOBO Report Series 3. RIVM, the Netherlands.

Martens, W.J., Niessen, L.W., Rotmans, J., Jetten, T.H. and McMichael, A.J. (1995) Potential impact of global climate change on malaria risk. *Environmental Health Perspectives* 103, 458–464.

Melville, A.R., Wilson, D.B., Glasgow, J.P. and Hocking, K.S. (1945) Malaria in Abyssinia. *East African Medical Journal* 22, 285–294.

New, M., Hulme, M. and Jones, P. (1999) Representing 20th century space-time climate variability. Part 1: Development of a 1961–90 mean monthly terrestrial climatology. *Journal of Climate* 12, 829–856.

Pascual, M. and Bouma, M.J. (2009) Do rising temperatures matter? *Ecology* 90, 906–912.

Pascual, M., Ahumada, J., Chaves, L., Rodo, X. and Bouma, M. (2006) Malaria resurgence in the East African highlands: temperature trends revisited. *Proceedings of the National Academy of Sciences* 103, 5829–5834.

Randolph, S.E. (2009) Perspectives on climate change impacts on infectious diseases. *Ecology* 90, 927–931.

Rogers, D.J. and Randolph, S.E. (2000) The global spread of malaria in a future, warmer world. *Science* 289, 1763–1766.

Siraj, A.S., Santos-Vega, M., Bouma, M.J., Yadeta, D., Carrascal, D.R., and Pascual, M. (2014) Altitudinal changes in malarial incidence in highlands of Ethiopia and Colombia. *Science* 343, 1154–1158.

Snow, R., Craig, M., Newton, C. and Steketee, R. (2003) *The Public Health Burden of P. falciparum in Africa: Deriving the Numbers.* The Disease Control Priorities Project, Working Paper No 11, Washington, DC.

Stern, D.I., Gething, P.W., Kabaria, C.W., Temperley, W.H., Noor, A.M., Okiro, E.A., *et al.* (2011) Temperature and malaria trends in highland East Africa. *PLoS ONE* 6(9), e24524.

Tanser, F.C., Sharp, B. and le Sueur, D. (2003) Potential effect of climate change on malaria transmission in Africa. *The Lancet* 362, 1792–1798.

Tulu, A. (1996) Determinants of malaria transmission in the highlands of Ethiopia: the impact of global warming on morbidity and mortality ascribed to malaria. PhD, London School of Hygiene and Tropical Medicine.

van Lieshout, M., Kovats, R.S., Livermore, M.T.J. and Martens, P. (2004) Climate change and malaria: analysis of the SRES climate and socio-economic scenarios. *Global Environmental Change* 14, 87–99.

von Humboldt, A. and Bonpland, A. (1805) *Essay sur la geographie des plantes.* Levrault, Paris.

WHO (World Health Organization) (1990) *Potential Health Effects of Climate Change.* WHO/PEP/90.10. WHO, Geneva, Switzerland.

Zhou, G., Minakawa, N., Githeko, A.K. and Yan, G. (2004) Association between climate variability and malaria epidemics in the East African highlands. *Proceedings of the National Academy of Sciences* 101, 2375–2380.

8 Dengue: Distribution and Transmission Dynamics with Climate Change

Hilary Bambrick[1] and Simon Hales[2]
[1]*Centre for Health Research, University of Western Sydney, New South Wales, Australia;*
[2]*University of Otago, Wellington, New Zealand*

8.1 Introduction

Dengue is a mosquito-borne disease (family: Flaviviridae) transmitted by the urban-dwelling mosquito *Aedes aegypti* and, less efficiently, by the cold-tolerant species *Aedes albopictus*. Approximately 390 million people living in tropical and subtropical regions are infected each year (Bhatt *et al.*, 2013). It affects mainly people living in urban areas and on the urban fringe because the primary vector, *Ae. aegypti*, is well adapted to urban environments and feeds preferentially on humans. Dengue manifests with a wide range of severity, from asymptomatic to life threatening. Dengue can only be treated symptomatically; there is no effective antiviral treatment. Current vaccines are only partially effective, and in any case may not be affordable for the majority.

Several factors must exist for dengue virus transmission to occur. Climatic conditions that are both appropriate for the biology and ecology of the vector and warm enough for virus replication are necessary – but not sufficient. Within these climatic constraints, biotic and societal factors determine whether or not the transmission of dengue actually occurs, and also its intensity. A viral source in a human is required for transmission to occur: the virus is not thought to 'overwinter' in the vector. In non-endemic regions with a suitable vector, the importation of the virus via, for example, infected tourists or returning workers is necessary for transmission. Human migration is not required if the virus is endemic. There also needs to be the presence of competent dengue vectors. Finally, there must be contact between these vectors and a susceptible (non-immune) human population. These factors are driven and shaped by social, economic and environmental circumstances.

8.2 Historical Distribution of Dengue and Dengue-like Illness

Determining dengue's historical distribution is difficult and somewhat speculative. Serological testing for the dengue virus became available in 1950, and some suspected outbreaks in preceding decades have been confirmed with retrospective testing. Many infectious diseases presenting with similar symptoms may be confused with dengue. These include: chikungunya, Rift Valley fever, Ross River fever, West Nile fever, yellow fever, malaria, leptospirosis, pappataci fever (sandfly fever), scarlet fever, typhoid fever, measles and influenza.

Africa and the Americas

Dengue or dengue-like outbreaks have been recorded throughout much of Africa since the

19th century, perhaps being less common in East Africa until the 1950s (Robinson, 1956). Dengue has been endemic in some African regions since the mid-20th century (van Rooyen and Rhodes, 1948), with only periodic outbreaks in others (Scott, 1943; Carey, 1971). In South and Central America, dengue cases were first reported in 1905, and it remains endemic throughout much of the region (Gratz and Knudsen, 1996). Transmission was interrupted in much of South America during an *Ae. aegypti* eradication campaign in the 1960s and 1970s, stimulated by fear of yellow fever, but by the 1980s, *Ae. aegypti* had returned (Aviles *et al.*, 1999, 2003; Vezzani and Carbajo, 2008). Uruguay was, until recently, the only South American country where locally transmitted (autochthonous) cases of dengue were absent. In the USA, the first reported outbreak of dengue-like illness occurred in 1780 in Philadelphia, Pennsylvania (Hirsch, 1883; Carey, 1971; Halstead, 1997; Rigau-Perez *et al.*, 1998; Rogers *et al.*, 2006). From the early 19th century, dengue-like illness appeared throughout the southern states (Hirsch, 1883; Gratz and Knudsen, 1996; Gubler, 1997; Kuno, 2009), with only sporadic cases reported from the central and northern states. Since 1940, most cases reported in the USA have been imported (with no local transmission), but there have been small, locally propagating outbreaks in several southern states (Gratz and Knudsen, 1996; Effler *et al.*, 2005; Trout *et al.*, 2010).

The Asia-Pacific

In the Asia-Pacific region, countries that were endemic for dengue in the mid-20th century included India, Sri Lanka, Thailand, Japan, China and Indonesia (van Rooyen and Rhodes, 1948), while periodic outbreaks were experienced across the Pacific islands (Gratz and Knudsen, 1996). In Australia, the first occurrence of dengue is reported to have occurred in 1873, imported by infected travellers from Mauritius, while the first local outbreaks occurred in 1879 and then 1885 in Townsville and Rockhampton in the state of Queensland. Queensland was, however, free from dengue for nearly three decades until 1981, but since then,

outbreaks have occurred almost annually (Kay *et al.*, 1984). A recent epidemic in Cairns and Townsville in 2008–2009 was much larger and more difficult to control than previous outbreaks (McCredie, 2009). In the state of New South Wales (NSW), the first cases are thought to have occurred in 1886 and again in 1898, possibly imported rather than locally transmitted. Several epidemics were reported in NSW between 1905 and 1942 (Lumley and Taylor, 1943; Hueston, 2004; Kuno, 2007). The Northern Territory and Western Australia are also reported as having experienced periodic outbreaks in the first half of the 20th century (Lumley and Taylor, 1943; Kuno, 2007).

Several countries that are not currently considered transmission areas have experienced dengue or dengue-like outbreaks in the past. For example, dengue appeared frequently in Japan from 1900 to 1945 (Kuno, 2007). The 1942–1945 outbreak was said to be one of the largest epidemics ever recorded in a temperate region, but has not recurred (Gratz and Knudsen, 1996).

Europe

In Europe, dengue occurred in Greece (Athens and Piraeus) in 1927–1928 and 1929–1933 (retrospectively serologically confirmed) (Kuno, 2009), and may have caused outbreaks in Spain and particularly in Seville (in 1784 and 1788) and Cadiz (in 1784 and 1785) (Hirsch, 1883; Gubler, 1997). Dengue was probably endemic in many eastern Mediterranean countries at the end of the 19th century, including in and around Lebanon, Turkey, Palestine and Syria (van Rooyen and Rhodes, 1948; Gratz and Knudsen, 1996).

8.3 Global Contraction and Expansion

Overall, at a global scale, and recognizing that exact boundaries of current and local transmission are difficult to establish, dengue appears to have undergone a geographical contraction over the past century (Fig. 8.1). However, over the past 50 years, the incidence of dengue has increased dramatically in many countries. Furthermore, local transmission has been reported for the first time in recent years in several new countries, including several in Africa, the Portuguese island of Madeira, Bhutan (Dorji *et al.*, 2009), Nepal (Pandey *et al.*, 2008), France (La Ruche *et al.*, 2010) and Croatia. Dengue has also recently re-emerged in Hawaii after many decades of absence (Effler *et al.*, 2005) and Florida, USA (Trout *et al.*, 2010).

There are several possible reasons to explain why dengue is occurring in new areas and re-emerging in others. These include increases in poor, urban populations, increased travel and trade, environmental disturbances (Gubler, 1997; Pinheiro and Corber, 1997; Patz *et al.*, 1998; Rigau-Perez *et al.*, 1998) and a breakdown in public health measures, including discontinuation of vector control programmes (Thai and Anders, 2011). This last possible reason may be particularly important because successful vector control over a number of years would have, in time, reduced population immunity through limiting viral exposure. If the vector subsequently returns, such populations are relatively serologically naïve and especially susceptible to an epidemic (Oki and Yamamoto, 2012). An already changing climate, bringing warmer and sometimes wetter weather, may also be contributing to geographic expansion

and areas with re-emerging risk (Kovats *et al.*, 2001; McMichael *et al.*, 2006).

8.4   Dengue, Climate and Climate Change

Climatic conditions provide an absolute constraint on the geographic area suitable for dengue transmission, and the climate change observed in recent decades may have contributed to increased incidence, expansion into new areas and viral re-emergence in others. However, debate remains about the potential role of climate change as a contributor to global trends in vector-borne diseases, both recently observed and potential future patterns (Kovats *et al.*, 2001; McMichael *et al.*, 2006). Among the multiple causal factors, proof of a role for climate change is challenging; but disproving it may be even harder.

Regardless of other factors, climate directly constrains vectorial capacity, the ability of a vector to transmit a virus. Generally speaking, and within limits, a warmer, wetter climate increases vectorial capacity, and thus the potential for disease transmission. Two main approaches are taken to estimate the effect of changing climate parameters on dengue transmission: either theoretical models or empirical models (Box 8.1). In order to estimate potential future transmission under climate change, these models include

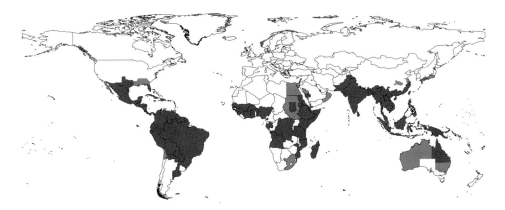

Fig. 8.1. Geographic distribution of dengue. Estimated regions of dengue transmission for 1975–present are shown in dark grey; additional regions of dengue transmission for 1900–1975 are shown in light grey. The map is based on a systematic review (Van Kleef *et al.*, 2010) and a recent map published by WHO (WHO, 2009). Where there were differences, the WHO map was assumed to be correct.

Box 8.1

Transmission of dengue is sensitive to climate. Temperature, rainfall and humidity affect the breeding cycle, survival and biting rate of the mosquito vectors. In general, higher ambient temperatures favour rapid vector development, increase the frequency of vectorial feeding and reduce the extrinsic incubation period (EIP) of the virus. The EIP is the time taken between the ingestion of an infective blood meal by the vector and the vector's capacity to transmit the virus in a later feed. (This is similar to malaria, in that the life cycle of the virus determines its capacity to be infective.) A shorter EIP increases the opportunities for virus transmission during the lifetime of an infected mosquito. If the ambient temperature is either too low or too high, then mosquitoes are unlikely to survive long enough to become infectious and transmit dengue (Gubler, 1997).

The findings of theoretical and empirical models of dengue and climate are broadly consistent: transmission of dengue is highly sensitive to climate, and relatively small increases in temperature (of the order of 1°C) can lead to substantial increases in transmission potential. While no studies have yet modelled the effect of long-term climate trends on dengue transmission over the past century, on theoretical grounds, observed changes in temperature will have increased the risk of dengue in some areas.

Theoretical models

Jetten and Focks (1997) modelled vectorial capacity (VC), defined as the average number of potentially infective contacts by a mosquito population per infectious person per year:

$$VC = mbca^2 P^n/-log^e(P)$$

where m is the number of female mosquitoes per person; b is the probability that an infectious mosquito transmits dengue while biting a susceptible human; c is the probability that a mosquito acquires a dengue infection while biting a viraemic human; a is the number of bites per person per day; P is the survival rate of the mosquito; and n is the EIP. Setting VC = 1, the epidemic potential (EP) is the reciprocal of the critical vector density (Jetten and Focks, 1997).

With a 2–4°C increase in temperature, both the range of transmission (latitude and altitude) and the length of the transmission season increase. This also decreases the average age of human infection, which could, in turn, increase cases of severe (haemorrhagic) dengue owing to a greater likelihood of multiple infections over a lifetime with different serotypes. This leads to an increased burden of the disease on individuals, and may also increase the number of deaths from dengue (Jetten and Focks, 1997).

Instead of a hypothetical temperature increase, Patz *et al.* used global climate model (GCM) projections of future climate to model the EP in a number of regions (Patz *et al.*, 1998) compared to baseline climate from 1931 to 1980. EP was assumed to be independent of rainfall where artificial water containers existed (i.e. urban areas). Temperature variability was not included in the model, but it was acknowledged that this might affect local breeding sites through evaporation. The model found an appreciable increase in EP at very small increases in temperature, up to a threshold of 40°C:

Among the three GCMs, the average projected temperature elevation was 1.16°C, expected by the year 2050. All three GCMs projected a temperature-related increase in potential seasonal transmission in five selected cities, as well as an increase in global EP, with the largest area change occurring in temperate regions. For regions already at risk, the aggregate EP across the three scenarios rose on average between 31 and 47% (range, 24–74%).

(Patz *et al.*, 1998)

the projected changes to future climate parameters under various scenarios.

8.5 Model Limitations

The theoretical models are useful in defining the outer climatic limits and potential rate of transmission in the absence of limiting social and other environmental factors. The empirical models are useful because they are derived from patterns of *actual* transmission occurrence and thus reflect, crudely, the numerous human and other environmental factors that contribute to dengue transmission, such as a region's dominant social, economic and microgeographic features. Further, each of the four serotypes of the virus vary in their transmissibility, and the dominant serotype may change over time within a given area (Oki and Yamamoto, 2012).

The main limitation to most models is that they project *climatic suitability* for transmission, rather than disease occurrence, which depends on several additional limiting or enhancing factors. Poverty is a particularly important non-climate determinant of dengue transmission risk, as it enhances population vulnerability in multiple ways. These include limiting capacity for public vector control programmes, promoting urban microenvironments that may be most conducive to vector breeding, i.e. intended or unintended water collection containers (such as those created by litter or broken structures), overcrowding, with a resultant high density of human hosts, inadequate health systems and poorer underlying population health. Culture can also play a role, such as the openness or resistance to biological controls for the vector, such as with copepods (Kay and Nam, 2005). Wealth, on the other hand, is to some degree protective against dengue transmission, and will become more so if vaccines are introduced.

Figure 8.2 illustrates the probability of dengue transmission in various example locations, based on climate suitability and population vulnerability, and shows how the risk changes over time as a result of projected climate changes. The areas which currently have zero risk of transmission because of unsuitable climate will shrink, while the areas where transmission probability is high will expand.

Figure 8.2 assumes, for the sake of simplicity, that population vulnerability remains static over coming decades as the climate changes, whereas in reality this is unlikely to be the case. India and China, for example, are currently experiencing rapid economic growth and would, in theory, shift to the left of the figure. But India, especially, has high rates of urbanization and domestic inequality, both of which are likely to hinder dengue reduction. Climate change and concurrent or interacting factors may increase existing wealth disparities, so that already vulnerable countries become even more vulnerable, through, for example, harm to already marginal agricultural systems (see Part IV on tertiary health effects). Studies that model the potential effects of future climate change on dengue predict increased climatic suitability for transmission and

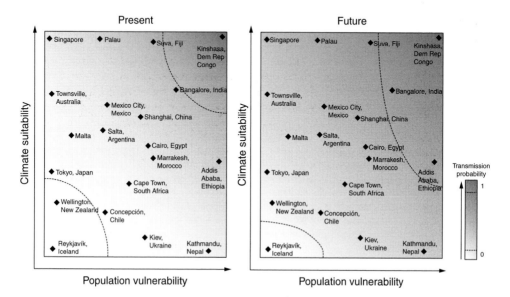

Fig. 8.2. Changes to transmission probability over time as a function of both climate suitability and population vulnerability. National per capita gross domestic product (GDP) rank was used as a proxy for population vulnerability (World Bank, 2013).

an expansion of the geographic regions at risk this century (Plate 4).

But, reductions in non-climatic population risk factors can partially, or even substantially, lower dengue risk otherwise heightened by climate change. Plate 4a, like most models, provides an estimate of future areas suitable for transmission under the assumption that today's relevant non-climate factors (e.g. housing quality and water supplies, levels of poverty, public health, surveillance and control measures) do not change over the coming century. Plate 4b illustrates the potential for one very important non-climate factor to limit geographic expansion from climate change by including estimates of future national per capita income in the model. In this model, the size of the areas subject to dengue transmission could be reduced to the point where increases in Gross Domestic Product (GDP) could, in some regions, completely offset geographic expansion that would otherwise be caused by a changing climate.

8.6   Conclusion: Dengue, Development and the Future

Increases in GDP and the infrastructure benefits this brings (piped water, improved building conditions, disease surveillance and vector control programmes) have contributed greatly to the geographic contraction of dengue in North America, China and Australia, despite overall temperature increases over the past century. Where dengue incidence has increased in recent decades, or where new areas have become transmission zones, increased urban poverty, increased human travel and trade and/or breakdown in public health measures are largely responsible, while climate trends may have exacerbated these effects. GDP is likely to remain a significant driver of dengue transmission in the future.

Novel developments in dengue control may play a significant role in the future distribution and transmission intensity of the virus. These may take the form of an effective vaccine, or involve biological control measures that reduce mosquito population densities or interrupt virus transmission. One such emerging arena is infection of the mosquito vectors with *Wolbachia* bacteria (see Box 8.2). While promising in the longer term, the effectiveness and sustainability of such technologies in poor country settings are still uncertain.

Mathematical models of dengue do not attempt to account for the effect of non-climate factors. Empirical models of spatial patterns of dengue *partly* account for these effects, because the geographic distribution of dengue on which they are based is determined and thus limited by the actual presence of all necessary factors. It thus follows that empirical models based on areas that are climatically suitable for dengue, yet lack sufficient populations of vectors, virus or humans for disease transmission to occur, will

---

Box 8.2

Wolbachia pipientis is an intracellular bacterium that naturally infects some insects. Female *Aedes* mosquitoes that are not infected die when mated with infected males, but infected females do not. *Wolbachia* affects the transmission of the dengue virus in several ways. First, infection reduces mosquito lifespan, reducing the opportunity for potential vectors to ingest infective blood meals. A shorter mosquito life reduces the time available for the dengue virus to complete its EIP, its life cycle within the mosquito, and lessens the chance that an infected mosquito will survive long enough to pass the virus on to a susceptible host in a later feed. Additionally, *Wolbachia* infection reduces the likelihood of viral transmission by reducing the ability of the virus to multiply within the mosquito, even if the mosquito acquires the virus through ingesting an infective blood meal.

There are multiple strains of *Wolbachia* bacteria. The strains that are most effective at reducing virus proliferation also have the greatest impact on mosquito lifespan, and this may pose a problem for establishing *Wolbachia* in vector populations. If mosquito lifespan is reduced too much, then the bacteria will not be passed on to subsequent generations. *Wolbachia* infection in mosquitoes is maternally inherited and can spread rapidly throughout a vector population. Recent field trials in Australia found that *Wolbachia* invaded wild populations of *Ae. aegypti* successfully, becoming established within just a few months (Hoffmann *et al.*, 2011).

underestimate the potentially suitable geographic region in which dengue may occur in future. Neither kind of model can account for the future global distribution of specific non-climatic factors that limit transmission (e.g. vaccination, vector eradication programmes), nor their effectiveness, though advances in this direction are occurring.

Dengue is re-emerging into areas that have been free of it for many decades. Disease control, such as through vector eradication programmes, has been difficult to sustain[1] and depends on the availability of resources, in other words, sufficient wealth. Global climate change is progressive and potentially irreversible, and to keep pace, dengue surveillance and control systems need to be strengthened, especially in poor countries and on the fringes of current endemic regions. In these regions, climate change will make dengue more difficult to control.

Acknowledgement

We thank an anonymous reviewer.

Note

[1] Vectorial resistance to insecticides, especially pyrethroids, is occurring. DDT (dichlorodiphenyltrichloroethane) remains effective; while there is social resistance to this, its use may be justified in some cases.

References

Åström, C., Rocklöv, J., Hales, S., Béguin, A., Louis, V. and Sauerborn, R. (2012) Potential distribution of dengue fever under scenarios of climate change and economic development. *EcoHealth* 9, 448–454.

Aviles, G., Rangeon, G., Vorndam, V., Briones, A., Baroni, P., Enria, D., *et al.* (1999) Dengue reemergence in Argentina. *Emerging Infectious Diseases* 5, 575–578.

Aviles, G., Paz, M.V., Rangeon, G., Ranaivoarisoa, M.Y., Verzeri, N., Roginski, S., *et al.* (2003) Laboratory surveillance of dengue in Argentina, 1995–2001. *Emerging Infectious Diseases* 9, 738–742.

Bhatt, S., Gething, P.W., Brady, O.J., Messina, J.P., Farlow, A.W., Moyes, C.L., *et al.* (2013) The global distribution and burden of dengue. *Nature*, doi:10.1038/nature12060.

Carey, D.E. (1971) Chikungunya and dengue: a case of mistaken identity? *Journal of the History of Medicine and Allied Sciences* 26, 243–262.

Dorji, T., Yoon, I.-K. and Holmes, E. (2009) Diversity and origin of dengue virus serotypes 1, 2, and 3, Bhutan. *Emerging Infectious Diseases* 15, 1630–1632.

Effler, P., Pang, L., Kitsutani, P. and Vorndam, V. (2005) Dengue fever, Hawaii, 2001–2002. *Emerging Infectious Diseases* 11, 742–749.

Gratz, N.G. and Knudsen, A.B. (1996) *The Rise and Spread of Dengue, Dengue Haemorrhagic Fever and Its Vectors: A Historical Overview (Up to 1995).* World Health Organization, Geneva, Switzerland.

Gubler, D.J. (1997) Dengue and dengue hemorrhagic fever: its history and resurgence as a global health problem. In: Gubler, D.J. and Kuno, G. (eds) *Dengue and Dengue Hemorrhagic Fever.* CAB International, Wallingford, UK, pp. 1–22.

Halstead, S.B. (1997) Epidemiology of dengue and dengue hemorrhagic fever. In: Gubler, D.J. and Kuno, G. (eds) *Dengue and Dengue Hemorrhagic Fever.* CAB International, Wallingford, UK, pp. 23–44.

Hirsch, A. (1883) *Handbook of Geographical and Historical Pathology – Volume I: Acute Infective Diseases.* The New Sydenham Society, London.

Hoffmann, A.A., Montgomery, B.L., Popovici, J., Iturbe-Ormaetxe, I., Johnson, P.H., Muzzi, F., *et al.* (2011) Successful establishment of *Wolbachia* in *Aedes* populations to suppress dengue transmission. *Nature* 476, 454–457.

Hueston, L. (2004) The increase in presentations of dengue fever in New South Wales. *NSW Public Health Bulletin* 15, 204–207.

Jetten, T.H. and Focks, D.A. (1997) Potential changes in the distribution of dengue transmission under climate warming. *American Journal of Tropical Medicine and Hygiene* 57, 285–297.

Kay, B. and Nam, V.S. (2005) New strategy against *Aedes aegypti* in Vietnam. *The Lancet* 365, 613–617.

Kay, B.H., Barker-Hudson, P., Stallman, N.D., Wiemers, M.A., Marks, E.N., Holt, P.J., *et al.* (1984) Dengue fever. Reappearance in northern Queensland after 26 years. *Medical Journal of Australia* 140, 264–268.

Kovats, R.S., Campbell-Lendrum, D.H., McMichel, A.J., Woodward, A. and Cox, J.S.H. (2001) Early effects of climate change: do they include changes in vector-borne disease? *Philosophical Transactions of the Royal Society of London. Series B: Biological Sciences* 356, 1057–1068.

Kuno, G. (2007) Research on dengue and dengue-like illness in East Asia and the Western Pacific during the first half of the 20th century. *Reviews in Medical Virology* 17, 327–341.

Kuno, G. (2009) Emergence of the severe syndrome and mortality associated with dengue and dengue-like illness: historical records (1890 to 1950) and their compatibility with current hypotheses on the shift of disease manifestation. *Clinical Microbiological Reviews* 22, 186–201.

La Ruche, G., Souarès, Y. and Armengaud, A. (2010) First two autochthonous dengue virus infections in metropolitan France, September 2010. *Eurosurveillance* 15(39), 2–6.

Lumley, G.F. and Taylor, F.J. (1943) *Dengue.* The School of Public Health and Tropical Medicine, University of Sydney, Australian Medical Publishing Company Limited, Sydney.

McCredie, J. (2009) Dengue fever epidemic hits northern Australia. *BMJ* 338, b967.

McMichael, A.J., Woodruff, R. and Hales, S. (2006) Climate change and human health: present and future risks. *The Lancet* 367, 859–869.

Oki, M. and Yamamoto, T. (2012) Climate change, population immunity, and hyperendemicity in the transmission threshold of dengue. *PloS One* 7, e48258.

Pandey, B.D., Morita, K. and Khanal, S.R. (2008) Dengue virus, Nepal. *Emerging Infectious Diseases* 14, 514–515.

Patz, J., Martens, W., Focks, D. and Jetten, T.H. (1998) Dengue fever epidemic potential as projected by general circulation models of global climate change. *Environmental Health Perspectives* 106, 147–153.

Pinheiro, F.P. and Corber, S.J. (1997) Global situation of dengue and dengue haemorrhagic fever, and its emergence in the Americas. *World Health Statistics Quarterly* 50, 161–169.

Rigau-Perez, J.G., Clark, G.G., Gubler, D.J., Reiter, P., Sanders, E.J. and Vorndam, A.V. (1998) Dengue and dengue haemorrhagic fever. *The Lancet* 352, 971–977.

Robinson, M.C. (1956) An epidemic of a dengue-like fever in the southern province of Tanganyika. *The Central African Journal of Medicine* 2, 394–396.

Rogers, D.J., Wilson, A.J., Hay, S.I. and Graham, A.J. (2006) The global distribution of yellow fever and dengue. *Advances in Parasitology* 62, 181–220.

Scott, H.H. (1943) The influence of the slave-trade in the spread of tropical disease. *Transactions of the Royal Society of Tropical Medicine and Hygiene* XXXVII, 169–188.

Thai, K.T. and Anders, K.L. (2011) The role of climate variability and change in the transmission dynamics and geographic distribution of dengue. *Experimental Biology and Medicine* 236, 944–954.

Trout, A., Baracco, G. and Rodriguez, M. (2010) Locally acquired dengue, Key West, Florida 2009–2010. *Morbidity and Mortality Weekly Report* 59, 577–581.

Van Kleef, E., Bambrick, H. and Hales, S. (2010) The geographic distribution of dengue fever and the potential influence of global climate change. *TropIKA.net* (http://journal.tropika.net/scielo.php?script=sci_pdf&pid=S2078-86062010005000001&lng=en&nrm=iso&tlng=en, accessed 15 March 2014).

van Rooyen, C. and Rhodes, A. (1948) *Virus Diseases of Man.* Thomas Nelson and Sons, New York.

Vezzani, D. and Carbajo, A.E. (2008) *Aedes aegypti, Aedes albopictus,* and dengue in Argentina: current knowledge and future directions. *Memórias do Instituto Oswaldo Cruz* 103, 66–74.

WHO (World Health Organization) (2009) *Dengue: Guidelines for Diagnosis, Treatment and Control.* WHO, Geneva, Switzerland.

World Bank, The (2013) Gross Domestic Product 2012 (http://databank.worldbank.org/data/download/GDP.pdf, accessed 18 March 2014).

9 Lyme Disease and Climate Change

Nick H. Ogden

*Zoonoses Division, Centre for Food-borne, Environmental
and Zoonotic Infectious Diseases, Public Health Agency
of Canada, Saint-Hyacinthe, Canada*

9.1 Introduction

Lyme disease, or rather Lyme borreliosis (LB), is caused by spirochetal bacteria of a species complex known as *Borrelia burgdorferi* sensu lato. There are four main species involved in human disease – *B. burgdorferi* sensu stricto, *Borrelia afzelii*, *Borrelia garinii* and *Borrelia bavariensis*. For simplicity, the whole species complex is termed *B. burgdorferi* in the following. Lyme disease spirochaetes are transmitted by hard-bodied ticks of the genus *Ixodes* and, as with other diseases transmitted by arthropod vectors, Lyme disease has been considered as potentially affected by climate change (Confalonieri *et al.*, 2007). The Lyme disease spirochaetes are maintained in nature by transmission cycles involving the ticks and a wide range of wild animal hosts (small and medium-sized mammals, birds and some reptiles), while humans (and some susceptible domesticated animal species) are accidental hosts, acquiring infection when they spend time in the woodland habitats where the ticks and wild animal hosts occur.

Climate and climate change may affect Lyme disease risk via four main outcomes: (i) changes to the geographic footprint of the occurrence of transmission cycles; (ii) changes in the abundance of infected ticks, and thus the level of risk, where Lyme disease transmission cycles already occur; (iii) changes to trajectories of bacterial evolution, with potential consequences for pathogenicity; and (iv) changes to human activities that alter the frequency with which humans are exposed to Lyme disease risk. In this chapter, the mechanisms underlying these possible outcomes are explored, while efforts to date to predict the effects of climate change as well as current evidence for the effects of climate change on Lyme disease risk are reviewed.

9.2 Ecology and Epidemiology of Lyme Disease

Lyme disease was first identified as a spirochaetal tick-borne disease following an outbreak in Lyme, Connecticut, USA, in the late 1970s (Burgdorfer *et al.*, 1982), but Lyme disease risk has clearly existed in Europe and North America for millennia (Margos *et al.*, 2011). Lyme disease is the most commonly reported vector-borne disease in the northern temperate zone, with an incidence of 100 cases per 100,000 population in the USA and some European countries (Lindgren and Jaenson, 2006; Bacon *et al.*, 2008). In the USA, risk of infection is highest from June to August, when infected ticks are active and when humans are more likely to be taking part in outdoor activities (Bacon *et al.*, 2008). Demonstrable risk factors for infection are leisure activities, location of residence and occupations that bring people into contact with the woodland habitats that support the tick vectors and animal hosts of *B. burgdorferi* (reviewed in Ogden *et al.*, 2014). There is currently no vaccine to protect humans from Lyme disease, but the disease is treatable with antibiotics (Ogden *et al.*, 2009).

The bacterium is maintained in natural transmission cycles between wild animal hosts and vector hard-bodied (Ixodid) ticks of the *Ixodes ricinus–Ixodes persulcatus* complex (see Fig. 9.1). A typical transmission cycle occurs globally for *B. burgdorferi* transmitted by *Ixodes pacificus* (western North America), *Ixodes scapularis* (central and eastern North America), *I. ricinus* (Western and Central Europe) and *I. persulcatus* (eastern Europe and Asia). In this cycle, infective nymphal ticks infect rodent, bird or medium-sized mammal reservoir hosts, which then transmit infection to uninfected larvae that feed on them. Infected engorged larvae maintain infection through the moult to become infective, host-seeking nymphs, completing the transmission cycle. Adult ticks feed mostly on reservoir-incompetent deer, so they do not participate in transmission cycles (Thompson *et al.*,

2001; Gern, 2008). There are, however, some variations, i.e. transovarial transmission and co-feeding transmission, to this 'typical' cycle that can be of ecological and public and animal health significance (Ogden *et al.*, 2014).

The existence of transmission cycles depends absolutely on the presence of competent tick vectors and reservoir hosts and, therefore, on the environmental conditions that support these. Within these broad bounds (and they are very broad: *I. ricinus* extends from North Africa to Scandinavia and *I. scapularis* is found from Texas in the USA to southern Canada), Lyme disease risk can vary due to a number of environmentally sensitive factors. These include the abundance and seasonal activity of ticks and the structure of the community of hosts for ticks and *B. burgdorferi* (Randolph, 2004; Tsao, 2009). The host community may include hosts that are incompetent as reservoirs, zooprophylactic or variable in their efficiency in maintaining and transmitting *B. burgdorferi* (Tsao, 2009), and these hosts may be efficient hosts for ticks or hosts that readily acquire resistance to

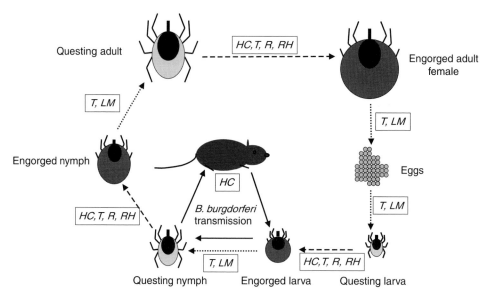

Fig. 9.1. The life cycle of the Ixodid tick vectors of Lyme disease, indicating also the cycle of *Borrelia burgdorferi* transmission from infected questing nymphs to vertebrate reservoir hosts, to larvae and then to nymphs through the moult. Dotted arrows indicate tick development processes, dashed lines indicate host-finding processes, while solid lines indicate the *B. burgdorferi* transmission cycle. Boxes indicate influences on these processes, including temperature (T), rainfall (R), air relative humidity (RH), soil/litter layer moisture (LM) and the abundance and species range of tick and reservoir hosts that comprise the host community (HC).

ticks (e.g. Dizij and Kurtenbach, 1995). In the following, the possible effects of climate change on each of these factors, as well as the effects of climate change on the dispersion of ticks and *B. burgdorferi*, are discussed.

However, it must be considered that a key feature of the natural ecology of the agent of Lyme disease is the tendency for key element species to be generalists. Species of *B. burgdorferi* sensu lato are mostly maintained by multiple species, and *B. burgdorferi* sensu stricto has efficient reservoir hosts that span several taxa (Margos *et al.*, 2011). The exophilic *Ixodes* spp. vectors are also usually generalists, being able to feed successfully on almost any terrestrial vertebrate (Kurtenbach *et al.*, 2006). Some key reservoir hosts, including *Peromyscus* spp. mice in North America and *Apodemus* spp. and *Myodes* spp. rodents in Europe, already have geographic ranges that span continents and have the capacity to persist in a wide range of woodland and woodland ecotone habitats (Amori *et al.*, 2008; Schlitter *et al.*, 2008; Linzey *et al.*, 2009). Deer are key hosts for adult ticks, even though they are generally incompetent as reservoirs of *B. burgdorferi*, and they too are generalists in terms of habitat use and have geographic ranges that extend across continents (Gallina and Lopez Arevalo, 2008). Consequently, the *B. burgdorferi* transmission cycle is, at its ecological core, generalist and capable of surviving much perturbation (as it is thought to have done in North America over the great land-use changes that occurred in the post-Columbian era), and, by inference, it has the capacity for successful invasion if climate change provides appropriate conditions (Kurtenbach *et al.*, 2006).

9.3 Possible Effects of Climate Change on Lyme Disease Risk

In this section, the possible direct and indirect effects of climate change are investigated. The potential effects of climate change are those projected according to the Intergovernmental Panel on Climate Change (IPCC) 4th report, which encompasses increasing temperatures, changes in rainfall patterns and increasing climate variability and incidence of extreme weather events (Parry *et al.*, 2007). It will be assumed throughout

that the degree or lack of efforts to mitigate greenhouse gas (GHG) emissions and atmospheric GHG concentrations will change the impact of climate change in some way proportional to their effects on the climate.

Direct effects of climate change on the tick vectors

Climate can impact the tick vectors in four ways that affect the survival of the tick. First, extremes of temperature outside a specific range will kill ticks directly when they are not parasitizing hosts (when they are relatively protected by their host's body temperature regulation). In nature, the direct effects of temperature on survival are modulated by the habitat in which ticks are found. In the laboratory, *I. scapularis* ticks die rapidly at temperatures below $-2°C$, but when ticks are placed in containers in the field in woodland habitats in northern Canada in winter, where air temperatures reach $-40°C$, their mortality rates are no greater than they are in summer in woodlands where the ticks are established (Lindsay *et al.*, 1999; Ogden *et al.*, 2004). Presumably, the litter layer or surface soil layers insulate the ticks from extremes of temperature, and indeed soil temperatures are usually an order of magnitude greater than air temperatures in northern regions in winter. Very high temperatures also reduce tick survival (Brownstein *et al.*, 2005). Second, low humidity is also lethal for ticks if they cannot find refugia in which to rehydrate, which they do by excreting hygroscopic saliva that acquires environmental moisture, which the ticks then ingest with the saliva (Bowman and Sauer, 2004). An indirect effect of relative humidity on survival is that if air relative humidity is low, the ticks must return more frequently to the litter layer to rehydrate, which depletes the finite energy resources available for host seeking and reduces the likelihood that they successfully find a host before dying of starvation (Randolph, 2004). Third, heavy rainfall generally inhibits host-seeking activity and again reduces the likelihood that they will successfully find a host (Randolph, 2004). Fourth, the duration of development of ticks from one stage to the next, the duration of pre-oviposition in engorged adult females and the duration of development of eggs from laying to hatching all depend to some extent on temperature (temperature-independent, daylength-dependent diapause mechanisms also operate, but these are very idiosyncratic for different species and populations of ticks; Ogden *et al.*, 2004; Randolph, 2004). Higher temperatures generally mean shorter life cycles, so, assuming constant per capita temperature-independent daily mortality rates, the higher ambient temperatures are, the higher is the likelihood that a larval tick survives to be an adult tick. In Canada, field and laboratory studies established three key features of the biology of *I. scapularis*: (i) the relationships between temperature and tick development and the contribution of temperature-independent diapause were elucidated; (ii) estimates of winter and summer mortality rates in suitable and suboptimal habitat were obtained from the field (Lindsay *et al.*, 1995; Ogden *et al.*, 2006); and (iii) measures of densities of questing ticks, host densities and tick infestations of hosts as well as approximate values for host-finding rates were obtained in field studies (e.g. Lindsay *et al.*, 1999). Using these data, a population model of *I. scapularis* was developed and, by simulations of tick populations in different locations in Canada, minimal temperature conditions for the survival of *I. scapularis* populations were identified. At this threshold condition, the life cycle length was such that total mortality of ticks was not greater than the number of offspring produced by each cohort (i.e. R_0 for the ticks was ≥ 1). Importantly, the most useful index of temperature conditions is annual cumulative degree days above 0°C because: (i) there is no tick development below 0°C, so only temperatures above this value impact development rates; and (ii) ticks reside in the litter layer during development, which buffers diurnal fluctuations in air temperature, while there appears to be a latency of several days in the ability of the tick's metabolism to respond to changing temperatures in terms of changes in development rates (Ogden *et al.*, 2004). Extreme weather events may have a lesser effect on ticks compared to mosquito vectors, because of the capacity of ticks to return to litter layer refugia, where they are protected from extreme temperature and rainfall; however, there will clearly be limits to this. A further effect of climate change is that the season of risk in locations where Lyme disease risk already occurs is likely to increase as ticks become active earlier in the year and remain active later in the year (Gray *et al.*, 2009).

Effects of climate change on hosts and their communities

Climate change is likely to impact the geographic range and local abundance of hosts of *B. burgdorferi* and tick vectors by direct effects and by indirect effects on the resources hosts need for their survival. As mentioned above, *B. burgdorferi*, and key vectors and hosts, are generalists, capable of surviving in a wide range of habitats and climatic conditions, so the core components of the transmission cycle will likely be resilient to long-term climatic changes and short-term climate variability. However, individual species will likely be affected by climate change, producing local idiosyncratic variations in Lyme disease risk within a general pattern of northward expansion and southern contraction of geographic ranges.

There has been much discussion in recent years of the 'dilution effect' of increased diversity of the host community, whereby increased biodiversity is thought to interrupt transmission from ticks to highly competent reservoirs. However, this is under debate because increased diversity may enhance transmission by boosting host and tick abundance (Ogden and Tsao, 2009). Whether or not increased biodiversity does correlate with reduced *B. burgdorferi* transmission, variations in host species range and abundance are likely very important in geographic variations in Lyme disease risk: incompetent lizard hosts are likely important in reducing Lyme disease risk in some ecosystems (Tsao, 2009), but how climate change will affect such complex systems remains to be assessed.

Effects on *B. burgdorferi* transmission cycles

Beyond the direct effects of climate and climate change on tick abundance, host range and abundance, and possibly survival of *B. burgdorferi* in ticks (high temperatures can reduce pathogen survival in ticks, although this has not been studied for *B. burgdorferi*), more subtle effects of climate change on transmission cycles have been suggested. In North America, different strains of *B. burgdorferi* sensu stricto are transmissible from infected rodent hosts for different periods of time (varying from a few weeks to the life of the rodent), and a key factor in the natural persistence of strains that are short-lived in their host is that the seasonal activity of nymphs infecting the hosts must be coincident with the larvae to which infected hosts must transmit infection for the transmission cycle to be perpetuated. Strains that

are long-lived in their hosts are more resilient to asynchronous seasonal activity of nymphal and larval ticks (Fig. 9.2). Seasonality of ticks is largely dependent on temperature effects on development and on host-seeking activity, and climate change is likely to affect seasonality, altering the geographic pattern of abundance of different strains (Fig. 9.2). A further consequence of this is the potential (both specifically and due to the general effects of climate change on the stability of *B. burgdorferi* transmission cycles) for

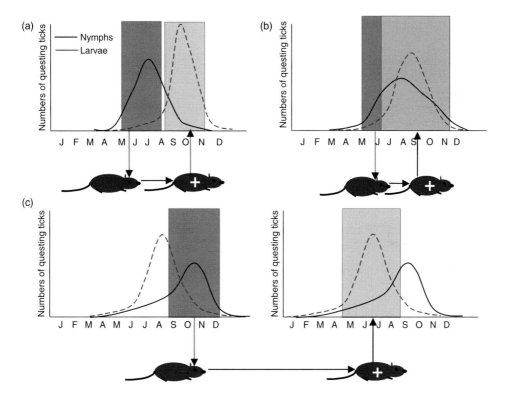

Fig. 9.2. Stylized seasonal activity patterns for larval and nymphal *I. scapularis* ticks under three climatic conditions are shown by the three graphs: (a) asynchronous nymphal and larval activity, with nymphal ticks active in spring before larvae, as occurs in north-eastern USA; (b) synchronous nymphal and larval activity as occurs in north midwestern USA; and (c) asynchronous nymphal and larval activity, with larvae active in spring before nymphs, as occurs in south-eastern USA. The dark shaded blocks indicate the main period of the year when nymphal ticks are active and capable of transmitting infection to rodents and other reservoirs, while the pale shaded blocks indicate the period when larvae are active and capable of acquiring infection from infected rodents and other reservoirs, and then maintaining infection through the moult to begin the transmission cycle again. Seasonal activity pattern (a) permits transmission of strains of *Borrelia burgdorferi* that cause long-lived infections in the reservoir hosts. Seasonal activity pattern (b) permits transmission of all strains of *B. burgdorferi*, whether they cause long- or short-lived infections in the reservoir hosts. Seasonal activity pattern (c) may permit only inefficient transmission of even strains of *B. burgdorferi* that cause long-lived infections in reservoir hosts because both infections and the rodents themselves may have a high chance of dying out before the infections can be transmitted to larvae.

evolutionary trajectories to change from specialism (via adaptive radiation) to one of generalism, which may have unpredictable effects on emergence of pathogenic strains (Kurtenbach et al., 2006). This provides just one example of how climate and climate change may have subtle effects on processes within the communities of hosts and vectors that directly and indirectly affect disease transmission cycles of infectious diseases in natural communities (Lafferty, 2009). In most cases, these effects are likely almost impossible to predict without more detailed knowledge of the ecological processes involved.

Effects of climate change on dispersion of ticks and *B. burgdorferi*

While a significant anticipated effect of climate change in risk assessments is northward range expansion of tick vectors and *B. burgdorferi*, actual changes in geographic range will depend on the capacity of hosts to disperse ticks and *B. burgdorferi*. Terrestrial hosts of *B. burgdorferi* and its tick vectors generally demonstrate limited capacity for dispersion. Maximal rodent home ranges are in the order of 1 km, and similar (<5 km) values have been suggested for deer in North America (reviewed in Leighton et al., 2012), even though deer may carry many thousands of ticks. Migratory birds are very likely an important method of long-distance dispersion of ticks in Europe and North America, because northward migration routes and seasons are coincident with the activity of nymphal *I. ricinus* (in Europe) and *I. scapularis* (in North America), as demonstrated by surveillance for ticks. Importantly, this means that ticks are continually being dispersed far into northern areas, where they can seed reproducing tick populations as the climate warms and becomes more suitable for tick survival. Studies to date suggest that this system may be resilient to the early effects of climate change, because while northward bird migration speed in spring increases with a warmer climate, the effect of this will likely initially be small (Ogden et al., 2008a). However, to what extent extreme changes in climate will affect this system is

unknown. Methods of dispersion and introduction of *B. burgdorferi* are less well studied. Migratory birds may vary in their efficiency as a route of introduction of *B. burgdorferi* for a number of reasons. The stress of migration may cause recrudescence of infection and transmissibility in migratory passerines, and this suggests that migratory birds could be an important route of dispersion (Gylfe et al., 2000). However, studies in North America suggest that northward migrating birds are infrequently infective for ticks (Ogden et al., 2008a). Furthermore, northward migrating passerines in north-eastern North America are more likely to acquire nymphal ticks, which will moult into adults that will feed on reservoir-incompetent deer. These ticks are, therefore, unlikely to be a frequent route of introduction of *B. burgdorferi*, and studies to date on northward expansion of ticks and *B. burgdorferi* support the notion that northward-moving hosts efficiently spread the ticks but inefficiently spread *B. burgdorferi* (leaving this role to terrestrial hosts), resulting in a significant (i.e. several years) time lag between tick and *B. burgdorferi* establishment (Ogden et al., 2010). How climate change may affect expansion or range change of *B. burgdorferi* per se rather than as a direct consequence of tick vector range change has not to date been studied.

Effects of climate change on human activities affecting Lyme disease incidence

The preceding sections have dwelt on the effects of climate and other factors that determine spatial and temporal variations in the environmental risk of Lyme disease. However, for the environmental risk to be realized as a real public health risk depends on the degree to which people come into contact with infected ticks by virtue of the ways in which people use the environment for their work, leisure, home building and other activities. In some parts of north-eastern USA, the design of residential areas means that there is peridomestic risk from Lyme disease (Jackson et al., 2006). In eastern Europe, the frequency of harvesting of mushrooms from forests, driven by socio-economic factors, has driven peaks in tick-borne disease risk that have been mistaken for climate change effects (Randolph and EDEN-TBD Sub-Project Team, 2010). In areas where Lyme disease risk already occurs, human activity and the likelihood of contacting ticks will most likely increase with climate change: seasons for work and leisure activities in woodland environments are likely to increase. In general, the relationship between future human behavior and future Lyme disease is relatively poorly understood and studied. As the climate changes, so does the risk, but changes in human behavior could act in ways to either magnify or dampen that evolving risk.

9.4   Assessments and Evidence for Effects of Climate Change on Lyme Disease Risk

To date, there have been a limited number of assessments of the potential effects of projected climate change on Lyme disease risk. Two in North America were roughly similar in their predictions for significant northward range expansion using statistical and simulation modelling approaches (respectively, Brownstein *et al.*, 2005; Ogden *et al.*, 2006), one suggested significant contraction of the southern range of risk (Brownstein *et al.*, 2005), two in Europe suggested northern expansion of the range of risk as well as expansion of the seasonal duration of risk (Lindgren and Jaenson, 2006; Gray *et al.*, 2009) and one suggested the possibility of increasing altitude of risk (Gilbert, 2010). Current evidence is clear: vector ticks and Lyme disease risk are expanding northwards in the northern hemisphere in a way that has been anticipated in the risk assessments based on the predictions of statistical or simulation modelling exercises. Evidence comes from the northward expansion of *I. ricinus* in Europe (Lindgren *et al.*, 2000), *I. scapularis* in Canada (Ogden *et al.*, 2008b, 2010; Leighton *et al.*, 2012), altitudinal increases in tick abundance (Moràn Cadenas *et al.*, 2007), increasing Lyme disease incidence in Scotland and Canada (Ogden *et al.*, 2009; Slack *et al.*, 2011) and evidence that new tick populations are setting up in areas where temperature

conditions are more favourable (Ogden *et al.*, 2008b, 2010; Leighton *et al.*, 2012), and synchrony of the northward spread of *I. scapularis* in Canada with a period of climate warming (Fig. 9.3). To date, however, none of the observed changes in tick and *B. burgdorferi* risk can be linked conclusively to climate change, even though this may be a very likely cause or cofactor. Conclusive proof of an effect of climate change will be difficult: climate change is globally universal, so there is no opportunity for meaningful case-control studies that may eliminate or estimate the relative importance of other drivers of changes in the complex ecology and sociobiology of tick-borne zoonoses such as Lyme disease.

9.5   Future Prospects

Time will tell as to whether the projections for Lyme disease risk are realized, whether or not the human and economic costs of Lyme and other vector-borne diseases are factored into the economic equation of decision making on efforts to mitigate climate change, and whether or not climate change already built into our climate system will bring significant change to Lyme disease risk globally. Further assessments of the impact of climate change on Lyme and other tick-borne zoonoses would be useful, given the evolving nature of the technology involved in predicting future climate. *B. burgdorferi* has existed for millennia, and there is possible evidence in phylogeographic data for past expansions and contractions in the range of ticks and *B. burgdorferi* (Hoen *et al.*, 2009). This raises the hypothesis that the study of the genetic structure of *B. burgdorferi* and its hosts and vectors may provide useful clues as to the extent of the effects of past climate changes to inform how current climate change may impact risk from Lyme disease. Even so, the establishment of surveillance to identify where Lyme disease risk is emerging or changing (e.g. Koffi *et al.*, 2012) would seem a prudent adaptation to the consequence of the effects of climate change or other drivers, so that communities newly at risk may be informed about how to protect themselves from this disease.

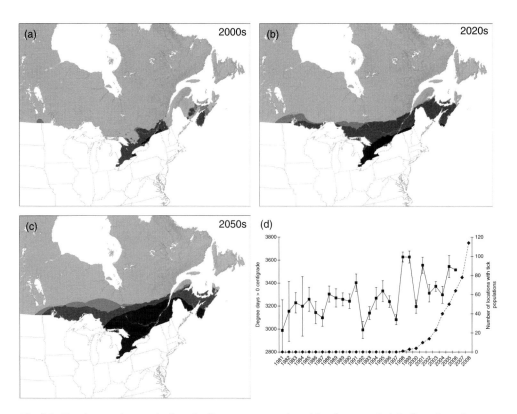

Fig. 9.3. Panels a–c show projections for the range expansion of *Ixodes scapularis* in Canada under current (panel a) and projected climate for the 2020s and 2050s (b and c, respectively) using the methodology in Ogden *et al.* (2008b). The solid graph with square symbols in panel d shows the mean and standard error of the annual cumulative degree-days above 0°C obtained from 14 meteorological stations in southern Quebec, Canada, for the period 1981–2006. The dotted graph with lozenge symbols indicates the estimated number of locations in Quebec where *I. scapularis* populations had established over the period up to 2008 (from the data used in Leighton *et al.*, 2012).

References

Amori, G., Hutterer, R., Kryštufek, B., Yigit, N., Mitsain, G., Palomo, L.J., *et al.* (2008) *Myodes glareolus.* In: IUCN (ed.) *IUCN Red List of Threatened Species. Version 2013.2.* International Union for Conservation of Nature (www.iucnredlist.org, accessed 17 March 2014).

Bacon, R.M., Kugeler, K.J. and Mead, P.S. (2008) Surveillance for Lyme disease – United States, 1992–2006. *Morbidity and Mortality Weekly Report Surveillance Summaries* 57, 1–9.

Bowman, A.S. and Sauer, J.R. (2004) Tick salivary glands: function, physiology and future. *Parasitology* 129, S67–S81.

Brownstein, J.S., Holford, T.R. and Fish, D. (2005) Effect of climate change on Lyme disease risk in North America. *Ecohealth* 2, 38–46.

Burgdorfer, W., Barbour, A.G., Hayes, S.F., Benach, J.L., Grunwaldt, E. and Davis, J.P. (1982) Lyme disease – a tick-borne spirochetosis? *Science* 216, 1317–1319.

Confalonieri, U., Menne, B., Akhtar, R., Ebi, K.L., Hauengue, M., Kovats, R.S., *et al.* (2007) Human health. In: Parry, M.L., Canziani, O.F., Palutikof, J.P., van der Linden, P.J. and Hanson, C.E. (eds) *Climate Change 2007: Impacts, Adaptation and Vulnerability.* Contribution of Working Group II to the Fourth

Assessment Report of the Intergovernmental Panel on Climate Change. Cambridge University Press, Cambridge, UK, pp. 391–431.

Dizij, A. and Kurtenbach, K. (1995) *Clethrionomys glareolus*, but not *Apodemus flavicollis*, acquires resistance to *Ixodes ricinus* L., the main European vector of *Borrelia burgdorferi*. *Parasite Immunology* 17, 177–183.

Gallina, S. and Lopez Arevalo, H. (2008) *Odocoileus virginianus*. In: IUCN (ed.) *IUCN Red List of Threatened Species. Version 2013.2.* International Union for Conservation of Nature (www.iucnredlist.org, accessed 17 March 2014).

Gern, L. (2008) *Borrelia burgdorferi sensu lato*, the agent of lyme borreliosis: life in the wilds. *Parasite* 15, 244–247.

Gilbert, L. (2010) Altitudinal patterns of tick and host abundance: a potential role for climate change in regulating tick-borne diseases? *Oecologia* 162, 217–225.

Gray, J.S., Dautel, H., Estrada-Peña, A., Kahl, O. and Lindgren, E. (2009) Effects of climate change on ticks and tick-borne diseases in Europe. *Interdisciplinary Perspectives on Infectious Diseases* 593232 (12), doi.org/10.1155/2009/593232.

Gylfe, A., Bergström, S., Lundström, J. and Olsen, B. (2000) Reactivation of *Borrelia* infection in birds. *Nature* 403, 724–725.

Hoen, A.G., Margos, G., Bent, S.J., Kurtenbach, K. and Fish, D. (2009) Phylogeography of *Borrelia burgdorferi* in the eastern United States reflects multiple independent Lyme disease emergence events. *Proceedings of the National Academy of Sciences* 106, 15013–15018.

Jackson, L., Hilborn, E. and Thomas, J. (2006) Towards landscape design guidelines for reducing Lyme disease risk. *International Journal of Epidemiology* 35, 315–322.

Koffi, J.K., Leighton, P.A., Pelcat, Y., Trudel, L., Lindsay, L.R., Milord, F., *et al.* (2012) Passive surveillance for *I. scapularis* ticks: enhanced analysis for early detection of emerging Lyme disease risk. *Journal of Medical Entomology* 49, 400–409.

Kurtenbach, K., Hanincová, K., Tsao, J., Margos, G., Fish, D. and Ogden, N.H. (2006) Key processes in the evolutionary ecology of Lyme borreliosis. *Nature Reviews Microbiology* 4, 660–669.

Lafferty, K.D. (2009) The ecology of climate change and infectious diseases. *Ecology* 90, 888–900.

Leighton, P.A., Koffi, J.K., Pelcat, Y., Lindsay, L.R. and Ogden, N.H. (2012) Predicting the speed of tick invasion: an empirical model of range expansion for the Lyme disease vector *Ixodes scapularis* in Canada. *Journal of Applied Ecology* 49, 457–464.

Lindgren, E. and Jaenson, T.G. (2006) *Lyme borreliosis in Europe: Influences of Climate and Climate Change, Epidemiology, Ecology and Adaptation Measures.* WHO Regional Office for Europe, Copenhagen, ISBN: 9289022914.

Lindgren, E., Tälleklint, L. and Polfeldt, T. (2000) Impact of climatic change on the northern latitude limit and population density of the disease-transmitting European tick *Ixodes ricinus*. *Environmental Health Perspectives* 108, 119–123.

Lindsay, L.R., Barker, I.K., Surgeoner, G.A., McEwen, S.A., Gillespie, T.J. and Robinson, J.T. (1995) Survival and development of *Ixodes scapularis* (Acari: Ixodidae) under various climatic conditions in Ontario, Canada. *Journal of Medical Entomology* 32, 143–152.

Lindsay, L.R., Mathison, S.W., Barker, I.K., McEwen, S.A. and Surgeoner, G.A. (1999) Abundance of *Ixodes scapularis* (Acari: Ixodidae) larvae and nymphs in relation to host density and habitat on Long Point, Ontario. *Journal of Medical Entomology* 36, 243–254.

Linzey, A.V., Matson, J. and Timm, R. (2009) *Peromyscus leucopus*. In: IUCN (ed.) *IUCN Red List of Threatened Species. Version 2013.2.* International Union for Conservation of Nature (www.iucnredlist.org, accessed 17 March 2014).

Margos, G., Vollmer, S.A., Ogden, N.H. and Fish, D. (2011) Population genetics, taxonomy, phylogeny and evolution of *Borrelia burgdorferi* sensu lato. *Infection, Genetics and Evolution* 11, 1545–1546.

Moràn Cadenas, F., Rais, O., Jouda, F., Douet, V., Humair, P.F., Moret, J., *et al.* (2007) Phenology of *Ixodes ricinus* and infection with *Borrelia burgdorferi* sensu lato along a north- and south-facing altitudinal gradient on Chaumont Mountain, Switzerland. *Journal of Medical Entomology* 44, 683–693.

Ogden, N.H. and Tsao, J.I. (2009) Biodiversity and Lyme disease: dilution or amplification? *Epidemics* 1, 196–206.

Ogden, N.H., Lindsay, L.R., Charron, D., Beauchamp, G., Maarouf, A., O'Callaghan, C.J., *et al.* (2004) Investigation of the relationships between temperature and development rates of the tick *Ixodes scapularis* (Acari: Ixodidae) in the laboratory and field. *Journal of Medical Entomology* 41, 622–633.

Ogden, N.H., Barker, I.K., Beauchamp, G., Brazeau, S., Charron, D., Maarouf, A., *et al.* (2006) Investigation of ground level and remote-sensed data for habitat classification and prediction of survival of *Ixodes scapularis* ticks in habitats of southeastern Canada. *Journal of Medical Entomology* 43, 403–414.

Ogden, N.H., Lindsay, L.R., Hanincová, K., Barker, I.K., Bigras-Poulin, M., Charron, D.F., *et al.* (2008a) The role of migratory birds in introduction and range expansion of *Ixodes scapularis* ticks, and *Borrelia burgdorferi* and *Anaplasma phagocytophilum* in Canada. *Applied and Environmental Microbiology* 74, 1780–1790.

Ogden, N.H., St-Onge, L., Barker, I.K., Brazeau, S., Bigras-Poulin, M., Charron, D.F. *et al.* (2008b) Risk maps for range expansion of the lyme disease vector, *Ixodes scapularis*, in Canada now and with climate change. *International Journal of Health Geographics* 7, 24.

Ogden, N.H., Lindsay, L.R., Morshed, M., Sockett, P.N. and Artsob, H. (2009) The emergence of Lyme disease in Canada. *Canadian Medical Association Journal* 180, 1221–1224.

Ogden, N.H., Bouchard, C., Kurtenbach, K., Margos, G., Lindsay, L.R., Trudel, L., *et al.* (2010) Active and passive surveillance, and phylogenetic analysis of *Borrelia burgdorferi* elucidate the process of Lyme disease risk emergence in Canada. *Environmental Health Perspectives* 118, 909–914.

Ogden, N.H., Artsob, H., Margos, G.M. and Tsao, J.I. (2014) Non-rickettsial tick-borne bacteria and the diseases they cause. In: Sonenshine, D. and Roe, M. (eds) *Ticks and Tick-Borne Diseases Volume 2*. Oxford University Press, Oxford, UK, pp. 278–312.

Parry, M.L., Canziani, O.F., Palutikof, J.P., van der Linden, P.J. and Hanson, C.E. (eds) (2007) *Climate Change 2007: Impacts, Adaptation and Vulnerability. Contribution of Working Group II to the Fourth Assessment Report of the Intergovernmental Panel on Climate Change*. Cambridge University Press, Cambridge, UK, New York.

Randolph, S.E. (2004) Tick ecology: processes and patterns behind the epidemiological risk posed by ixodid ticks as vectors. *Parasitology* 129, S37–S65.

Randolph, S.E. and EDEN-TBD Sub-Project Team (2010) Human activities predominate in determining changing incidence of tick-borne encephalitis in Europe. *Euro Surveillance* 15, 24–31.

Schlitter, D., van der Straeten, E., Amori, G., Hutterer, R., Kryštufek, B., Yigit, N., *et al.* (2008) *Apodemus sylvaticus*. In: IUCN (ed.) *IUCN Red List of Threatened Species. Version 2013.2*. International Union for Conservation of Nature (www.iucnredlist.org, accessed 17 March 2014).

Slack, G.S., Mavin, S., Yirrell, D. and Ho-Yen, D.O. (2011) Is Tayside becoming a Scottish hotspot for Lyme borreliosis? *Journal of the Royal College of Physicians of Edinburgh* 41, 5–8.

Thompson, C.A., Spielman, A. and Krause, P.J. (2001) Coinfecting deer-associated zoonoses: Lyme disease, babesiosis, and ehrlichiosis. *Clinical Infectious Diseases* 33, 676–685.

Tsao, J.I. (2009) Reviewing molecular adaptations of Lyme borreliosis spirochaetes in the context of reproductive fitness in natural transmission cycles. *Veterinary Research* 40, 36, doi:10.1051/vetres/2009019.

10 Climate Change and Human Parasitic Disease

Haylee J. Weaver

*School of Science, University of the Sunshine Coast,
Maroochydore DC, Queensland, Australia*

10.1 Stoll's 'Wormy World' Today

While humans and parasites have been inexorably linked throughout history, it was only in the 20th century that the first attempt was made to understand just how prevalent human parasites were, and their effects on health. After World War II, many American service personnel returned from the theatres of war in the Pacific with new parasitic infections. This prompted the first quantification of the global prevalence of human helminthiases. American parasitologist, Norman Stoll estimated that in the 1940s there were over two billion infections of human parasites (Stoll, 1947). This figure, achieved by extrapolating the rates of parasitism for the main human helminthiases (i.e. metazoan helminth 'worm' parasites), did not include protozoan parasites. Stoll remarked that there was '... work still to be done', and while rates of helminth infections have reduced dramatically during the 20th century as a result of mass drug administration and the implementation of control programmes, the most recent estimates indicate that, currently, there are still at least one billion people infected with one or more parasitic disease (Stoll, 1947; Hurlimann *et al.*, 2011). Climate change adds a new dimension and urgency to the task of combating human parasitic diseases. The ways in which parasites react to climate change will affect the conduct of treatment and control programmes, both those in current use and new approaches under development. This chapter provides an overview of the main mechanisms by which climate change will affect parasites and the effects on human health.

10.2 The Burden of Disease of Human Parasites

While the prevalence of infection with parasites can be quantified, estimates of health impacts and morbidity are usually confounded by non-specific symptoms that are often under-reported (Brooker, 2010). Estimates of the burden of parasitic diseases and its effect on health are provided by the 2002 Global Burden of Disease project. These estimates neither consider people with co-infections (i.e. simultaneous infection with more than one species of parasite or pathogen) nor include those who are asymptomatic and unaware they have an infection. Deaths due to parasites are usually relatively low, often with high morbidity instead of mortality. For example, approximately 15,000 people die yearly from schistosomiasis, but more than 250 million people are infected. This contrasts with malaria, the annual death toll of which was perhaps two orders of magnitude higher at its peak, though considerable recent progress has been made in lowering this (Snow and Marsh, 2010). This highlights that for parasites other than malaria, the real risk to health is from acute and chronic morbidity as opposed to mortality.

Parasitic infections affect the health of hosts in different ways, depending on the species of parasite and the intensity of infection. For example, heavy infections of soil-transmitted helminths (STHs) cause undernutrition, growth wasting and stunting, which in turn lead to cognitive and educational deficits in children (Bethony et al., 2006; Weaver et al., 2010). Infection with the STH hookworm also causes anaemia, with severity increasing proportional to the intensity of infection. This is particularly important for pregnant women (Hotez and Kamath, 2009). Similarly, infections of *Schistosoma* spp. cause chronic inflammation and anaemia, fatigue, decreased physical fitness in adults and growth wasting, stunting and cognitive deficits in children (Hotez and Kamath, 2009).

10.3 How Will Climate Change Affect Parasites?

Different methods of transmission of species mean that climate change will affect parasites in two distinctly different ways. While some parasites reproduce and are transmitted exclusively between humans (e.g. STHs), most are zoonotic, i.e. transmitted from animals to humans. Environmental disturbance will alter biological and ecological aspects of the development of both parasites and their intermediate hosts and vectors. Alterations to life cycles, rates of larval development, availability of intermediate hosts and changes to distribution may result in the increased ability of parasites to infect new hosts in new areas or, conversely, reduce the prevalence and distribution of some parasite species (Weaver et al., 2010). The other way in which climate change will influence parasites is via human social and economic change. The prevalence of parasites within a population relies just as heavily on the practices and lifestyles of their human hosts as on the changing environmental factors.

Climate and environment

Research on the relationships between climate, environmental change and helminth ecology is increasing, but is at this stage biased toward parasites of wildlife and livestock (see examples in Kutz et al., 2005; van Dijk et al., 2010), although zoonotic parasites are gaining more attention (see, for example, Mas-Coma et al., 2008; Jenkins et al., 2011). Models using remote sensing and geographic information systems for potential transmission scenarios under climate change conditions have been used since the early 2000s, including for schistosomiasis in the Peoples' Republic of China (Yang et al., 2006; Zhou et al., 2008) and in parts of Africa (Kabatereine et al., 2004). Despite increasing precision on the predicted outcomes of changes in temperatures and other environmental parameters, rainfall and precipitation changes are more difficult to include in biological models. Advances are being made to integrate stochastic events into models. However, large-scale ecological manipulation in the form of the development of water infrastructure (e.g. the Three Gorges Dam project and the south–north water transfer project in China) will alter the distribution patterns of intermediate host snails and may facilitate increased transmission of schistosomiasis in ways unable to be predicted by modelling (Zhou et al., 2008). To date, however, other control measures appear to have prevented this (Gray et al., 2012).

Not all parasites are affected by environmental factors to the same extent. Changes in environmental factors, for example increased desiccation resulting from prolonged high temperatures, or conversely, increased availability of standing water caused by changes in rainfall patterns, will exert an effect both on the development of parasites and on the intermediate hosts (or vectors) they depend on for completion of the life cycle (Mas-Coma et al., 2009) (Fig. 10.1). Those that have a free-living larval stage (e.g. STHs) are heavily reliant on the external environment being stable and are susceptible to desiccation in hot and/or dry temperatures. Intermediate hosts of parasites, such as the snails that host larval stages of *Schistosoma* spp., are equally susceptible to changes in external environments. Increases in temperature will alter the developmental rates of the hosts and vectors, as rates are modulated strongly by high temperatures, and invertebrate intermediate hosts or vectors are usually unable to regulate their body temperature (Reiter, 2001), thus

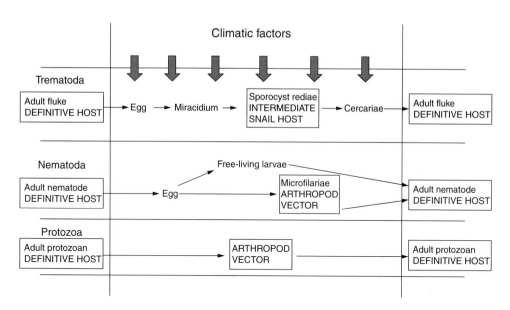

Fig. 10.1. Examples of the susceptibilities of the parasites of humans to climate and environmental factors. (Modified from Mas-Coma *et al.*, 2008.)

reducing the overall survival of parasites. However, local changes in precipitation can increase the availability of standing water (Mas-Coma *et al.*, 2009), which, in the example of *Schistosoma* spp., may increase habitat for snail intermediate hosts.

Socio-economic aspects

Parasitic infections in humans are biased heavily to the poorest people, and to tropical areas. Most of the highly prevalent human parasitic diseases fall into the category of Neglected Tropical Diseases, as listed by the World Health Organization (http://www.who.int/neglected_diseases/diseases/en/). The combination of poverty and neglect, coupled with complex factors such as social, political, economic and cultural aspects, create vulnerability to parasitic diseases (Manderson *et al.*, 2009) (see Fig. 10.2). Climate change will exert an effect on each of these factors, which will potentially destabilize the current situation for parasite control.

Polyparasitism and co-infection is often common. People are often infected with more than one species of parasite. This can create a feedback loop, in which poor health impairs work

capacity, hindering appropriate nutrition, therapy and preventative measures, increasing the risk of additional parasitic loads. Co-infection with other pathogens (e.g. HIV or *Mycobacterium tuberculosis*) alongside parasitic infection is often synergistic. It follows that the socio-economic value placed on mono-infection underestimates the full health burden in polyparasitized populations.[1]

Governments have often been responsible for administering treatment or elimination programmes, but political instability, conflict and corruption can erode the efficacy of these approaches. For example, outbreaks of zoonotic diseases increased during the Balkans war in the 1990s, due to famine, movement of populations to new areas and the breakdown of public health and medical infrastructure (Cascio *et al.*, 2011). In China, schistosomiasis prevention and control programmes conducted in vulnerable areas have led to a substantial reduction in infection (Li *et al.*, 2010), reducing infection rates from over 11 million people in the 1950s to less than 700,000 by 2000 (Zhou *et al.*, 2005). However, more recent changes in agricultural practices (such as large-scale water infrastructure projects) and increasing levels of development have allowed *Oncomelania hupensis*, the intermediate

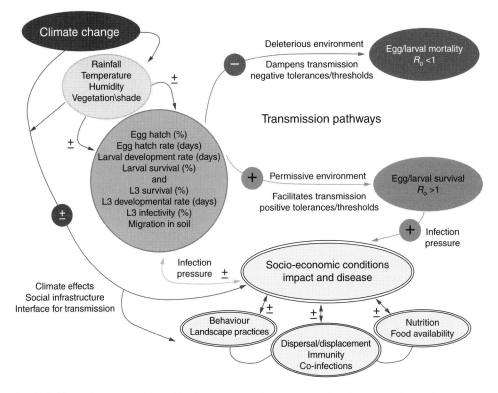

Fig. 10.2. Examples of the relationships between climate, environment and social factors on the distribution and abundance of STHs (Weaver *et al.*, 2010). (Reprinted with permission from Elsevier.)

host of *Schistosoma japonicum*, to expand its distribution.

In contrast to China, where the number of human infections of schistosomiasis has reduced by 90% (despite population increase), the number of cases in the Philippines has remained almost constant. This is despite the rate of population growth in recent decades, the Philippines being approximately double that of China. The number of Filipinos infected with schistosomiasis (*c*.560,000) now approaches the number living in China with the same condition (Bergquist and Tanner, 2010).

Human migration is also an important mechanism by which parasitic diseases spread to new areas, or recolonize former territories. Urbanization is increasing as people seek employment and other opportunities. In large cities, densely populated slums, characterized by inadequate, and in some cases deteriorating, social, health and economic conditions, increase the likelihood of parasite transmission (Desjeux, 2004).

Urban sprawl, commercial development, land-use change including deforestation, agricultural change and intensification, irrigation and dams illustrate ways in which human-induced environmental change can increase parasite transmission risk (Patz *et al.*, 2000). Exposure to parasites and their hosts and vectors is altered, often adversely, by these activities, including by the importation of large workforces (potentially infected) for construction projects. Immunologically naïve populations are especially vulnerable to the health effects of (to them) novel parasitic infections. Conversely, immigrants can introduce parasites to previously uninfected regions, with adverse effects. Visceral leishmaniasis in South America has increased in some areas where native forests are being replaced by plantation species. In one example, plantation species altered the optimal thresholds for survival and increased the population of foxes, which are a natural reservoir host for *Leishmania* sp., thus increasing the prevalence of infection in workers at the plantation or others newly arrived in the area (Cascio *et al.*, 2011). Leishmaniasis is also becoming more prevalent in urban areas (e.g. Manaus in Brazil) via the increased migration of rural populations to city areas (Desjeux, 2004). Control of Chagas disease (American trypanosomiasis) has also been complicated by extensive plantations of palm oil, native to Africa (Guhl *et al.*, 2009; Nóbrega *et al.*, 2009). Therefore, the relationships between climate, environmental factors and social factors must all be considered when investigating how climate change will affect parasites. Subtle environmental changes have the potential to affect parasite abundance. At the same time, changes to the way humans move around and manipulate the environment will also influence parasite abundance and should be considered alongside environmental parameters.

10.4 Main Groups of Human Parasites

This section will include a summary of the main human parasites and outlines the main mechanisms by which climate change will affect them. It is not possible to provide an exhaustive list of parasites and their susceptibility to climate change. Instead, this serves as a starting point for those interested in learning more about climate change and human parasites.

10.5 Protozoan Parasites

Many species of parasitic protozoan are transmitted to humans via arthropod vectors. Two indicative examples are given below for *Leishmania* spp. and *Trypanosoma* spp.

Leishmania spp.

Leishmaniasis, caused by one of many species of the protozoan genus, *Leishmania*, occurs in tropical and subtropical regions of Asia, Latin America, the Middle East and Africa. Its main forms are cutaneous and visceral, but their actual prevalence is unknown due to under-reporting of the disease (Hotez *et al.*, 2012). Cutaneous

leishmaniasis is zoonotic and transmitted by phlebotomine sandflies that bite animal reservoir hosts (usually rodents) and in turn transmit the protozoans to humans, but anthroponotic (human–phlebotomine–human) transmission also occurs, in parts of South Asia (Hotez *et al.*, 2012). Most cases of cutaneous leishmaniasis occur in a belt stretching from Morocco to Afghanistan (Hotez and Kamath, 2009; Hotez *et al.*, 2012).

Visceral leishmaniasis is also both zoonotic and anthroponotic, caused by *Leishmania infantum* and *Leishmania donovani*, respectively. Its zoonotic forms occur in northern parts of Africa, Southeast Asia (e.g. Thailand) and in parts of South America, including Brazil and (recently) northern Argentina (Hotez and Kamath, 2009; Colwell *et al.*, 2011). In the case of *L. infantum*, dogs are the main reservoir host, and control relies largely on drug treatment as cases arise (Hotez *et al.*, 2012). Visceral leishmaniasis is often an opportunistic infection and is associated with HIV/AIDS co-infection (Hotez and Kamath, 2009).

Trypanosoma spp.

Chagas disease, caused by the vector-borne protozoan, *Trypanosoma cruzi*, affects over 7 million people in Latin America (Colwell *et al.*, 2011). Trypanosomes circulate in wildlife (e.g. armadillos and monkeys) and are transmitted to humans via triatominid bugs, but can also be acquired by ingesting orally, such as from contaminated sugarcane or fruit juices (Colwell *et al.*, 2011). The acute febrile phase of Chagas disease, if untreated, can lead to a chronic stage affecting the heart, cardiovascular system and/or the digestive system (Nóbrega *et al.*, 2009; Rodrigues Coura and Borges-Pereira, 2010). Of particular concern is the role of changing land-use patterns, deforestation, plantations and increasing urbanization in bringing people into closer contact with the bug vectors of *T. cruzi* (Guhl *et al.*, 2009; Colwell *et al.*, 2011).

Human African trypanosomiasis (HAT), also called sleeping sickness, is caused by *Trypanosoma brucei*, and is thought to infect approximately 50,000–70,000 people, with 17,000 new cases occurring annually (Hotez and Kamath, 2009). HAT occurs in central

Africa (i.e. Angola, Republic of Congo, Sudan and Democratic Republic of Congo). It is transmitted by the tsetse fly and causes a range of immunological and neuropsychiatric symptoms in affected people (Malvy and Chappuis, 2011). The complexity of symptoms means that HAT can be difficult to diagnose and, subsequently, to treat.

Box 10.1 Effects of Climate Change on Vector-borne Protozoa

- The arthropod vectors of *Leishmania* spp. and *Trypanosoma* spp. are highly susceptible to climatic disturbance. Alterations to ecosystems or habitats may result in expansions or contractions of distribution.
- Increasing urban expansion and changing land-use patterns will increase people's risk of exposure to arthropod vectors.
- Movements of populations as a result of climate change related activities (work, natural disasters, conflict, etc.) will also alter the distribution of the parasites, in accordance with the fluctuation in distribution of vectors.

10.6 Platyhelminth Parasites (Trematodes)

Trematode parasites (also called flukes) have a complex life cycle that usually involves intermediate stages being present in a gastropod (or other mollusc) and an adult stage in a vertebrate definitive host. Some species have an additional intermediate stage in a smaller vertebrate that is consumed by the final (definitive) host. The best-known trematode parasites of humans are the species of *Schistosoma* that cause schistosomiasis. Other zoonotic trematode parasites include *Fasciola* spp.

Schistosoma spp.

More than 200 million people in sub-Saharan Africa (equating to over 90% of the total global infection) are infected with urinary or intestinal forms of schistosomiasis, caused by *Schistosoma haematobium* and *Schistosoma mansoni*, respectively (Hotez and Kamath, 2009). Infection with

these species also occurs in parts of the Middle East and South America. In Asia, three species exist, all of which cause intestinal forms, namely *S. japonicum* in China and the Philippines, *Schistosoma mekongi* in southern Laos and eastern Cambodia and *Schistosoma malayensis* in peninsular Malaysia (Conlan *et al.*, 2011).

People suffering intestinal schistomiasis can develop bowel ulceration, liver fibrosis and portal hypertension (Hotez and Kamath, 2009). Signs and symptoms of the urinary form include haematuria, dysuria and bladder wall pathology, and there is a significant association between this disease and squamous cell bladder carcinoma (Hotez and Kamath, 2009). Urinary schistosomiasis has also been associated with promoting horizontal transmission of HIV/AIDS in sub-Saharan Africa (Hotez and Kamath, 2009; McMichael *et al.*, 2009). In addition, chronic infection with schistosomes is an important cause of anaemia, fatigue and undernutrition, with its sequelae of growth retardation, and cognitive impairment (Hotez and Kamath, 2009). Polyparasitism has a compounding adverse health impact, as described above.

Considerable evidence exists that communities living near watercourse and large-scale irrigation projects in many parts of the world, especially sub-Saharan Africa, have a high risk of infection. This is due not only to the risk of exposure to schistosome parasites and gastropod intermediate hosts from wading in water but also from inadequate health education and provision of toilets, so that urination of infected people in rivers and dams is common and perpetuates infection. The effects of climate change (e.g. changes to rainfall patterns) may aggravate this risk (Mangal *et al.*, 2008; Hotez and Kamath, 2009). In Asia, people working in and around watercourses or standing water also experience the highest risk of infection, with approximately 1 million people in China infected with *S. japonicum* (Conlan *et al.*, 2011).

In Egypt, mass drug administration via the National Schistosomiasis Control Project has resulted in a decrease of *S. haematobium* infection from 7% in 1993 to 1% in 2006, and for *S. mansoni*, infection has reduced from 15% in 1993 to 1.5% in 2006 (Salem *et al.*, 2011). Unfortunately, however, reuse of needles during an earlier parenteral phase of this programme led to a high transmission of viral hepatitis, with transmission occurring via this route until the mid-1980s (Frank *et al.*, 2000).

Fasciola spp.

Zoonotic fascioliasis is caused by either *Fasciola hepatica* or *Fasciola gigantica*. The former occurs in Europe, the Americas and Oceania, while both species are present in Africa and Asia (Mas-Coma *et al.*, 2009). Definitive hosts include humans and livestock, and intermediate hosts are freshwater snails. Fascioliasis is emerging or re-emerging in parts of Latin America, Africa, Asia and Europe (Mas-Coma *et al.*, 2009). Increased rainfall and sustained rain events have been implicated as contributing factors in outbreaks of human fascioliasis in Iran, along with local culinary practices involving wild aquatic plants that may harbour encysted meta-cercariae (larval forms) (Salahi-Moghaddam *et al.*, 2011). Changes in temperature, rainfall and evapotranspiration rates affect fluke development, and changes to these may accelerate the fluke life cycle.

Box 10.2 Effects of Climate Change on Trematode Parasites

- Larval stages of flukes and intermediate host snails are entirely dependent on the availability of standing water for survival. This may increase or decrease, depending on regional weather patterns.
- Development rates of larval stages of flukes may change as optimal survival conditions are subject to strict temperature thresholds and may reach a peak, beyond which survival is compromised as temperatures increase.
- Movement of populations and habitat/environmental modification via development (particularly water infrastructure) and agriculture will affect distribution and potential exposure to flukes.

10.7 Nematodes (Roundworms)

Two groups of nematodes are represented here – those with a direct life cycle (STHs) and those that are vector-borne via arthropods (filiariae).

Soil-transmitted helminths: *Ascaris lumbricoides, Trichuris trichiura, Ancylostoma duodenale* and *Necator americanus*

STHs follow malaria as the second most common human parasitic diseases. The term STH includes four different nematode species occurring in the gastrointestinal tract of humans: *Ascaris lumbricoides, Trichurus trichiura, Ancylostoma duodenale* and *Necator americanus*. The latter two are both often referred to by the generic term 'hookworm', as their eggs are indistinguishable (identification of STH infection is commonly achieved through examination of faecal samples for egg presence). Neither of these species involves an intermediate host. For *A. lumbricoides* and *T. trichiura*, infection is sustained via egg ingestion, whereas for hookworm, infective larvae can penetrate the skin (most typically a bare foot) in addition to being ingested. In sub-Saharan Africa, rates of infection for STHs have remained fairly steady over the past 60 years, while these rates have declined in other regions (Hotez and Kamath, 2009). The prevalence of STH infections in China has declined significantly following the introduction of control programmes, with approximately 129 million infections recorded in 2003, down from over half a billion in 1993 (Li *et al.*, 2010).

Hookworm infection is the most common STH in sub-Saharan Africa, infecting an estimated 198 million people (Hotez and Kamath, 2009). Almost 40 million people are infected with hookworm in China, mainly in its subtropical and tropical areas (Li *et al.*, 2010). Warm, humid coastal regions are hotspots for hookworm infection (Hotez and Kamath, 2009), parameters that overlap with the distribution of other parasites, especially malaria.

Children bear the highest burden of *A. lumbricoides* infection, which is estimated to affect about 175 million people in sub-Saharan Africa. Infection with *T. trichiura* is also higher in children than in adults, with approximately 162 million affected in sub-Saharan Africa alone (Hotez and Kamath, 2009). Both ascariasis and trichuriasis are more prevalent in urban areas, a pattern which is the reverse of that seen for hookworm. Foci of ascariasis and trichuriasis occur in equatorial and southern Africa, including in Madagascar (Hotez and Kamath, 2009).

Within northern Africa and the Middle East, 23 million people are estimated to be infected with ascariasis and 9 million with trichuriasis (Hotez *et al.*, 2012). Many of these people are likely to be co-infected, and some will have at least other chronic infection with a parasite or other pathogen.

Vector-borne filarial nematodes

Wuchereria bancrofti and *Brugia malayi*

Lymphatic filariasis (LF) is caused by parasitic nematodes transmitted by culicid or anopheline mosquitoes. Unlike malaria, no development occurs in the vectors. Microfilarial larvae hatch in the human host and migrate to the epidermis, where they are ingested by mosquitoes as they draw blood. The mosquitoes then mechanically transmit the microfilariae when biting a new host. In Africa, LF is caused by *Wuchereria bancrofti*, but in Asia the causative species is either *W. bancrofti* or *Brugia malayi*. In Africa, LF is endemic in Egypt, Nigeria, Democratic Republic of Congo, Tanzania, Ethiopia and Kenya (Hotez and Kamath, 2009; Hotez *et al.*, 2012). It also occurs in parts of the Middle East, including Yemen, with small foci of infections occurring in Djibouti and Saudi Arabia (Hotez *et al.*, 2012). Generally, these parasites do not cause high mortality but can cause significant morbidity, disability and disfigurement, via hydrocele and lymphoedema. Most other countries in northern Africa and the Middle East have established treatment regimes and elimination programmes to break the cycle of transmission of vector-borne filarial nematodes (Hotez *et al.*, 2012). A comprehensive drug-treatment campaign in PR China has resulted in the elimination of LF (Utzinger *et al.*, 2010).

Box 10.3 Effects of Climate Change on Nematode Parasites

Environmental factors such as temperature, precipitation and relative humidity will affect the development rates of larval STHs. High temperatures and/or low precipitation will exceed the threshold for survival and may result in a reduction in parasite abundance. However, changes to environmental parameters will not be uniform across all geographic areas, and this may enable parasites to survive in new areas.

Vectors are highly susceptible to changes in environmental parameters – LF prevalence may reduce in areas where mosquitoes are unable to survive. Conversely, LF prevalence may rise in new areas that mosquitoes infiltrate due to the changing climate envelope.

Movement of human populations will alter the distribution of nematodes. Environmental upheaval (e.g. drought, flooding, sea level rise) will displace communities, forcing immigration to new areas. Conflict and political instability, as a result of the effects of climate change, will also force populations to move.

10.8 Conclusion

Climate change will alter the distribution of parasite species, both directly via environmental factors and indirectly via factors linked to human behaviour. These factors will vary in both space and time, meaning that the 21st century's wormy world is likely to comprise mosaics of parasite prevalence and distribution not seen before, ultimately driven by a changing climate. It is essential that parasites be considered on their own terms when understanding how climate change will affect human health. Identification of future risk areas for parasites is key to the development of appropriate targeted treatment and control programmes to combat the disease caused by parasites of humans.

Note

[1] If the burden of a single parasite is *x* and the burden of a non-parasitic pathogen is *y*, the total burden in that polyparasitized person will exceed *x* + *y*.

References

Bergquist, R. and Tanner, M. (2010) Controlling schistosomiasis in Southeast Asia: a tale of two countries. *Advances in Parasitology* 72, 109–144.

Bethony, J., Brooker, S., Albonico, M.D., Geiger, S.M., Loukas, A., Diemert, M.D., *et al.* (2006) Soil-transmitted helminth infections: ascariasis, trichuriasis and hookworm. *The Lancet* 367, 1521–1532.

Brooker, S. (2010) Estimating the global distribution and disease burden of intestinal nematode infections: adding up the numbers. *International Journal for Parasitology* 40, 1137–1144.

Cascio, A., Bosilkovski, M., Rodriguez-Morales, A.J. and Pappas, G. (2011) The socio-ecology of zoonotic infections. *Clinical Microbiology and Infection* 17, 336–342.

Colwell, D.D., Dantas-Torres, F. and Otranto, D. (2011) Vector-borne parasitic zoonoses: emerging scenarios and new perspectives. *Veterinary Parasitology* 182, 14–21.

Conlan, J.V., Sripa, B., Attwood, S. and Newton, P.N. (2011) A review of parasitic zoonoses in a changing Southeast Asia. *Veterinary Parasitology* 182, 22–40.

Desjeux, P. (2004) Leishmaniasis: current situation and new perspectives. *Comparative Immunology, Microbiology and Infectious Diseases* 27, 305–318.

Frank, C., Mohamed, M.K., Strickland, G.T., Lavanchy, D., Arthur, R.R., Magder, L.S., *et al.* (2000) The role of parenteral antischistosomal therapy in the spread of hepatitis C virus in Egypt. *The Lancet* 355, 887–891.

Gray, D.J., Thrift, A.P., Williams, G.M., Zheng, F., Li, Y.-S., Guo, J., *et al.* (2012) Five-year longitudinal assessment of the downstream impact on Schistosomiasis transmission following closure of the Three Gorges Dam. *PLoS Neglected Tropical Diseases* 6, e1588.

Guhl, F., Pinto, N. and Aguilera, G. (2009) Sylvatic triatominae: a new challenge in vector control transmission. *Memórias do Instituto Oswaldo Cruz Rio de Janeiro* 104, 71–75.

Hotez, P. and Kamath, A. (2009) Neglected tropical diseases in sub-Saharan Africa: review of their prevalence, distribution and disease burden. *PLoS Neglected Tropical Diseases* 3, e412.

Hotez, P., Savioli, L. and Fenwick, A. (2012) Neglected tropical diseases of the Middle East and Northern Africa: review of their prevalence, distribution and opportunities for control. *PLoS Neglected Tropical Diseases* 6, e1475.

Hurlimann, E., Schur, N., Boutsika, K., Stensgaard, A.-S., Laserna de Himpsl, M., Ziegelbauer, K., *et al.* (2011) Toward and open-access global database for mapping, control and surveillance of neglected tropical diseases. *PLoS Neglected Tropical Diseases* 5, e1404.

Jenkins, E.J., Schurer, J.M. and Gesy, K.M. (2011) Old problems on a new playing field: helminth zoonoses transmitted among dogs, wildlife and people in a changing northern climate. *Veterinary Parasitology* 182, 54–69.

Kabatereine, N.B., Brooker, S., Tukahebwa, E.M., Kazibwe, F. and Onapa, A.W. (2004) Epidemiology and geography of *Schistosoma mansoni* in Uganda: implications for planning control. *Tropical Medicine and International Health* 9, 372–380.

Kutz, S.J., Hoberg, E.P., Polley, L. and Jenkins, E.J. (2005) Global warming is changing the dynamics if Arctic host–parasite systems. *Proceedings of the Royal Society of London B Biological Sciences* 272, 2571–2576.

Li, T., He, S., Zhao, H., Zhao, G. and Zhu, X.-Q. (2010) Major trends in human parasitic diseases in China. *Trends in Parasitology* 26, 264–270.

McMichael, A.J., Butler, C.D. and Weaver, H.J. (2009) Climate change and AIDS: a joint working paper. UNEP/UNAIDS, Nairobi.

Malvy, D. and Chappuis, F. (2011) Sleeping sickness. *Clinical Microbiology and Infection* 17, 986–995.

Manderson, L., Aagaard-Hansen, J., Allotey, P., Gyapong, M. and Sommerfeld, J. (2009) Social research on neglected diseases of poverty: continuing and emerging themes. *PLoS Neglected Tropical Diseases* 3, e332.

Mangal, T.D., Paterson, S. and Fenton, A. (2008) Predicting the impact of long-term temperature changes on the epidemiology and control of schistosomiasis: a mechanistic model. *PLoS One* 3, e1438.

Mas-Coma, S., Valero, M.A. and Bargues, M.D. (2008) Effects of climate change on animal and zoonotic helminthiases. *Review Scientific and Technical Review Office International des Epizooties* 27, 443–452.

Mas-Coma, S., Valero, M.A. and Bargues, M.D. (2009) Climate change effects on trematodiases, with emphasis on zoonotic fascioliasis and schistosomiasis. *Veterinary Parasitology* 163, 264–280.

Nóbrega, A.A., Garcia, M.H., Tatto, E., Obara, M.T., Costa, E., Sobel, J., *et al.* (2009) Oral transmission of Chagas disease by consumption of Açaí palm fruit, Brazil. *Emerging Infectious Diseases* 15, 653–655.

Patz, J.A., Graczyk, T.K., Geller, N. and Vittor, A.Y. (2000) Effects of environmental change on emerging parasitic diseases. *International Journal for Parasitology* 30, 1395–1405.

Reiter, P. (2001) Climate change and mosquito-borne disease. *Environmental Health Perspectives* 109 (Suppl 1), 141–161.

Rodrigues Coura, J. and Borges-Pereira, J. (2010) Chagas disease: 100 years after its discovery. A systematic review. *Acta Tropica* 115, 5–13.

Salahi-Moghaddam, A., Habibi-Nokhandam, M. and Fuentes, M.V. (2011) Low-altitude outbreaks of human fascioliasis related with summer rainfall in Gilan province, Iran. *Geospatial Health* 6, 133–136.

Salem, S., Mitchell, R.E., El-Alim El-Dorey, A., Smith, J.A. and Barocas, D.A. (2011) Successful control of schistosomiasis and the changing epidemiology of bladder cancer in Egypt. *BJU International* 107, 206–211.

Snow, R.W. and Marsh, K. (2010) Malaria in Africa: progress and prospects in the decade since the Abuja Declaration. *The Lancet* 376, 137–139.

Stoll, N. (1947) This wormy world. *Journal of Parasitology* 33, 1–18.

Utzinger, J., Bergquist, R., Olveda, R. and Zhou, X.-N. (2010) Important helminth infections in southeast Asia: diversity, potential for control and prospects for elimination. *Advances in Parasitology* 72, 1–30.

van Dijk, J., Sargison, N.D., Kenyon, F. and Skuce, P.J. (2010) Climate change and infectious disease: helminthological challenges to ruminants in temperate regions. *Animal* 4, 377–392.

Weaver, H.J., Hawdon, J.M. and Hoberg, E.P. (2010) Soil-transmitted helminthiases: implications of climate change and human behavior. *Trends in Parasitology* 26, 574–581.

Yang, G.-J., Vounatsou, P., Tanner, M., Zhou, X.-N. and Utzinger, J. (2006) Remote sensing for predicting potential habitats of *Oncomelania hupensis* in Hongze, Baima and Gaoyou lakes in Jiangsu province, China. *Geospatial Health* 1, 85–92.

Zhou, X.-N., Wang, L.-Y., Chen, M.-G., Wu, X.-H., Jiang, Q.-W., Chen, X.-Y., *et al.* (2005) The public health significance and control of schistosomiasis in China: then and now. *Acta Tropica* 96, 97–105.

Zhou, X.-N., Yang, G.-J., Yang, K., Wang, X.-H., Hong, Q.-B., Sun, L.-P., *et al.* (2008) Potential impact of climate change of schistosomiasis transmission in China. *American Journal of Tropical Medicine and Hygiene* 78, 188–194.

11 Impacts of Climate Change on Allergens and Allergic Diseases: Knowledge and Highlights from Two Decades of Research

Paul J. Beggs

Department of Environment and Geography, Faculty of Science, Macquarie University, New South Wales, Australia

11.1 Introduction

One of the important secondary health effects of climate change is its impacts on allergens and resulting allergic diseases such as asthma and allergic rhinitis. These diseases are of great global significance, for a number of reasons. First, they are very common, with the global average prevalence of current asthma and rhinoconjunctivitis symptoms in 13- to 14-year-olds at 14.1% and 14.6%, respectively, and in some countries the prevalence is as high as 32.6% and 45.1%, respectively (Aït-Khaled *et al.*, 2009; Lai *et al.*, 2009). Second, they are on the increase, with, for example, the prevalence of current asthma symptoms increasing in Africa, Latin America and parts of Asia (Pearce *et al.*, 2007). And third, they impose a significant burden on both individuals and societies. For example, one study found hay fever alone was associated with impairment in at-work performance and lost productivity attributable to reduced on-the-job effectiveness of 11–40% (Vandenplas *et al.*, 2008).

While research on the relationships between climate per se and allergens, particularly aeroallergens such as pollen and mould spores, has been conducted for many decades, research focused on the impacts of climate change on allergens has largely been restricted to the last two decades or so. We now know that climate change impacts four distinct groups of allergens, these being respiratory allergens, contact allergens, plant food allergens and insect allergens. While research on the latter three categories is extremely limited, and in the case of plant food allergens, largely theoretical, the former category (respiratory allergens) has now had a large body of work devoted to it.

For this reason, the impacts of climate change on aeroallergens and allergic respiratory diseases have, in recent times, started to receive the recognition and prominence they deserve. They were highlighted as one of only seven key health effects that supported the US Environmental Protection Agency's (EPA's) determination that current and future concentrations of greenhouse gases endangered public health, in the EPA's 'Endangerment and Cause or Contribute Findings for Greenhouse Gases under the Clean Air Act' of 7 December 2009 (US EPA, 2009). Similarly, they were the focus of one of just eight climate change health effects chapters in the recent *Health Effects of Climate Change in the UK 2012* report (Vardoulakis and Heaviside, 2012).

These indirect and, to many, less obvious or even hidden, unknown or unacknowledged impacts of climate change on human health are an intriguing and multifaceted story, the likes of which have been revealed through the tireless efforts of a relatively small number of researchers scattered around the world. The remainder of this chapter summarizes what we know about the impacts of climate change on allergens and allergic diseases and then explores the development of this topic and the resulting body of research.

11.2   Respiratory Allergens

Plants, a source of many clinically important allergens including pollen, are particularly sensitive to climate change as a result of not only their response to changes in temperature, rainfall and other climate variables but also their response to changes in atmospheric carbon dioxide (CO_2) concentration. The impacts of climate change on aeroallergens, and in particular pollen, include impacts on pollen production and atmospheric pollen concentration, pollen season, plant and pollen spatial distribution and pollen allergenicity, and similar impacts on mould spores.

Experimental research where ragweed (*Ambrosia artemisiifolia*) has been grown at pre-industrial, current and potential future atmospheric CO_2 concentrations has found that pollen production in this species is increased significantly at current compared to pre-industrial atmospheric CO_2 concentration and also at potential future compared to current atmospheric CO_2 concentration (Wayne et al., 2002; Ziska and Caulfield, 2000). Similar research, comparing pollen production in ragweed grown in the higher temperature and atmospheric CO_2 concentration of an urban environment to that grown in the adjacent rural environment, also found increased production in this species in the conditions of the urban environment: analogous to a future global atmosphere (Ziska et al., 2003). Many complementary studies of atmospheric pollen concentrations over several decades show increases in response to local warming and/or rising atmospheric CO_2 concentration (Ziello et al., 2012). As just one example, Frei and Gassner have associated increasing temperatures with increasing birch pollen quantities

in Switzerland over the 38-year period, 1969–2006 (Frei and Gassner, 2008).

Consistent with much climate change impacts research showing that spring events are occurring earlier with warming temperatures, many pollen studies have now demonstrated an earlier start to, and in some cases a lengthening of, the pollen season over recent decades. Two noteworthy studies serve to illustrate this phenomenon. Van Vliet et al. analysed daily pollen counts for 14 plant species or families from 1969 to 2000 in the western part of the Netherlands and found an advance of the mean start of the pollen season of 3–22 days for the decade of the 1990s compared to the decade of the 1970s (van Vliet et al., 2002). More recent research by Ziska et al. shows that the duration of the ragweed (*Ambrosia* spp.) pollen season has been increasing in recent decades as a function of latitude in central North America, by as much as 13–27 days at latitudes above ~44°N (Ziska et al., 2011).

Research by Singer et al. found that ragweed grown at a potential future atmospheric CO_2 concentration produced pollen that was significantly more allergenic than pollen produced by plants grown at both current and pre-industrial atmospheric CO_2 concentrations (Singer et al., 2005). There is also some evidence to suggest that higher air temperature can increase birch pollen allergenicity (Hjelmroos et al., 1995; Ahlholm et al., 1998). Similarly, Wolf et al. studied *Alternaria alternata* (a ubiquitous allergenic fungus) experimentally and found that when it was grown on timothy plants grown at potential future atmospheric CO_2 concentrations, it produced nearly three times more spores and more than twice the total antigenic protein per plant than at lower atmospheric CO_2 concentrations (Wolf et al., 2010).

Relationships also exist between indoor allergens, such as mould and house dust mite, and climate. Although there are clear links between indoor and outdoor climate, the impacts of climate change on indoor allergens have received very little consideration. A recent and substantial exception to this is the report prepared for the US EPA on public health consequences and the cost of climate change impacts on indoor environments (Mudarri, 2010), which considers indoor allergens, particularly indoor mould. One of the key points of the report is that 'increased relative humidity from climate change will increase the moisture content of materials indoors and thus increase the risk for mould growth', and that 'these conditions will be exacerbated by heavy periodic rainfalls that will likely stress the ability of buildings of all types to adequately manage excess water flow' (Mudarri, 2010).

11.3   Contact Allergens

Only a single study has contemplated the impacts of climate change on allergic skin disease. Mohan *et al.* examined the responses of poison ivy (*Toxicodendron radicans*) to elevated atmospheric CO_2 concentration (Mohan *et al.*, 2006). They found poison ivy grown at elevated atmospheric CO_2 concentration (570 ppm) was more allergenic, or toxic, than plants grown at ambient atmospheric CO_2 concentration, containing a significantly increased concentration of the unsaturated triene congener of urushiol (the relative amount of which is related to the human contact dermatitis response).

11.4   Plant Food Allergens

In 2008, Beggs and Walczyk suggested, for the first time, the potential for global climate change, and in particular increased atmospheric CO_2 concentration and temperature, to have an impact on the allergenicity of plant food allergens such as groundnut (Beggs and Walczyk, 2008). While some research in this area is currently under way, and more is planned for the future, there remains no experimental research published to verify this theory, despite the call for this to be done urgently.

11.5   Insect Allergens

In another unique study, this time related to the potential impacts of climate change on stinging insects and associated reactions, Demain *et al.* examined documented insect reactions in Alaska over the period 1992–2007 and found up to a fourfold increase in patients during this period that had occurred after increases in annual and winter regional temperatures (Demain *et al.*, 2009). It was suggested that these warmer temperatures might have caused a poleward shift in the range of these insects.

11.6   Development of Climate Change and Allergen Research

After two decades or so of research on the impacts of climate change on allergens and allergic diseases, now is an opportune time to reflect on the development of this research, and to identify its highlights and leaders. While such an assessment inevitably draws on the experience and expertise, and therefore the subjectivity, of the author of this chapter, an analysis of the literature through research database searches provides an appropriate level of objectivity. For this purpose, the following discussion draws on two searches in Thomson Reuters' Web of Knowledge, Web of Science. Both searches were over all years available (1985–2012), and were conducted on 9 March 2012 on the '2012-03-07' update. All databases available in Web of Science were included in the searches, and it should be noted that coverage was variable according to database, extending from 1992 to the present for the three subject-based citation indexes and 1990 to the present for the two conference proceedings citation indexes. The first of these searches was for items published including the search topic terms "climate change"[1] and "allerg*".[2] For comparison, the second search was for items published including the search topic terms "climate change" and "human health". Results were sorted alphabetically by the first author's surname in order to facilitate an assessment of the number of items by each lead author. Citation reports for both searches were created in Web of Science in order to produce data on the number of published items each year and the number of citations of these items each year.

As shown in Fig. 11.1a, "climate change" and "allerg*" literature, specifically, started in the late 1990s and was limited to just one to five articles per year until 2004, after which there was in most years a considerable increase in articles compared to the previous year. There were, of course, articles discussing climate change and allergens prior to 1998. Among the very first were those by: John Last – specifically, Last

(a)

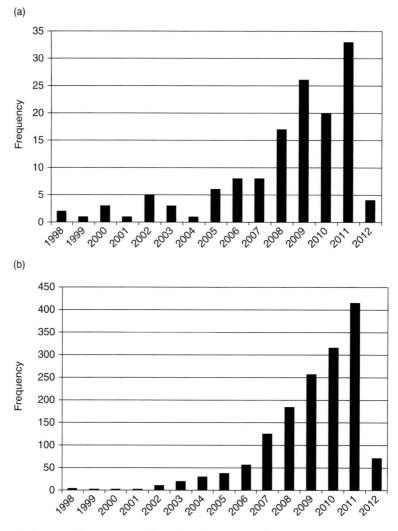

(b)

Fig. 11.1. (a) Number of items published including the search topic terms "climate change" and "allerg*" in each year from 1985 to 2012 and (b) number of citations of these items, from Thomson Reuters Web of Knowledge (as of 9 March 2012, updated 2012-03-07).

and Guidotti, and Last, which suggested that there would be significant changes in the distribution of vegetation, including many allergens (Last and Guidotti, 1990/91; Last, 1993); and Janice Longstreth, who suggested 'changes in the prevalence or intensity of asthma and hay fever episodes in affected individuals' could come about through likely 'quantitative and/or qualitative changes in the airborne concentration of allergens' such as mould spores and pollens (Longstreth, 1991).

The growth of research on climate change and allerg* reflects the overall growth in climate change and human health research, with the latter shown in Fig. 11.2a. Of interest is the similar variability in the number of articles per year in the initial years, and that the period 2005–2008 saw a marked upturn in the amount of research in both, which continues to this day. The notable differences are, of course, the far smaller number of climate change and allerg* articles (total of 138 versus 707), with

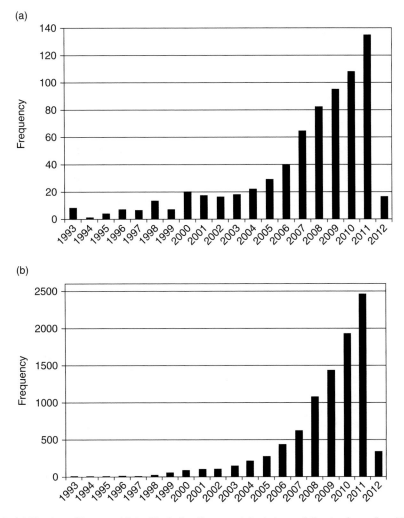

Fig. 11.2. (a) Number of items published including the search topic terms "climate change" and "human health" in each year from 1985 to 2012 and (b) number of citations of these items, from Thomson Reuters Web of Knowledge (as of 9 March 2012, updated 2012-03-07).

these articles in many respects being a subset of the literature in the climate change and human health field, and that the climate change and human health literature starts earlier (1993) than the climate change and allerg* literature.

The impact of climate change and allerg* research, as measured through citation of it (Fig. 11.1b), started in 1998 at understandably very low levels and continued at very low levels until 2002, after which it continued to grow year after year, exponentially. A particularly large relative increase in citations of articles in this area was seen in 2007, more than doubling

the previous year's total. Citations in each year to climate change and human health articles are shown in Fig. 11.2b for comparison. Again, the growth of citations in the two areas has been very similar, with the same exceptions as were noted for articles themselves – climate change and allerg* citations being fewer (a total of 1530 versus 9376) and starting later.

Again, using citations as a measure of impact, the leading climate change and allerg* articles are ranked in Table 11.1. Topping the list is a review article by Gennaro D'Amato and eight of his colleagues, including Lorenzo Cecchi.

Table 11.1. Top ten climate change and allerg* publications by total times cited, and lead authors by number of publications, from 1985 to 2012 from Thomson Reuters Web of Knowledge (as of 9 March 2012). Lead authors with the same number of publications are listed in alphabetical order.

	Publications			Lead authors	
Rank	Reference	Total citations	Rank	Name	Number of publications
1	D'Amato *et al.*, 2007	120	1	Frei, T.	8
2	Gilmour *et al.*, 2006	106	2	Beggs, P.J.	7
3	Beggs, 2004	93	3	D'Amato, G.	6
4	Wayne *et al.*, 2002	86	3	Ziska, L.H.	6
5	Bernard *et al.*, 2001	82	5	Emberlin, J.	3
6	Emberlin *et al.*, 2002	77	5	Makra, L.	3
7	Ziska *et al.*, 2003	73	7	Ariano, R.	2
8	Beggs and Bambrick, 2005	52	7	Behrendt, H.	2
9	Shea *et al.*, 2008	48	7	Perez-Badia, R.	2
10	Rogers *et al.*, 2006[a]	42	7	Shea, K.M.	2

[a]Norval *et al.* had 44 citations but did not discuss climate change and allerg* (Norval *et al.*, 2007). Reference to allerg* in this article is restricted to nickel and experimental allergic encephalomyelitis. M. Norval is also lead author of another climate change and allerg* publication, which refers largely to sun allergy (polymorphic light eruption), with only a fleeting mention to climate change and allergic diseases.

The review, while extensive (citing 190 references), included climate change as just one of two subsections of one of the three main sections of the article, with the main focus of the article, as reflected in its title, being allergenic pollen and pollen allergy in Europe (D'Amato *et al.*, 2007). Similarly, in the review by Ian Gilmour and his colleagues (2006), the impact of climate change (on aeroallergens) was just the fourth and final section of the article (Gilmour *et al.*, 2006).

A number of authors have been particularly prolific in the climate change and allergen area (Table 11.1). Twenty-seven of the 138 climate change and allerg* articles (~20%) have been lead authored by just four people. Most prolific has been Thomas Frei, who has written extensively on the impacts of climate change on airborne pollen quantities in Switzerland over the latter decades of the last century and the first few years of this century.

Figure 11.3 is an attempt to bring all of this together – a synthesis of the relatively short history of climate change and allergen research, and in particular its highlights. Not surprisingly, a number of the studies and authors in Table 11.1 are included, but not all. The three single papers on contact allergens, plant food allergens and insect allergens are included, for they are truly

novel and without them this timeline would exclude any mention of these areas. All the other studies are experimental or observational studies on respiratory allergens – pollen or mould spores. These are largely the first few studies that have documented the impacts of climate change (whether it be increasing temperatures and/or increasing CO_2 concentrations) on the seasonality, allergenicity or amount of these allergens. Noteworthy is the fact that Lewis Ziska led three of the studies highlighted but was also a co-author of three others. Together with his appearance as the lead author of a highly cited paper in this field and a prolific author in this field (Table 11.1), it would be difficult to argue that he is not the visionary international leader of this field.

11.7 Limitations

This analysis has included only articles using the term 'climate change'. There will be a number of relevant articles that use only alternate terms such as 'global warming', 'greenhouse effect', etc. Searches for "greenhouse effect" and "allerg*" and "global warming" and "allerg*" resulted in far fewer items, many of which were indeed found in the "climate change" search, because this term was also used. Similarly, while the search term "allerg*"

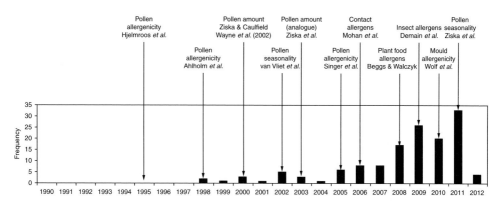

Fig. 11.3. Highlights in the development of climate change and allergen research from 1990 to 2012. Data from Fig. 11.1a are also provided for comparison. Highlighted studies may or may not be included in the Fig. 11.1a studies.

found all forms of this word root, such as allergen, allergens, allergic, allergy, allergies, allergenic, etc., there were likely to be some articles omitted from the search because they used words such as pollen, mould spore, etc., instead of allerg*. The latter searches have not been included here because the number of missed relevant articles is likely to be small, and because a "climate change" and "pollen" search results in of the order of 1700 results, most of which focus on long-term past (geological) climate change and changes in 'fossil' pollen as a reflection of changes in past plant life and as a climate proxy. Finally, some of the above analysis is based on the assumption that the first-named author of an article is the main or leading author. While it is believed that this is generally the case, it is not necessarily universally so.

11.8   Conclusion

Climate change is having, and will continue to have, a range of impacts on allergens. The evidence to date suggests that these impacts will have adverse outcomes for allergic diseases.

The basic responses to this particular significant climate change and global health issue, like any other, are mitigation of climate change and adaptation to its impacts. With the global failure to mitigate climate change fully or even significantly, we are now left having both to continue mitigation efforts and to establish and implement adaptation options.

At the same time, considerable further research on the impacts of climate change on allergens and allergic diseases is desperately required. This presents emerging and established researchers alike, from a range of disciplines, with an interesting, challenging and rewarding opportunity. The diversity of relatively recently discovered indirect impacts of climate change on allergic diseases, via impacts on pollen, mould spores, poison ivy, stinging insects and possibly plant food allergens such as groundnut, also serve as both a reminder of how intimately we are a part of the biosphere, and indeed the climate system, and a warning that there are likely to be impacts of climate change on global health that are unknown and have not even been contemplated.

Notes

[1] In Web of Science, double quotation marks around the topic search terms finds only exact matches for the terms.

[2] In Web of Science, an asterisk (*) retrieves zero or more characters. Therefore the "allerg*" search term retrieves any word starting with allerg, such as allergy, allergen, allergic, allergenic, allergenicity, etc.

References

Ahlholm, J.U., Helander, M.L. and Savolainen, J. (1998) Genetic and environmental factors affecting the allergenicity of birch (*Betula pubescens* ssp. *czerepanovii* [Orl.] Hämet-Ahti) pollen. *Clinical and Experimental Allergy* 28, 1384–1388.

Aït-Khaled, N., Pearce, N., Anderson, H.R., Ellwood, P., Montefort, S., Shah, J. and ISAAC Phase Three Study Group (2009) Global map of the prevalence of symptoms of rhinoconjunctivitis in children: The International Study of Asthma and Allergies in Childhood (ISAAC) Phase Three. *Allergy* 64, 123–148.

Beggs, P.J. (2004) Impacts of climate change on aeroallergens: past and future. *Clinical and Experimental Allergy* 34, 1507–1513.

Beggs, P.J. and Bambrick, H.J. (2005) Is the global rise of asthma an early impact of anthropogenic climate change? *Environmental Health Perspectives* 113, 915–919.

Beggs, P.J. and Walczyk, N.E. (2008) Impacts of climate change on plant food allergens: a previously unrecognized threat to human health. *Air Quality, Atmosphere and Health* 1, 119–123.

Bernard, S.M., Samet, J.M., Grambsch, A., Ebi, K.L. and Romieu, I. (2001) The potential impacts of climate variability and change on air pollution-related health effects in the United States. *Environmental Health Perspectives* 109, 199–209.

D'Amato, G., Cecchi, L., Bonini, S., Nunes, C., Annesi-Maesano, I., Behrendt, H., *et al.* (2007) Allergenic pollen and pollen allergy in Europe. *Allergy* 62, 976–990.

Demain, J.G., Gessner, B.D., McLaughlin, J.B., Sikes, D.S. and Foote, J.T. (2009) Increasing insect reactions in Alaska: Is this related to changing climate? *Allergy and Asthma Proceedings* 30, 238–243.

Emberlin, J., Detandt, M., Gehrig, R., Jaeger, S., Nolard, N. and Rantio-Lehtimäki, A. (2002) Responses in the start of *Betula* (birch) pollen seasons to recent changes in spring temperatures across Europe. *International Journal of Biometeorology* 46, 159–170 [Erratum published (2003) 47, 113–115].

Frei, T. and Gassner, E. (2008) Climate change and its impact on birch pollen quantities and the start of the pollen season an example from Switzerland for the period 1969–2006. *International Journal of Biometeorology* 52, 667–674.

Gilmour, M.I., Jaakkola, M.S., London, S.J., Nel, A.E. and Rogers, C.A. (2006) How exposure to environmental tobacco smoke, outdoor air pollutants, and increased pollen burdens influences the incidence of asthma. *Environmental Health Perspectives* 114, 627–633.

Hjelmroos, M., Schumacher, M.J. and van Hage-Hamsten, M. (1995) Heterogeneity of pollen proteins within individual *Betula pendula* trees. *International Archives of Allergy and Immunology* 108, 368–376.

Lai, C.K.W., Beasley, R., Crane, J., Foliaki, S., Shah, J., Weiland, S. and ISAAC Phase Three Study Group (2009) Global variation in the prevalence and severity of asthma symptoms: Phase Three of the International Study of Asthma and Allergies in Childhood (ISAAC). *Thorax* 64, 476–483.

Last, J.M. (1993) Global change: ozone depletion, greenhouse warming, and public health. *Annual Review of Public Health* 14, 115–136.

Last, J. and Guidotti, T.L. (1990/91) Implications for human health of global ecological changes. *Public Health Reviews* 18, 49–67.

Longstreth, J. (1991) Anticipated public health consequences of global climate change. *Environmental Health Perspectives* 96, 139–144.

Mohan, J.E., Ziska, L.H., Schlesinger, W.H., Thomas, R.B., Sicher, R.C., George, K., *et al.* (2006) Biomass and toxicity responses of poison ivy (*Toxicodendron radicans*) to elevated atmospheric CO_2. *Proceedings of the National Academy of Sciences* 103, 9086–9089.

Mudarri, D. (2010) *Public Health Consequences and Cost of Climate Change Impacts on Indoor Environments.* Prepared for: The Indoor Environments Division, Office of Radiation and Indoor Air, Environmental Protection Agency, Washington, DC.

Norval, M., Cullen, A.P., de Gruijl, F.R., Longstreth, J., Takizawa, Y., Lucas, R.M., *et al.* (2007) The effects on human health from stratospheric ozone depletion and its interactions with climate change. *Photochemical and Photobiological Sciences* 6, 232–251.

Pearce, N., Aït-Khaled, N., Beasley, R., Mallol, J., Keil, U., Mitchell, E., *et al.* and ISAAC Phase Three Study Group (2007) Worldwide trends in the prevalence of asthma symptoms: Phase III of the International Study of Asthma and Allergies in Childhood (ISAAC). *Thorax* 62, 758–766.

Rogers, C.A., Wayne, P.M., Macklin, E.A., Muilenberg, M.L., Wagner, C.J., Epstein, P.R., *et al.* (2006) Interaction of the onset of spring and elevated atmospheric CO_2 on ragweed (*Ambrosia artemisiifolia* L.) pollen production. *Environmental Health Perspectives* 114, 865–869.

Shea, K.M., Truckner, R.T., Weber, R.W. and Peden, D.B. (2008) Climate change and allergic disease. *The Journal of Allergy and Clinical Immunology* 122, 443–453.

Singer, B.D., Ziska, L.H., Frenz, D.A., Gebhard, D.E. and Straka, J.G. (2005) Increasing Amb a 1 content in common ragweed (*Ambrosia artemisiifolia*) pollen as a function of rising atmospheric CO_2 concentration. *Functional Plant Biology* 32, 667–670.

US EPA (Environmental Protection Agency) (2009) EPA's Endangerment Finding: Health Effects; date created and modified: 4 December, 2009 (http://epa.gov/climatechange/Downloads/endangerment/ EndangermentFinding_Health.pdf, accessed 16 March 2014).

van Vliet, A.J.H., Overeem, A., de Groot, R.S., Jacobs, A.F.G. and Spieksma, F.T.M. (2002) The influence of temperature and climate change on the timing of pollen release in the Netherlands. *International Journal of Climatology* 22, 1757–1767.

Vandenplas, O., D'Alpaos, V. and van Brussel, P. (2008) Rhinitis and its impact on work. *Current Opinion in Allergy and Clinical Immunology* 8, 145–149.

Vardoulakis, S. and Heaviside, C. (eds) (2012) *Health Effects of Climate Change in the UK 2012: Current Evidence, Recommendations and Research Gaps*. Health Protection Agency, London.

Wayne, P., Foster, S., Connolly, J., Bazzaz, F. and Epstein, P. (2002) Production of allergenic pollen by ragweed (*Ambrosia artemisiifolia* L.) is increased in CO_2-enriched atmospheres. *Annals of Allergy, Asthma, and Immunology* 88, 279–282.

Wolf, J., O'Neill, N.R., Rogers, C.A., Muilenberg, M.L. and Ziska, L.H. (2010) Elevated atmospheric carbon dioxide concentrations amplify *Alternaria alternata* sporulation and total antigen production. *Environmental Health Perspectives* 118, 1223–1228.

Ziello, C., Sparks, T.H., Estrella, N., Belmonte, J., Bergmann, K.C., Bucher, E., *et al.* (2012) Changes to airborne pollen counts across Europe. *PloS One* 7, e34076.

Ziska, L.H. and Caulfield, F.A. (2000) Rising CO_2 and pollen production of common ragweed (*Ambrosia artemisiifolia*), a known allergy-inducing species: implications for public health. *Australian Journal of Plant Physiology* 27, 893–898.

Ziska, L.H., Gebhard, D.E., Frenz, D.A., Faulkner, S., Singer, B.D. and Straka, J.G. (2003) Cities as harbingers of climate change: common ragweed, urbanization, and public health. *The Journal of Allergy and Clinical Immunology* 111, 290–295.

Ziska, L., Knowlton, K., Rogers, C., Dalan, D., Tierney, N., Elder, M.A., *et al.* (2011) Recent warming by latitude associated with increased length of ragweed pollen season in central North America. *Proceedings of the National Academy of Sciences* 108, 4248–4251.

12 Wildfires, Air Pollution, Climate Change and Health

Haruka Morita and Patrick Kinney
Department of Environmental Health Sciences,
Mailman School of Public Health, Columbia University, New York, USA

12.1 Introduction

Wildfires have important implications for human health, including through injury, mental health effects and air pollution on a global scale (Johnston *et al.*, 2012; Marlier *et al.*, 2012). Wildfire occurrence and behaviour are influenced strongly by temperature, precipitation and other meteorological factors, suggesting that climatic change could result in their changing patterns. Over the past century, an increase in the frequency and impacts of wildfires has been observed (Stocks *et al.*, 2002; Spracklen *et al.*, 2009; Weinhold, 2011). This changed trend has been noted in Canada (Amiro *et al.*, 2009) and Russia, where, in recent years, millions of hectares have been burned annually (Flannigan *et al.*, 2008; Shvidenko, 2011). The consequences of fires include health-harming air pollution and the release of additional greenhouse gases (GHG), further exacerbating climate change. The feedbacks linking climate, wildfire activity, air pollution and health are shown graphically in Fig. 12.1.

Interactions among climate variables, air pollution emissions and human health are well illustrated by recent events in Russia, where, in the summer of 2010, over 32,000 wildfires burned over 2 million hectares (Mha) of forests from mid-July to mid-August (Barriopedro *et al.*, 2011a; Bondur, 2011; Williams *et al.*, 2011). Thousands of lives were affected and 53 people were directly killed (Barriopedro *et al.*, 2011a). Impacts were enhanced by the fires' geographic location and high accompanying temperatures, which were record breaking in the weeks preceding and during the wildfires (Konovalov *et al.*, 2011). In July and August, 25 days exceeded the previous daily high temperature records, with many days above 30°C (Grumm, 2011). The heatwave was also estimated to have been responsible for about 55,000 excess deaths (Barriopedro *et al.*, 2011b). In the past, the vast majority of Russian wildfires have occurred in its Asian region; however, the 2010 wildfires burned the central area of European Russia (Shvidenko, 2011), home to roughly 71% of the 143 million Russian people (The Ministry of Foreign Affairs of Russia, 2012). Many homes were burned, but far more people were exposed to plumes of harmful emissions from the fires (Bondur, 2011). Air pollution concentrations considerably exceeded permissible levels in August 2010 (see the following) (Konovalov *et al.*, 2011; Zvyagintsev *et al.*, 2011). In another recent example, 172 civilians died from burns in bushfires near Melbourne, Australia, in 2009 (Handmer *et al.*, 2010).

A growing body of research is linking more intense and frequent episodes of wildfire with the environmental changes wrought by climate change. While fires play an important role in some ecosystems, such as by renewing tree stands in circumboreal forests (Flannigan *et al.*, 2008),

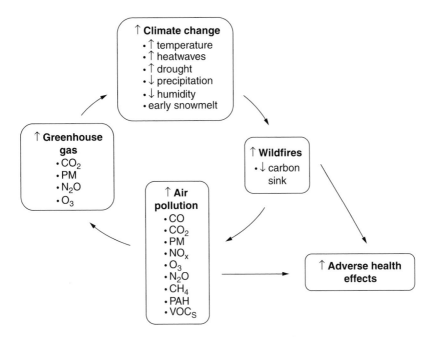

Fig. 12.1. Feedback loop showing the impacts of climate change, wildfires and their emissions, on each other as well as on human health.

their prevention and management is becoming critical, as their scale and consequences appear to be escalating. This chapter explores the positive (reinforcing) feedback loop created by the impact of climate change on wildfires and their consequent harmful emissions, including GHGs and other compounds which harm human health.

12.2 Meteorological Factors Affecting Wildfire Risk

Fire activity is influenced by many intertwined elements, including meteorology, climate, fuel type and dryness and human activity, including fire prevention and suppression efforts (Flannigan *et al.*, 2008). Some fires are also lit accidentally or deliberately; this aspect of human behaviour by a small minority of people, especially adolescents, appears inevitable. Meteorological factors include temperature, precipitation, relative humidity, wind speed (Liu *et al.*, 2010), probability of lightning and atmospheric stability (Groisman *et al.*, 2007). In northern latitudes,

studies narrow these to two critical variables: temperature and precipitation (Gillett *et al.*, 2004; Spracklen *et al.*, 2009; Liu *et al.*, 2010). Higher surface temperatures increase water demand by vegetation; if this is not met by precipitation, then the resultant drier land in the summer creates conditions favourable for wildfire (Groisman *et al.*, 2007).

During the summer of 2010, Russia suffered its most severe drought since 1972 (Grumm, 2011), characterized by anomalous temperature increases as well as low precipitation and humidity (Barriopedro *et al.*, 2011a). Decreased soil and fuel moisture, with increased evapotranspiration (Gedalof *et al.*, 2005; Westerling *et al.*, 2006) boosted the flammability of vegetation. At high altitudes and latitudes, summer drought also affects snowmelt (Westerling *et al.*, 2006). Earlier snowmelt, extended dry seasons and further soil and vegetation drying lengthen the period of potential biomass ignition and increase wildfire potential (Westerling *et al.*, 2006). Summer anticyclones in northern latitudes promote drought. A persistent anticyclone, as well as early spring thaw and low precipitation between January and July 2010, were associated with the 2010 Russian heat-wave (Barriopedro et al., 2011a). The 'blocking' anticyclone during the Russian wildfire was particularly persistent, effectively diverting cooler and damper air around it (McKenzie et al., 2004; Gedalof et al., 2005; Grumm, 2011). There are varying perspectives on whether the Russian heatwave was due to anomalously hot and dry conditions from natural variability in climate or climate change. One study said that the persistent anticyclone formation was due to 'internal atmospheric dynamical processes' or natural causes (Dole et al., 2011). However, this does not deny the possibility that past and future heatwaves were and will be influenced strongly by climate change. Another study concluded that the hot and dry conditions were due to external influences that were outside the scope of atmospheric processes. This study points to sea surface temperatures (SST) in the Northern Indian Ocean, which affected the anomalously persistent atmospheric circulations formed over Russia (Trenberth and Fasullo, 2012). The hot and dry conditions conducive for wildfire ignitions resulted from the anticyclone, but also were likely influenced by climate change. Increasing concentrations of long-living GHGs such as CO_2 and the accumulation of them in the atmosphere feeds back to intensify the warming of the surface. The anticyclone and global warming therefore may have had a synergistic effect that intensified the heatwave and facilitated the development of hot and dry conditions to exacerbate the wildfires. The previous study also assessed the effects of SST, but did not find significant patterns that could be associated with the anticyclone circulation formation.

Summer anticyclones are also associated with anomalous increases in the 500 hPa geopotential height[1] (Gedalof et al., 2005), and some simulations have shown that future scenarios of increased GHGs will increase the 500 hPa height in the USA, potentially leading to more intense heatwaves (Meehl and Tebaldi, 2004). Such anomalies were observed for the heatwaves in Chicago in 1995 and in Paris in 2003, although without catastrophic wildfires (Meehl and Tebaldi, 2004). However, greater areas experiencing large-scale wildfires have also been associated with 500 hPa height increases in North America (Gedalof et al., 2005).

12.3   Health Effects of Smoke Exposure

Wildfire emissions contribute significantly to particulate and gaseous air pollution concentrations in downwind areas. Emissions from wildfires include carbon dioxide (CO_2), nitrous oxide (N_2O), ozone (O_3), particulate matter (PM), carbon monoxide (CO), methane (CH_4), nitrogen oxides (NO_x), organic compounds, aldehydes, volatile organic compounds (VOCs), free radicals and polycyclic aromatic hydrocarbon (PAH) (Amiro et al., 2009). NO_2 and O_3 are irritants, carbon monoxide is an asphyxiant, PAHs are carcinogens, aldehydes are both irritants and carcinogens and PM can cause inflammation and oxidative stress, as well as increasing deaths both acutely and chronically (Naeher et al., 2007). Van Donkelaar et al. assessed the mortality impacts from the 2010 Russian wildfire, specifically from PM10. Excess deaths were quantified using PM10 data from various monitoring stations in Moscow and mortality data from a World Health Organization (WHO) meta-study of 33 European cities. Based on published evidence from prior studies, they assumed a 0.6% increase in excess death associated with a 10 µg/m³ increase in PM10. Assuming a daily average of 400 deaths in Moscow, this amounts to 434 excess deaths from PM10 between 2 and 10 August (van Donkelaar et al., 2011). Because of such excess health burdens, technology to track emissions and estimate air pollution concentrations is critical in assessing wildfire health impacts.

National Aeronautics and Space Administration (NASA) satellites can measure emissions-related variables including aerosol optical thickness (AOT),[2] UV aerosol index (AI), single-scattering albedo (SSA) and total column CO (CO_{TC}) (Witte et al., 2011). AOT is a measure of how much light transmission is inhibited by absorption or scattering of light. AI measures aerosols such as dust and soot that absorb UV radiation. SSA is a measure of light extinction via scattering compared to absorption. CO_{TC} captures the atmospheric concentrations of CO. CO is not a direct GHG but it hastens global warming indirectly by facilitating reactions for O_3 formation; its rate is determined by the degree of vertical mixing of CO in the atmosphere. Inversions, characterized by a positive vertical temperature gradient, limit vertical mixing and contain fire emissions near the surface, as opposed to distributing them vertically in the atmosphere (Zvyagintsev *et al.*, 2011). CO_{TC}, AOT and AI were very high during the 2010 wildfires in Russia, indicative of very high emission levels. SSA decreased during this period, consistent with the increase in AI. These anomalies diminished after 18 August, as the fires subsided (Witte *et al.*, 2011). In another study, aerosol optical depth (AOD) data from the NASA satellite estimated mean PM2.5 concentrations ranging from 13.1 to 35.7 $\mu g/m^3$ during the pre-fire period between 7 and 21 July. These data correlated well with the observed mean PM2.5 range of between 16.7 and 32.1 $\mu g/m^3$, derived from various monitoring stations in Moscow during the pre-fire period using a 66.8% PM2.5/PM10 ratio. Derived mean PM2.5 concentrations during the fire period between 3 and 10 August ranged from 223 to 333 $\mu g/m^3$ (van Donkelaar *et al.*, 2011).

Back trajectories from Moscow during the wildfires revealed an anticyclonic pattern over the city, which eventually dissipated with the return of the four variables (i.e. AOT, AI, CO_{TC} and SSA) to average readings (Witte *et al.*, 2011). Thick plumes of harmful smoke were transported to Moscow from the south and east, and the anticyclone circulations trapped these wildfire emissions over the city and surrounding regions (Plate 5) (Witte *et al.*, 2011). CO is emitted in the greatest quantity compared to all other pollutants from wildfires, and satellite observations showed that the CO concentrations were elevated significantly during the wildfire (Konovalov *et al.*, 2011). A maximum daily CO concentration of 37 mg/m^3 was reported, more than seven times the maximum permissible concentration (MPC) in Russia of 5 mg/m^3 (Zvyagintsev *et al.*, 2011). Other compounds were also observed at significantly elevated concentrations. The maximum daily concentration of PM10 was 1.7 mg/m^3 (MPC 0.5 mg/m^3) (Zvyagintsev *et al.*, 2011). The maximum daily concentration of nitrogen dioxide (NO_2) exceeded 500 $\mu g/m^3$, while the MPC was 200 $\mu g/m^3$ (Konovalov *et al.*, 2011). These are primary air pollutants with direct health impact, and they are also precursors of secondary air pollutants including O_3 (Fowler, 2003). During the 2010 wildfire, maximum daily O_3 levels above 300 $\mu g/m^3$ were observed, almost double the Russian air-quality

standard of 160 $\mu g/m^3$ (Konovalov *et al.*, 2011). The highest O_3 concentrations were associated with a narrow temperature gradient, indicative of weak vertical mixing (Zvyagintsev *et al.*, 2011). These four pollutants are among the six that are regulated by the National Ambient Air Quality Standard set by the US EPA, due to harmful effects on human health and the environment (Fowler, 2003).

Health impacts from wildfire emissions are experienced not only by local populations but also by populations of neighbouring regions and countries, as smoke plumes have been observed to travel long distances. Wildfire emissions from Quebec, Canada, were tracked to Washington, DC (Colarco *et al.*, 2004) and Baltimore, Maryland, USA (Sapkota *et al.*, 2005). NASA satellites showed that the emissions from a 2003 Russian wildfire circled the world in 17 days (Damoah *et al.*, 2004). Understanding the extent of pollution dispersion and the various human health consequences is critical in anticipating the magnitude of, and in responding to, the health effects from wildfire emissions.

Emmanuel measured impacts of haze on respiratory health experienced in Singapore from the 1997 fires in Indonesia that occurred between July and early October (Emmanuel, 2000). Deliberately lit fires to clear land occur annually in Indonesia, but their scale was worsened by the drought that year associated with an intense El Niño event (Page *et al.*, 2002). The haze from the fire affecting Singapore between August and November 1997 increased PM10 from its monthly average of 30–50 $\mu g/m^3$ to 60–100 $\mu g/m^3$. In Singapore, health surveillance found increases of 12% in upper respiratory tract illnesses, 19% in asthma and 26% in rhinitis associated with the increase in PM10 concentration. However, there was no detected increase in hospital admissions or mortality.

Rappold *et al.* evaluated the health effects associated with the North Carolina (USA) peat bog wildfire in 2008 by linking AOD measurements and data on emergency department visits from syndromic surveillance. Emergency department visits were significantly higher in exposed counties on and after high-exposure days compared to the referent non-exposed counties. Respiratory outcomes (i.e. asthma, obstructive pulmonary disorder, pneumonia and acute bronchitis) increased significantly. Cardiovascular outcomes (i.e. heart failure, cardiac dysrhythmia and myocardial infarction) also rose significantly (Rappold *et al.*, 2011). Peat fires are common in boreal forests and contain high fuel loads. Their emissions vary as compared with flaming fires of trees, as peat burns by smouldering (Flannigan *et al.*, 2008; Langmann *et al.*, 2009). Smouldering fires emit greater quantities of CO than flaming fires (Amiro *et al.*, 2009; Shvidenko, 2011). However, peat fires are difficult to detect remotely because they are often hidden by trees (Konovalov *et al.*, 2011). Peat can also burn underground for months (Amiro *et al.*, 2009), posing challenges to future projections.

Health impact assessments (HIAs) have proven to be a useful tool for evaluating the health effects from fire events (e.g. van Donkelaar *et al.*, 2011). However, there are several challenges in conducting HIAs for wildfires. Baseline health statuses are important to forecast susceptibility to air pollution; however, such data are often lacking. HIAs in regions with low background pollution levels, such as in Singapore, may uncover a heightened health effect compared to regions with high background pollution levels, like Russia. In addition, some people are exposed to greater emission concentrations during wildfire events, such as firefighters, while others, such as with pre-existing respiratory conditions, are susceptible to even tiny increases in air pollution concentration (Naeher *et al.*, 2007). Naeher *et al.* cited various studies on occupational exposures and found that most reported increases in respiratory outcomes such as lower lung function. Greater airway hyperresponsiveness was observed among firefighters.

To date, there has been limited information on human exposures to wildfire emissions. The majority of the studies have been ecological, where wildfire-related air pollution levels are matched geographically to potentially affected populations. A study on associations of cardiorespiratory hospital admissions with PM10 from vegetation fires pointed out the limitation of derived exposure measures (Hanigan *et al.*, 2008). Wildfire emissions combine numerous toxic compounds, and it is difficult to tease out the health effects of their components (van Donkelaar *et al.*, 2011). Further, fire events can emit different mixtures, depending on the fuel type (Spracklen *et al.*, 2009), limiting the generalizability of studies (Naeher *et al.*, 2007) and making attribution of individual compounds with specific health effects difficult. To complicate HIAs further, wildfire events are often accompanied by heatwaves that can amplify adverse health effects.

12.4 Wildfire Impacts on Climate

Wildfires are significant sources of air pollution and GHGs, including CO_2, N_2O, O_3 and black carbon, that absorb infrared radiation and trap heat in the lower atmosphere. This contributes to the global warming projected for this century (Alley *et al.*, 2007). Forests store large amounts of carbon, which, if released through combustion, can have a large impact on the global carbon balance. Vegetation often regrows after wildfires, recreating sinks for the carbon that was released. However, the frequency and magnitude of increased fire means that forests may not renew fast enough to replenish depleted carbon sinks. Further, the charred surfaces from wildfires decrease albedo, facilitating greater heat absorption, also contributing to global warming (Flannigan *et al.*, 2008). In effect, GHG emissions from wildfires contribute to climate change and further promote conditions conducive to wildfires, potentially of greater intensity (see Fig. 12.1).

12.5 Relevant Spatial and Temporal Climate Change Patterns

Atmospheric CO_2 concentrations were 280 ± 10 parts per million (ppm) during the pre-industrial era before 1750 (Prentice *et al.*, 2001), and today are about 400 ppm. The Intergovernmental Panel on Climate Change (IPCC) reported an increase in global mean temperature of $0.74 \pm 0.18°C$ in the 100 years between 1906 and 2005 (Alley *et al.*, 2007). To date, various evidence of climatic change has been observed. Northern Eurasia experienced the most consistent and substantial increase in surface air temperature during the 20th century, with a greater prominence in the past 50 years (Groisman *et al.*, 2007). The incidence of abnormally cold nights and days decreased, whereas the incidence of abnormally warm nights and days increased. Days with thaw, or snowmelt, occurred earlier during the latter half of the 20th century, and the severity of cold seasons has diminished significantly.

Climatic changes in the northern latitudes are of particular importance to wildfires, for several reasons. Temperature changes are most marked in these higher latitudes (Groisman *et al.*, 2007). Hence, these regions are experiencing a greater impact from climate change than are tropical regions. Boreal forests that serve as fuel for wildfires are mostly aggregated in the northern latitudes and equatorial regions (Stocks *et al.*, 1998) (see Plate 6). The three countries with the largest forest area are Russia, Brazil and Canada, with 809, 478 and 310 Mha, respectively (Food and Agriculture Organization, 2005). Russia and Canada together account for 28.3% of the world's forests.

The more pronounced climate change and the large percentage of boreal forests in the northern latitudes render mitigation and adaptation strategies critical in preventing intense wildfires in these regions. Reductions in wildfire emissions would reduce their impacts on global warming while also yielding health co-benefits through reductions in harmful air pollutants.

12.6 Future Climate Change Impacts on Wildfire Activity

The IPCC projects that, by the end of the 21st century, CO_2 concentration may increase to between 540 and 970 ppm, and global surface temperature may rise by 1.1–6.4°C (Alley *et al.*, 2007; Parry *et al.*, 2007). Environmental consequences from climate change that are already observed are likely to accelerate during this century with the projected increases in CO_2 concentrations and temperature. The climate and health impacts of wildfires highlight the importance of estimating future wildfires under changing climate conditions, including the application of future wildfire scenarios to assess their impact on air pollution. Particularly informative variables for projecting future air pollution emissions include area burned, changes in fuel load and fuel consumption.

Fire risk projection

Among the various methods available, the ones used more widely are the Canadian Forest Fire

Weather Index (FWI) System and the Canadian Forest Fire Behaviour Prediction (FBP) System. The FWI System combines daily temperature, relative humidity, wind speed and 24-h rainfall measurements to yield outputs for six components – Fine Fuel Moisture Code, Duff Moisture Code, Drought Code, Initial Spread Index, Build-up Index and Fire Weather Index – to predict wildfire risk based on weather variables (Amiro *et al.*, 2009). The FBP System estimates fuel consumption, rate of spread, fire intensity at the head of the fire, crown fraction burned and fire type (surface or crown), by using various inputs on fuel, weather, topography, foliar moisture content and duration of prediction (Natural Resources Canada, 2012). These systems emphasize the impact of specific variables based on a set of assumptions to predict future wildfires, and have been employed by various studies to predict future fire risk around the world.

Wildfire risk projections for Canada and Russia comparing a $1 \times CO_2$ baseline concentration from the 1980–1989 decade with outputs of doubled CO_2 scenarios from four General Circulation Models (GCMs) revealed a consistent increase in wildfire risk in central Canada and Siberia, especially for June and July. Risk was quantified as daily, monthly and seasonal severity rating (DSR, MSR and SSR, respectively) using the FWI System (Stocks *et al.*, 1998). Wildfire risk projections for the USA comparing a $1 \times CO_2$ baseline concentration during the 1980–1994 period to $2 \times CO_2$ scenario for the year 2060 revealed an increase of between 10% and 50% in SSR across the country, with the greatest SSR in Alaska (Flannigan *et al.*, 2000). Although the projected magnitude of wildfire risk varies between studies, the predictions uniformly project increased risk in the northern latitudes with climate change. Projections of wildfire risk and wildfire emissions are each critical to inform wildfire prevention strategies in the context of climate change.

Emissions projection

Emissions projections are derived from area burned, fuel load, fuel consumed and emission factors (French *et al.*, 2007). An emissions factor represents the mass of a gaseous or particulate species emitted per kilogram of dry matter burned. Knowing these quantities allows calculation of the total emissions produced (Amiro *et al.*, 2009; Langmann *et al.*, 2009). Fuel load is dependent on vegetation type, climate, soil composition and time interval between fires or other disturbance (Langmann *et al.*, 2009). Boreal (e.g. Canada, Russia) and tropical forests (e.g. Brazil) contain the greatest fuel loads and hence are of great concern (Langmann *et al.*, 2009).

Increases in area burned and fuel consumed have been projected with future climate scenarios. In numerous studies, the results indicate an increase in future emissions, although the degree of these increases varies between studies. Using the FWI and FBP systems, the predicted increase in the area burned in Canada under a $2 \times CO_2$ scenario (representing the 2040–2060 decades) and a $3 \times CO_2$ scenario (the 2080–2100 decades) (Amiro *et al.*, 2009) were compared to a baseline CO_2 concentration in the period between 1975 and 1995. Mean annual area burned was projected to be one-third times greater with the doubled CO_2 concentration and two times greater with $3 \times CO_2$ concentration (see Fig. 12.2). Fuel consumption was also predicted to increase, with 116.8 terragrams (Tg) of dry fuel consumed under the $3 \times CO_2$ scenario compared to 60.3 Tg

under the $1 \times CO_2$ baseline. With these outputs, calculations of future emissions in Canada for various air pollutants, such as CO_2, CO, NO_x and particulate matter less than 2.5 microns (PM2.5), approximately doubled for all pollutants measured when comparing the $3 \times CO_2$ to the baseline. CO_2 and CO emissions from wildfires were predicted to be 183.3 Tg/year and 12.5 Tg/year, respectively, in the $3 \times CO_2$ scenario compared to $1 \times CO_2$ baseline emissions of 94.6 Tg/year of CO_2 and 6.4 Tg/year of CO. PM2.5 and NO_x emissions were predicted to be 1519 gigagramme (Gg)/year and 350 Gg/year, respectively, compared to a baseline concentration of 784 Tg/year of PM2.5 and 181 Gg/year of NO_x (Amiro *et al.*, 2009). It is important to note that fuel load is the most challenging factor to estimate in projecting future emissions (French *et al.*, 2007). These calculations assumed no change in fuel load and assessed the sole impact of CO_2 concentration increase on future emissions (Amiro *et al.*, 2009). However, public health impacts can be estimated with wildfire and emissions projections that inform the development of climate change policies and public health strategies.

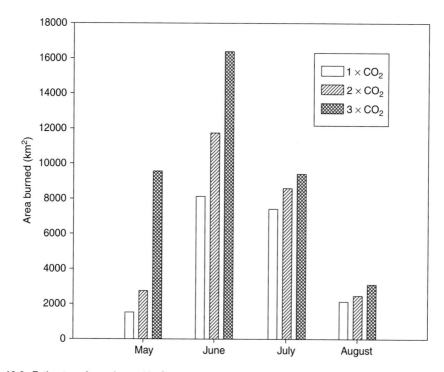

Fig. 12.2. Estimates of area burned in Canada for three future climate scenarios (Amiro *et al.*, 2009). (© 2008 Canadian Science Publishing or its licensors. Reproduced with permission.)

12.7 Conclusion

There remain many shortcomings and challenges in accurately measuring past and projecting future wildfire events and their consequences, especially for human health. However, ecological studies convincingly show that increased air pollution levels are associated with adverse health effects. Wildfire emissions include several pollutants that are commonly experienced in urban areas such as PAH and PM, and it is thus reasonable to carry out preliminary HIAs based on epidemiological findings from urban pollution studies.

In addition, wildfires emit GHGs which exacerbate global climate change and increase the chance of more fires. They also burn property, livestock and crops, and can directly kill people and traumatize survivors. The consequences of climate change such as abnormally high temperature, early snowmelt and increased dryness all predispose to wildfires.

Both adaptation and mitigation strategies are needed to address these challenges. Climate change mitigation strategies that aim to slow the development of adverse environmental conditions that promote wildfires are necessary. In the short to medium term, the capacity to spread knowledge, such as through weather advisories and warning systems in anticipation of extreme heat and fire-inducing weather, will help public institutions and populations prepare for and respond more effectively to potential wildfires, and also reduce their exposure to emissions, thus lowering adverse health effects. These climate and public health strategies will be key in disrupting the feedback loop to prevent future wildfires; they will also yield health co-benefits.

Notes

[1] Geopotential height is the approximate height above sea level of a pressure level. For example, if a station reports that the 500 hPa height at its location is 5600 m, it means that the level of the atmosphere over that station at which the atmospheric pressure is 500 hPa is 5600 m above sea level. This is an estimated height based on temperature and pressure data (http://w1.weather.gov/glossary/index.php?letter=h).
[2] Or AOD.

References

Alley, R., Berntsen, T., Bindoff, N.L., Chen, Z., Chidthaisong, A., Friedlingstein, P., *et al.* (2007) Summary for policymakers. In: Solomon, S., Qin, D., Manning, M., Chen, Z., Marquis, M., Avery, K.B., Ignor, M. and Miller, H.L. (eds) *Climate Change 2007: The Physical Science Basis. Contribution of Working Group I to the Fourth Assessment Report of the Intergovernmental Panel on Climate Change.* Cambridge University Press, Cambridge, UK, pp. 1–18.

Amiro, B.D., Cantin, A., Flannigan, M.D. and de Groot, W.J. (2009) Future emissions from Canadian boreal forest fires. *Canadian Journal of Forest Research* 39, 383–395.

Barriopedro, D., Fischer, E.M., Luterbacher, J., Trigo, R.M. and García-Herrera, R. (2011a) The hot summer of 2010: redrawing the temperature record map of Europe. *Science* 332, 220–224.

Barriopedro, D., Fischer, E.M., Luterbacher, J., Trigo, R.M. and García-Herrera, R. (2011b) Supporting online material for the hot summer of 2010: redrawing the temperature record map of Europe. *Science* [Online] (http://www.sciencemag.org/content/suppl/2011/03/15/science.1201224.DC1/Barriopedro-SOM.pdf, accessed 16 March 2014).

Bondur, V. (2011) Satellite monitoring of wildfires during the anomalous heat wave of 2010 in Russia. *Izvestiya Atmospheric and Oceanic Physics* 47, 1039–1048.

Colarco, P.R., Schoeberl, M.R., Doddridge, B.G., Marufu, L.T., Torres, O. and Welton, E.J. (2004) Transport of smoke from Canadian forest fires to the surface near Washington, DC: injection height, entrainment, and optical properties. *Journal of Geophysical Research* 109, D06203.

Damoah, R., Spichtinger, N., Forster, C., James, P., Mattis, I., Wandinger, U., *et al.* (2004) Around the world in 17 days – hemispheric-scale transport of forest fire smoke from Russia in May 2003. *Atmospheric Chemistry and Physics* 4, 1311–1321.

Dole, R., Hoerling, M., Perlwitz, J., Eischeid, J., Pegion, P., Zhang, T., *et al.* (2011) Was there a basis for anticipating the 2010 Russian heat wave? *Geophysical Research Letters* 38, L06702.

Emmanuel, S.C. (2000) Impact to lung health of haze from forest fires: the Singapore experience. *Respirology* 5, 175–182.

Flannigan, M., Stocks, B., Turetsky, M. and Wotton, M. (2008) Impacts of climate change on fire activity and fire management in the circumboreal forest. *Global Change Biology* 15, 549–560.

Flannigan, M.D., Stocks, B.J. and Wotton, B.M. (2000) Climate change and forest fires. *Science of the Total Environment* 262, 221–229.

Food and Agriculture Organization (2005) Ten countries with largest forest area (million ha) (http://www.fao.org/forestry/12959-0a21d605dcfef9d94c63d8f474f56dc65.jpg, accessed 10 October 2012).

Fowler, C.T. (2003) Human health impacts of forest fires in the southern United States: a literature review. *Journal of Ecological Anthropology* 7, 39–63.

French, N.H.F., Kasischke, E.S., Turetsky, M.R., de Groot, W., Honrath, R.E. and Ottmar, R.D. (2007) Carbon, trace gas, and particulate emissions from wildfires in the Boreal regions of North America. 16th Annual International Emission Inventory Conference. Emission Inventories: 'Integration, Analysis, and Communications', 2007 Raleigh, North Carolina (http://www.google.com.au/url?sa=t&rct=j&q=&esrc=s&source=web&cd=1&ved=0CCkQFjAA&url=http%3A%2F%2Fwww.epa.gov%2Fttnchie1%2Fconference%2Fei16%2Fsession10%2Ffrench.pdf&ei=NkUlU6ujG8rqkgXs44HoCQ&usg=AFQjCNFbMH2uXAReYKcBtBGvTo_NK4bwGA&sig2=DSzO-hvxD_RDuo5UGNNAmg&bvm=bv.62922401,d.dGI, accessed 16 March 2014).

Gedalof, Z., Peterson, D.L. and Mantua, N.J. (2005) Atmospheric, climatic, and ecological controls on extreme wildfire years in the northwestern United States. *Ecological Applications* 15, 154–174.

Gillett, N.P., Weaver, A.J., Zwiers, F.W. and Flannigan, M.D. (2004) Detecting the effect of climate change on Canadian forest fires. *Geophysical Research Letters* 31, L18211.

Groisman, P.Y., Sherstyukov, B.G., Razuvaev, V.N., Knight, R.W., Enloe, J.G., Stroumentova, N.S., *et al.* (2007) Potential forest fire danger over Northern Eurasia: changes during the 20th century. *Global and Planetary Change* 56, 371–386.

Grumm, R.H. (2011) The central European and Russian heat event of July–August 2010. *Bulletin of the American Meteorological Society* 92, 1285–1296.

Handmer, J., O'Neil, S. and Killalea, D. (2010) *Review of Fatalities in the February 7, 2009, Bushfires.* Bushfire CRC and Centre for Risk and Community Safety, RMIT University, Melbourne, Australia.

Hanigan, I.C., Johnston, F.H. and Morgan, G.G. (2008) Vegetation fire smoke, indigenous status and cardio-respiratory hospital admissions in Darwin, Australia, 1996–2005: a time-series study. *Environmental Health* 7, 42, doi:10.1186/1476-069X-7-42.

Johnston, F.H., Henderson, S.B., Chen, Y., Randerson, J.T., Marlier, M., DeFries, R.S., *et al.* (2012) Estimated global mortality attributable to smoke from landscape fires. *Environmental Health Perspectives* 120, 695–701.

Konovalov, I.B., Beekmann, M., Kuznetsova, I.N., Yurova, A. and Zvyagintsev, A.M. (2011) Atmospheric impacts of the 2010 Russian wildfires: integrating modelling and measurements of the extreme air pollution episode in the Moscow megacity region. *Atmospheric Chemistry and Physics* 11, 10031–10056.

Langmann, B., Duncan, B., Textor, C., Trentmann, J. and van der Werf, G.R. (2009) Vegetation fire emissions and their impact on air pollution and climate. *Atmospheric Environment* 43, 107–116.

Liu, Y., Stanturf, J. and Goodrick, S. (2010) Trends in global wildfire potential in a changing climate. *Forest Ecology and Management* 259, 685–697.

McKenzie, D., Gedalof, Z., Peterson, D.L. and Mote, P. (2004) Climatic change, wildfire, and conservation. *Conservation Biology* 18, 890–902.

Marlier, M.E., DeFries, R.S., Voulgarakis, A., Kinney, P.L., Randerson, J.T., Shindell, D.T., *et al.* (2012) El Niño and health risks from landscape fire emissions in Southeast Asia. *Nature Climate Change* 3, 131–136.

Meehl, G.A. and Tebaldi, C. (2004) More intense, more frequent, and longer lasting heat waves in the 21st century. *Science* 305, 994–997.

Ministry of Foreign Affairs of Russia, The (2012) Population data. The Embassy of the Russian Federation to the United Kingdom of Great Britain and Northern Ireland (http://www.rusemb.org.uk/russianpopulation/, accessed 20 September 2012).

Naeher, L.P., Brauer, M., Lipsett, M., Zelikoff, J.T., Simpson, C.D., Koenig, J.Q., *et al.* (2007) Woodsmoke health effects: a review. *Inhalation Toxicology* 19, 67–106.

Natural Resources Canada (2012) Background Information: Canadian Forest Fire Behavior Prediction (FBP) System (http://cwfis.cfs.nrcan.gc.ca/background, accessed 16 March 2014).

Page, S.E., Siegert, F., Rieley, J.O., Boehm, H.D.V., Jaya, A. and Limin, S. (2002) The amount of carbon released from peat and forest fires in Indonesia during 1997. *Nature* 420, 61–65.

Parry, M.L., Canziani, O.F. and Palutikof, J.P. (2007) Technical summary. In: Parry, M.L., Canziani, O.F., Palutikof, J.P., van der Linden, P.J. and Hanson, C.E. (eds) *Climate Change 2007: Impacts, Adaptation and Vulnerability. Contribution of Working Group II to the Fourth Assessment Report of the Intergovernmental Panel on Climate Change.* Cambridge University Press, Cambridge, UK, pp. 23–78.

Prentice, I.C., Farquhar, G.D., Fasham, M.J.R., Gouldenm, M.L., Heimann, M., Jaramillo, V.J., *et al.* (2001) The carbon cycle and atmospheric carbon dioxide. In: Houghton, J.T., Ding, Y., Griggs, D.J., Noguer, M., van der Linden, P.J., Dai, X., Maskell, K. and Johnson, C.A. (eds) *Climate Change 2001: The Scientific Basis.* Cambridge University Press, Cambridge, UK, pp. 183–237.

Rappold, A.G., Stone, S.L., Cascio, W.E., Neas, L.M., Kilaru, V.J., Carraway, M.S., *et al.* (2011) Peat bog wildfire smoke exposure in rural North Carolina is associated with cardiopulmonary emergency department visits assessed through syndromic surveillance. *Environmental Health Perspectives* 119, 1415–1420.

Sapkota, A., Symons, J.M., Kleissl, J., Wang, L., Parlange, M.B., Ondov, J., *et al.* (2005) Impact of the 2002 Canadian forest fires on particulate matter air quality in Baltimore City. *Environmental Science and Technology* 39, 24–32.

Shvidenko, A.Z. (2011) Impact of wildfire in Russia between 1998–2010 on ecosystems and the global carbon budget. *Doklady Earth Sciences* 441, 1678–1682.

Spracklen, D.V., Mickley, L.J., Logan, J.A., Hudman, R.C., Yevich, R., Flannigan, M.D., *et al.* (2009) Impacts of climate change from 2000 to 2050 on wildfire activity and carbonaceous aerosol concentrations in the western United States. *Journal of Geophysical Research* 114, D20301.

Stocks, B.J., Fosberg, M.A., Lynham, T.J., Mearns, L., Wotton, B.M., Yang, Q., *et al.* (1998) Climate change and forest fire potential in Russian and Canadian boreal forests. *Climatic Change* 38, 1–13.

Stocks, B.J., Mason, J.A., Todd, J.B., Bosch, E.M., Wotton, B.M., Amiro, B.D., *et al.* (2002) Large forest fires in Canada, 1959–1997. *Journal of Geophysical Research* 107, 8149, doi:10.1029/2001JD000484.

Trenberth, K.E. and Fasullo, J.T. (2012) Climate extremes and climate change: the Russian heat wave and other climate extremes of 2010. *Journal of Geophysical Research: Atmospheres* 117, D17103.

van Donkelaar, A., Martin, R.V., Levy, R.C., da Silva, A.M., Krzyzanowski, M., Chubarova, N.E., *et al.* (2011) Satellite-based estimates of ground-level fine particulate matter during extreme events: a case study of the Moscow fires in 2010. *Atmospheric Environment* 45, 6225–6232.

Weinhold, B. (2011) Fields and forests in flames: vegetation smoke and human health. *Environmental Health Perspectives* 119, a386.

Westerling, A.L., Hidalgo, H.G., Cayan, D.R. and Swetnam, T.W. (2006) Warming and earlier spring increase western US forest wildfire activity. *Science* 313, 940–943.

Williams, J., Albright, D., Hoffmann, A.A., Eritsov, A. and Leonard, M. (2011) Findings and implications from a coarse-scale global assessment of recent selected mega-fires. 5th International Wildland Fire Conference, 2011 Sun City, South Africa, pp. 9–13.

Witte, J.C., Douglass, A.R., da Silva, A., Torres, O., Levy, R.C. and Duncan, B.N. (2011) NASA A-Train and Terra observations of the 2010 Russian wildfires. *Atmospheric Chemistry Physics Discussions* 11, 19113–19142.

Zvyagintsev, A., Blum, O., Glazkova, A., Kotel'nikov, S., Kuznetsova, I., Lapchenko, V., *et al.* (2011) Air pollution over European Russia and Ukraine under the hot summer conditions of 2010. *Izvestiya Atmospheric and Oceanic Physics* 47, 699–707.

13 Famine, Hunger, Society and Climate Change

Colin D. Butler

Faculty of Health, The University of Canberra, Australia, and National Centre for Epidemiology and Population Health, The Australian National University, Australia, and Benevolent Organisation for Development, Health & Insight (BODHI)

And he gave it for his opinion that he that could make two ears of corn or two blades of grass to grow upon a spot of ground where only one grew before would deserve better of mankind and do more essential service to his country than the whole race of politicians put together.

(Swift, 1729)

13.1 Introduction: Famine, Population Growth and the Cold War

In 1966, a famine was emerging in the chronically impoverished, caste-ridden northern Indian state of Bihar. Its ecological causes combined drought and flood with limited use of the then emerging technology to extract groundwater using tube wells (Brass, 1986).

India was an important Asian democracy. It was non-aligned during the Cold War, a prize ally for either of that undeclared conflict's main protagonists. Winning the hearts and minds of the people of the Third World had long been part of US policy, including through a programme established in the 1950s but christened as 'food for peace' in 1960 by John F. Kennedy, then campaigning for the presidency. On hearing of the Bihar famine, senior US state officials recommended emergency shipments of American grain. However, Kennedy's successor, President Lyndon

Johnson, prevaricated, instead authorizing shipments only in increments, in exchange for evidence that India's family planning programme was being strengthened.

India is the birthplace of Amartya Sen, justly famous for his work on famine (Sen, 1981). In 1998, Sen was awarded the Nobel Prize in Economics, partly for his concept of economic 'entitlement'. India has a long familiarity with devastating famine, particularly in the 19th century, as described vividly, for example, by Mike Davis in his book, *Late Victorian Holocausts: El Niño Famines and the Making of the Third World* (Davis, 2001).

Across the Himalayas, China, the land that had long called itself the Middle Kingdom, had an even longer experience of recorded famine. Chinese bureaucrats are reported to have documented 1823 famines between 108 BCE and 1911 CE (Mallory, 1926). Famines did not end in China in 1911. In Communist China, between 1958 and 1961, Mao's catastrophic 'great leap forward' culminated in a famine kept secret at the time, killing at least 30 million people (Becker, 1996).

At the time that America was providing grain relief to India its officials may not even have known of the then recent Chinese famine. Lloyd Boyd Orr, founding director of the United Nation's Food and Agricultural Organization (FAO), was one of the rare Western visitors to go to China during the famine. Orr (then aged about 80), along with other dignitaries, including the President of the Royal Society, reportedly accepted at face value Chinese assurances of adequate food (Becker, 1996).

In 1949, Orr was awarded the Nobel Peace Prize for his pioneering work in nutrition. He was also intensely concerned with the links between food insecurity and possible conflict. As FAO Director, he had tried to promote world peace and food security through the establishment of a World Food Board and expanded agricultural productivity (Staples, 2003). Two decades later, the Nobel Prize committee reaffirmed their recognition of the link between food security and the avoidance of conflict by awarding its highest peace honour to the agricultural scientist, Norman Borlaug. Borlaug is the main parent of the 'Green Revolution', a term adopted because it was thought greater food supply could lower the attraction of Communism and thus avert the Red Revolution that many policy makers feared.

13.2 The Apparent Conquest of Famine by the Green Revolution

The Green Revolution relied on unparalleled international and interdisciplinary cooperation (Boyer *et al.*, 2013). Achieved by the long-established techniques of selective plant breeding, it enabled far more food to be grown in a given area, provided there was adequate fertilizer and water. Its dependency on monocultures and fossil fuels, especially as the basic source of pesticides and nitrogen fertilizer, has been criticized by many activists, most notably Vandana Shiva (Shiva, 1991). In many countries, such industrialized agriculture favoured large farmers over small farmers, and helped to cement or intensify inequality. However, few if any scholars have credibly proposed how the widespread famines, perceived by many as otherwise plausible, could have been averted without the increased yield facilitated by the Green Revolution, the origins of which can be traced to the 1930s. It was well under way by the mid-1960s (a decade of intense concern about impending global food insecurity), though on a scale which at the time was little recognized.

There are many interlocking reasons for inequality and poverty, but most of these are social or 'natural' (e.g. geographical or environmental). Attributing poverty and inequality to the Green Revolution is simplistic, unless its meaning is widened to include the human institutions who apply and control these agricultural technologies.

In the following two decades, the Green Revolution was spectacularly successful in putting the global plough (food supplies) ahead of the stork (human numbers) (see Fig. 13.1). World hunger also declined, not only as a proportion but also even in absolute terms. A few famines occurred from 1970 to 2000 (including in Bangladesh in 1974 and Ethiopia in 1984), but their scale was modest compared to what had been feared and predicted.

In his 1970 Nobel Prize acceptance speech, Borlaug warned:

> The Green Revolution has won a temporary success in man's war against hunger and deprivation; it has given man a breathing space. If fully implemented, the revolution can provide sufficient food for sustenance during the next three decades. But the frightening power of human reproduction must also be curbed; otherwise the successes of The Green Revolution will be ephemeral only.
>
> (Tribe, 1994)

This warning was quickly forgotten. In 1984, US President Ronald Reagan became the first of a series of US presidents to declare explicitly that the importance of human population size had been overestimated (Butler, 2004). Since then, the plough has no longer been gaining on the stork. Although some food experts, including the former head of the World Food Programme, Josette Sheeran, have claimed that food production per person has continued to increase in recent decades and is now at a record high (Sheeran, 2011), the reality is less comforting. Already, more than one-third of the US maize crop is diverted to biofuels. While a fraction of its food value is recovered and fed to cattle as 'distiller's grain', this quantity of maize could feed millions of people if fed directly to them. If grain and soy lost from the human diet because of diversion to biofuel[1] is subtracted, the residual amount of grain and soy (and probably total food) per person is stable or declining (see Fig. 13.1). Compounding this is the problem of the increasing quantity of calories in food edible by humans that is diverted to feed livestock, including farmed fish. At the same time, the global population

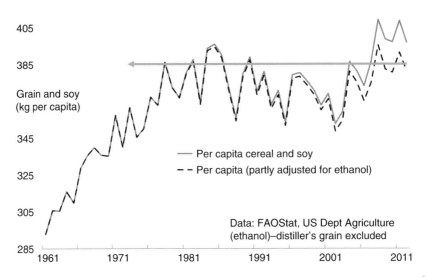

Fig. 13.1. Between 1960 and 1985, grain production increased far faster than population, due to the Green Revolution. Global hunger also declined substantially. Since 1985, increasing quantities of soy (not a grain, but a food that shares important similar properties, especially its ease of storage and shipping) have been grown, especially as feed. In more recent years, however, substantial soy and grain (especially maize) has been diverted from food, or feed, to fuel. Sufficient grain is still grown to alleviate hunger, were it distributed more equitably.

continues to rise by over 70 million per annum, an annual increment probably exceeding that in 1968, when global population increase peaked as a percentage.

The incidence of famine again appears to be increasing. Several famines have occurred recently in North Africa, including Kenya, Somalia, Sudan and Niger. The Islamist group, Al Shabaab, which exerts much control over modern Somalia, has worsened the human toll of the Somali famine greatly by tactics such as intimidating, kidnapping and killing famine relief workers (Maxwell *et al.*, 2012). Yet, in nearby Somaliland, a quasi-independent breakaway nation, little, if any, famine occurred at all, even though it too has experienced severe drought. An Asian country, the Democratic People's Republic of Korea, has also experienced severe food insecurity and frank starvation since the late 1990s.

13.3 Inequality, Vegetarianism and Other Forms of Food Waste

It is true that a complete global sharing of food resources, with the total abandonment of feeding grain and other crops such as soy to livestock as feed, would have lessened world famine considerably without any Green Revolution, as was advocated in 1971 in the book, *Diet for a Small Planet* (Lappé and Collins, 1971). However, as with alternatives to the Green Revolution, I am unaware of scholars then or now who have viewed complete vegetarianism as a viable or total solution to world hunger. Vegetarian diets can be healthy for many people,[2] but are neither culturally preferred nor ecologically possible in many places. Nor, probably, are they physiologically adequate for many populations, including those forced by ecological factors to rely largely on livestock for food consumption. Genetic variations such as haemochromatosis are advantageous for people with limited iron intake (Naugler, 2008). Those lacking this or similar genes may be at a disadvantage if forced to be vegetarian.

Arguments for a completely equal distribution of global food resources are also problematic. While excessive inequality cannot be tolerated indefinitely in either human society or many animal groups, all human and most animal groups are, none the less, characterized by many forms of inequality, such as of power, experience, strength and ability. A large reduction of human inequality, on a scale sufficient to generate adequate food for all, may be partly achievable, not least as being overfed is harmful to health and sustainability. But such large-scale redistribution also seems unrealistic, or even utopian.

However, much food is lost from human ingestion due to pests and poor storage. The custom of eating offal has vanished among most affluent populations, leading to a waste of nutrients. In rich countries, enormous quantities of edible food are thrown out of supermarkets, restaurants and people's homes (Parfitt et al., 2010).

13.4 Famines Have Ecological and Social Causes

Opinions vary concerning the causality of famine. At one extreme are experts influenced by Sen, who argue that the dominant cause of all or most recent famines is social (Patel, 2012). Others are more nuanced, conceptualizing famine as 'ecosocial'. Mike Davis recounts how more powerful populations, whether in India, China or parts of Africa, have thrived while others have starved. India's colonial ruler regularly appropriated Indian-grown food for consumption in Britain, even during severe Indian famines.[3] However, Davis implicitly recognizes an ecoclimatic contribution by including the cyclic climatic event, El Niño, in his book's subtitle (Davis, 2001).

Sen's seminal work on 'effective' entitlement, defined as the capacity to gain food by cultivation or purchase, was shaped by his childhood in Calcutta (today Kolkata). Sen witnessed the catastrophe during World War II now known as the Bengal Famine, which claimed 2–3 million lives (Sen, 1981). He showed that harvests in the year of the greatest death toll (1942) exceeded that of 1941. However, as in the famines described by Davis, I argue that an interaction between ecoclimatic and social factors was at play. If a bumper harvest had occurred in Bengal in 1941, then no famine would likely have occurred in either that year or the one following.

Others support the view that most famines have mixed causation. In 1926, Mallory wrote: 'There are some famines that are due almost certainly to "natural" causes (i.e. drought, locusts, plant disease), and there are scarcely any to which

natural phenomena do not contribute' (Mallory, 1926). Brass, discussing the Bihari famine of 1966–1967, describes proximate and remote causes, both ecological and social (Brass, 1986).

Even the prolonged famine in the Democratic People's Republic of Korea has ecological as well as social causes. While the hermit kingdom's cruel and repressive regime has numerous policies which repel foreign aid and assistance, a series of poor harvests, some associated with torrential rains and others with unusually high temperatures, have also contributed.

However, the lesser-known famine in North Vietnam during 1944–1945, which killed up to 2 million people, had little, if any, contributing climatic cause. Instead, the direct reason for this was the requisitioning of locally grown food by the occupying Japanese army. The role of Vietnam's colonial master, France, has also been blamed. Earlier in the 20th century, the Ukrainian famine may also illustrate an exclusively socially caused event, resulting from the deliberate suppression of the Ukrainian people by the policies of Stalinist Russia (see Fig. 13.2).

Fig. 13.2. The plaque reads '7,000,000. Dedicated to victims of the 1932–1933 enforced famine – genocide in Ukraine', Canberra, ACT, Australia (photo C.D. Butler).

At a finer scale, the social and environmental factors that can generate abundant locally available food do not guarantee either local ingestion or good utilization of ingested nutrients. In the agriculturally rich, populous Indian state of Uttar Pradesh, up to 60% of children are chronically undernourished (Black *et al.*, 2008), and thus stunted physically and mentally (Dillingham and Guerrant, 2004). This dismal loss of human potential arises through a combination of being underfed on a background of poor hygiene, dirty water and limited sanitation. This leads to excess disease, including diarrhoea, fevers and malabsorption, including from parasites (see Fig. 13.3). Not only the ingestion of food but also access to toilets and clean water requires 'entitlement'.

Furthermore, even if a population has sufficient entitlement to obtain and ingest food of adequate quality, quantity and variety, adequate nutrition for less favoured members of that group is not automatic. Many groups discriminate against females, or preferentially feed earners rather than dependents, even if dependents are ill.

Crucially, discrimination against less powerful individuals, and even entire populations, is particularly likely if food supplies are stressed by environmental factors such as drought, flooding or, as in parts of the Sudan, fear of random attack by bombers operated by the national government. In parts of Africa, physical insecurity has favoured the planting of the less nutritious root crop, cassava, which remains edible below ground for up to 36 months and is easier to hide from marauding enemies than, for example, maize (Rosenthal and Ort, 2011).

In summary, attributing famine *solely* to social causes is overly simplistic. Even if humans were as egalitarian and cooperative as ants, equal ingestion of food and sufficient nutrients to avoid famine is not always possible during times of extreme generalized food scarcity. But no human society has been completely egalitarian; young children, the poor and the elderly are at special risk. If food scarcity occurs because of either social or environmental causes, then intensified nutritional inequality is very likely. The rationing of food and other forms of entitlement which occurred during World War II in the UK is an exception.

There is evidence that Australian Aborigines maintained low population numbers in

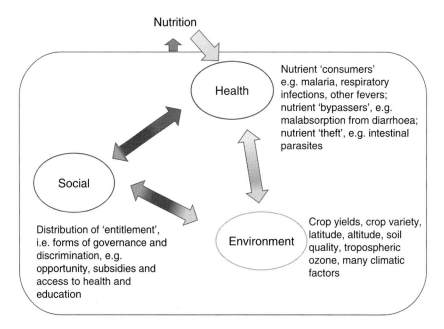

Fig. 13.3. Population nutrition can be conceptualized as arising from the interaction of social, health and environmental factors, in a complex system. Many relationships are bidirectional. For example, poor population health status can impair social function (e.g. by impaired education).

part so they could avoid famine or other forms of undue hardship even if food supplies became scarce (Butler, 2012). The Australian climate, dominated by the El Niño Southern Oscillation that swings from the drought of El Niño to the floods of La Niña, is very variable in Eastern Australia.

This section has argued that, more often than not, famines are caused by an interaction of ecoclimatic causes (e.g. drought, flood and plant diseases such as potato blight) and social causes (such as poor governance, conflict and discrimination against minorities and vulnerable individuals). This interplay between ecoclimatic and political events is a characteristic of tertiary events, as outlined elsewhere in this book. Famines and large-scale food insecurity, even without frank starvation, also have the potential to harm health on a far larger scale than any of the primary or secondary effects discussed in previous sections of this book.

The next part of this chapter outlines the evidence of an effect of climate change on crop growth. It is plausible that, as the other tertiary effects of climate change worsen

(i.e. large-scale population dislocation and conflict), increasingly governance could fail, also contributing to famine.

13.5 Rising Food Prices and Climate Change

For centuries, the long-term trend of food prices has been in decline (Naylor and Falcon, 2010). Since 1961, the FAO has calculated an index of food prices (see Fig. 13.4). The increase in this index in 2008 surprised many observers, even though it was consistent with the warnings given by Borlaug and the World Scientist's Warning to Humanity.

The most plausible proximate cause of the food price increase in 2008 is that year's rise in energy prices, especially oil, which reached over US$140/barrel (Piesse and Thirtle, 2009). Higher energy prices also increased the price of fertilizer. That price spike was not long sustained, due to the global financial crisis and recession. However, in December 2010, the global price index of food approached the 2008 peak and remained

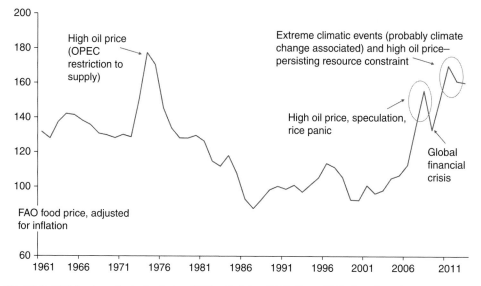

Fig. 13.4. FAO food nominal price index, 1961 to October 2013. Since 2005, there have been two pronounced spikes in food prices. The first one (2008) was driven by high oil prices, speculation and restrictions on rice exports. However, extreme weather events, themselves possibly influenced by climate change, are likely causes of the second rise, which started at the end of 2010. Prices declined slightly in mid-2013, but are still historically high. The diversion of crops to biofuels is a background factor, but cannot fully explain either rise.

unusually high until mid-2013. Yet, the price of energy was higher at its 2008 peak.

To date, few published studies have analysed the cause for this second price rise. Two factors seem important in addition to fairly high energy prices. One is the apparently relentless rise in the fraction of global food crops used for biofuels, which itself is largely in response to the growing scarcity of easily recoverable crude oil, essential for transport using current technology.

But, the second reason may be a non-linear response in the price of food to extreme climatic events. The most notable of these were the Russian and Ukrainian heatwave of 2010 and the Pakistani flood in the same year. The Russian heatwave was especially significant to the global food price because it affected the traditional 'bread bowl' of the former Soviet Union. It led to a 66% increase in the global price of wheat within 2 months and a lesser rise in other grain prices (Naylor and Falcon, 2010). The floods in Pakistan did not exert a very large death toll, but displaced over 20 million people, some of them for months. Both events reduced crop production in 2010, sufficiently to contribute to a decline in per capita global grain production (see Table 13.1). Previously, the extraordinary European heatwave of 2003 also lowered crop yields. However, this event did not appreciably affect global food prices. The USA then experienced a severe drought and heatwave in 2011 and 2012. Although the dryness has apparently occurred at other times in the last millennium (Boyer et al., 2013), its conjunction with such heat may be unprecedented, and thus attributable in part to elevated greenhouse gas levels.

There is debate as to whether climate change contributed to the Russian heatwave. However, there is increasing support for the proposition that the burden of evidence concerning the contribution of climate changes to such events should be reversed. That is, some climate scientists are now arguing that climate change should be accepted as a contributor to these extreme events until proved otherwise (Trenberth, 2011).

The recent Somali famine may also be related causally to climate change. In the past three decades, the Indian Ocean has warmed especially fast, in association with increased precipitation over the tropical Indian Ocean. Williams and Funk claim that, since 1980, this has suppressed convection over tropical eastern Africa, decreasing precipitation during the 'long rains' season of March–June. Unfortunately, their attempts to alert Somali authorities to the likelihood of famine were unsuccessful, but may have benefitted other, better-governed parts of the Horn of Africa, including Somaliland and Ethiopia (Funk, 2011; Williams and Funk, 2011).

Crop production is vulnerable to many factors associated with climate change, apart from extreme weather events such as those listed in Table 13.2. Some regions and crops are expected to benefit, especially in high latitudes. Like humans, plants have evolved to cope best in a given temperature range. For example, the increased yield for potatoes in Scotland has been attributed to warmer temperatures there (Gregory and Marshall, 2012).

Overall, evidence suggests that the effects of climate change on land-based agriculture are increasingly adverse, especially for wheat and maize, but not (yet) for rice and soy (Lobell et al., 2011). Nocturnal warming in tropical areas has been found to harm the yields of both wheat (Prasad et al., 2008) and rice (Peng et al., 2004), and these adverse effects seem likely to be increasingly influential. None the less, the magnitude of detected decline in land-based agricultural production due to increasing temperatures and changes in rainfall is small compared to increased harvests due to improved farming

Table 13.1. Two particularly extreme weather events occurred in 2010, the heatwave in Russia and Ukraine and the Pakistani floods. In 2009, Russia and Ukraine grew 12.1% of the global wheat production, but in 2010, this proportion fell to 9% (of a smaller total). Paddy rice production declined markedly in Pakistan in 2010, but this had only a small impact on total global production. In 2009, Pakistan grew only 1.5% of the world's paddy rice.

	Wheat				Paddy rice		
	Russia	Ukraine	Global	Per capita	Pakistan	Global	Per capita
2010 (as per cent of 2009)	67	81	95	94	70	98	97

Table 13.2. Since 2003, several extreme weather events have occurred which have reduced crop production. Some of these have been linked to climate change. At present, there is no comprehensive data set of 'extreme agricultural events', and the apparent trend in the severity of these events may partly reflect recall and reporting biases. Nevertheless, other factors (e.g. declining crop yield growth, falling soil fertility, competition from feed and biofuels and ongoing population growth) mean that any reduction in crop growth due to extreme weather events is increasingly likely to have an adverse effect on global food prices, per capita global food production and, thus, on global health.

Event	Year	Location	CC attribution?	Significant crop production effect?
Heatwave	2003	Europe	4/5	Yes
Fires and heat	2009	Australia	4/5	Yes
Typhoon	2009	Taiwan	2/5	Yes
Heatwave	2010	Russia and Ukraine	3/5	Yes (significant global price rise at time)
Flood	2010	Pakistan	2/5	Yes
Flood	2011	Thailand	0/5	Likely (no FAO data yet)
Drought	2011	North-east Africa	3/5	Yes (famine)
Heat and drought	2011–2012	USA	4/5	Likely (no FAO data yet), significant price rise in maize, soy
Drought	2012	Niger	?	Yes (famine)

knowledge and technology. It is also trivial compared to the amount of food fed to livestock, used for biofuels, consumed beyond baseline needs by the overnourished and wasted in other ways (Foley *et al.*, 2011).

13.6 Modelling Climate Change and Famine

The modelling of past and future agroclimatic effects is a formidable challenge, even without considering their health impacts. Existing agroclimate models are excessively simple and biased toward the optimistic (Butler, 2010; Gornall *et al.*, 2010). The decline detected in food production ascribed to climate change is already likely to be understated, and that of future climate change even more so.

Concern over future climate change and crops is amplified by increasing doubt over the benefits and strength of the carbon fertilization effect (CFE), especially for C_4 plants such as maize and sugar (Long *et al.*, 2006). In response, agroclimatic models increasingly incorporate positive and more neutral CFE effects. However, the CFE may also enhance the growth of pests

(Ziska *et al.*, 2009) and damage some crops, including cassava, a staple for about 750 million mostly poor people (Gleadow *et al.*, 2009). These effects to date are not incorporated into models.

Current agroclimatic models also poorly incorporate increased extremes, including rainfall intensity, sea level rise, saline intrusion, glacial melting and the possibilities of monsoon weakening and intensification of the El Niño Southern Oscillation and other ocean currents and atmospheric oscillations. They also omit the effect of climate change on mycotoxins, and on crop and animal diseases (Butler, 2010).

The capacity of ingested food to provide adequate nutrition is influenced by other factors, such as the level of physical activity (not just paid work, but unpaid labour such as hauling water, sometimes over hilly terrain) and by states of health and disease that lower appetite (Mangili *et al.*, 2006), nutrient absorption and metabolic rates.

13.7 Fisheries

Climate change is also predicted to have complex effects on fisheries, including by changing the pattern of ocean currents, thus redistributing marine productivity, especially to higher latitudes. A declining trend in the global phytoplankton concentration since 1899, in eight out of ten ocean regions, has been linked with warming sea surface temperatures. Increasing ocean acidity and climate change associated deoxygenated zones will also harm future marine productivity. Furthermore, ocean acidification associated with increased carbon dioxide concentrations interferes with the development of a wide range of aquatic species. It is already harming coral reef systems and further stressing fish stocks already in decline.

13.8 Solutions

Increased famines, consequent in part to climate change, appear likely. If left unaddressed, climate change threatens increasingly profound negative effects on food security, and hence on health and nutrition. Rising food prices stimulate social unrest and can contribute to the overthrow of governments. Conflict can occur easily due to the scarcity of raw materials, including food and fertile soil. While climate change requires adaptation, mitigation is an even more important priority.

Much can be done to reduce the enormous waste of food, whether preharvest (such as losses on the field), postharvest (e.g. eaten by rodents) or postprocessing (e.g. thrown out from supermarkets or wasted after purchase). Food fed to people suffering from chronic diarrhoea can be utilized better if their illnesses can be reduced by better health care, adequate sanitation and clean water. Overfed people, especially if consuming animal products raised on grain and soy, can and should reduce their consumption of these products.

There are calls for a new agricultural revolution, to be fostered by further investment. While this might help at the margins, too much hope may already be invested in this strategy, though much could be done to bring the Green Revolution to Africa (Ejeta, 2010). The World Bank in 2013 warned that if the average temperature rose by four degrees, then the most alarming effect was likely to be on food production (World Bank, 2013). However, food production is still likely to be affected greatly by climate change, even if global temperature increase can be held at a level far lower than this.

It is important that the determinants of fertility in low-income populations are altered in ways that will increase child survival and the effective demand for education and lower population growth. The final chapter of this book will discuss other general principles of solution. Adapting crops to cope with drought, flood and inundation from the sea is feasible to a point, but only to thresholds which may soon be exceeded. A holistic, global approach is required, which integrates society, education, health care, equity and technology.

Notes

[1] Not only from maize but also from other important crops such as sugarcane and palm oil.
[2] If supplemented with eggs, dairy products or at least vitamin B12 supplements.
[3] On the other hand, the rapid spread of the Indian rail network – a then recent British invention – did much to relieve Indian famine.

References

Becker, J. (1996) *Hungry Ghosts: China's Secret Famine*. Henry Holt, New York.
Black, R.E., Allen, L.H., Bhutta, Z.A., Caulfield, L.E., de Onis, M., Ezzati, M., *et al.* (2008) Maternal and child undernutrition: global and regional exposures and health consequences. *The Lancet* 371, 243–260.
Boyer, J.S., Byrne, P., Cassman, K.G., Cooper, M., Delmer, D., Greene, T., *et al.* (2013) The U.S. drought of 2012 in perspective: a call to action. *Global Food Security* 2, 139–143.

Brass, P.R. (1986) The political uses of crisis: the Bihar famine of 1966–1967. *The Journal of Asian Studies* 45, 245–267.

Butler, C.D. (2004) Human carrying capacity and human health. *Public Library of Science Medicine* 1, 192–194.

Butler, C.D. (2010) Climate change, crop yields, and the future. *SCN News* 38, 18–25.

Butler, C.D. (2012) Population trends and the environment. In: Friis, R.H. (ed.) *Praeger Handbook of Environmental Health.* Praegar, Westport, Connecticut, pp. 215–231.

Davis, M. (2001) *Late Victorian Holocausts: El Niño Famines and the Making of the Third World.* Verso, London.

Dillingham, R. and Guerrant, R.L. (2004) Childhood stunting: measuring and stemming the staggering costs of inadequate water and sanitation. *The Lancet* 363, 94–95.

Ejeta, G. (2010) African Green Revolution needn't be a mirage. *Science* 327, 831–832.

Foley, J.A., Ramankutty, N., Brauman, K.A., Cassidy, E.S., Gerber, J.S., Johnston, M., *et al.* (2011) Solutions for a cultivated planet. *Nature* 337, 337–342.

Funk, C. (2011) We thought trouble was coming. *Nature* 476, 7.

Gleadow, R.M., Evans, J.R., McCaffery, S. and Cavagnaro, T.R. (2009) Growth and nutritive value of cassava (*Manihot esculenta Cranz.*) are reduced when grown in elevated CO_2. *Plant Biology* 11, 76–82.

Gornall, J., Betts, R., Burke, E., Clark, R., Camp, J., Willett, K., *et al.* (2010) Implications of climate change for agricultural productivity in the early twenty-first century. *Philosophical Transactions of the Royal Society B* 365, 2973–2989.

Gregory, P.J. and Marshall, B. (2012) Attribution of climate change: a methodology to estimate the potential contribution to increases in potato yield in Scotland since 1960. *Global Change Biology* 18, 1372–1388.

Lappé, F.M. and Collins, J. (1971) *Diet for a Small Planet.* Institute for Food and Development Policy, San Francisco, California.

Lobell, D.B., Schlenker, W. and Costa-Roberts, J. (2011) Climate trends and global crop production since 1980. *Science* 333, 616–620

Long, S.P., Ainsworth, E.A., Leakey, A.D.B., Nösberger, J. and Ort, D.R. (2006) Food for thought: lower-than-expected crop yield stimulation with rising CO_2 concentrations. *Science* 312, 1918–1921.

Mallory, W.H. (1926) *China: Land of Famine.* American Geographical Society, New York.

Mangili, A., Murman, D.H., Zampini, A.M., Wanke, C.A. and Mayer, K.H. (2006) Nutrition and HIV infection: review of weight loss and wasting in the era of highly antiretroviral therapy from the nutrition for healthy living cohort. *Clinical Infectious Diseases* 42, 836–842.

Maxwell, D., Haan, N., Gelsdorf, K. and Dawe, D. (2012) The 2011–12 famine in Somalia: introduction to the special edition. *Global Food Security* 1, 1–4.

Naugler, C. (2008) Hemochromatosis: a Neolithic adaptation to cereal grain diets. *Medical Hypotheses* 70, 691–692.

Naylor, R.L. and Falcon, W.P. (2010) Food security in an era of economic volatility. *Population and Development Review* 36, 693–723.

Parfitt, J., Barthel, M. and Macnaughton, S. (2010) Food waste within food supply chains: quantification and potential for change to 2050. *Philosophical Transactions of the Royal Society B: Biological Sciences* 365, 3065–3081.

Patel, R.C. (2012) Food sovereignty: power, gender, and the right to food. *PLoS Medicine* 9, e1001223.

Peng, S., Huang, J., Sheehy, J.E., Laza, R.C., Visperas, R., Zhong, X., *et al.* (2004) Rice yields decline with higher night temperature from global warming. *Proceedings of the National Academy of Sciences* 101, 9971–9975.

Piesse, J. and Thirtle, C. (2009) Three bubbles and a panic: an explanatory review of recent food commodity price events. *Food Policy* 34, 119–129.

Prasad, P., Pisipati, S., Ristic, Z., Bukovnik, U. and Fritz, A. (2008) Impact of night-time temperature on physiology and growth of spring wheat. *Crop Sciences* 48, 2372–2380.

Rosenthal, D.M. and Ort, D.R. (2011) Examining cassava's potential to enhance food security under climate change. *Tropical Plant Biology* 5, 30–38.

Sen, A.K. (1981) *Poverty and Famines: An Essay on Entitlement and Deprivation.* Clarendon Press, Oxford, New Delhi.

Sheeran, J. (2011) Preventing hunger: sustainability not aid. *Nature* 479, 469–470.

Shiva, V. (1991) *The Violence of the Green Revolution. Third World Agriculture, Ecology and Politics.* Third World Network and with ZED Books, Penang and London.

Staples, A.L.S. (2003) To win the peace: the Food and Agriculture Organization, Sir John Boyd Orr, and the
    World Food Board proposals. *Peace and Change* 28, 495–523.
Swift, J. (1729) A Modest Proposal. For Preventing the Children of Poor People in Ireland from Being a
    Burden to their Parents or Country, and for Making Them Beneficial to the Public (http://art-bin.com/
    art/omodest.html#hit, accessed 16 March 2014).
Trenberth, K.E. (2011) Attribution of climate variations and trends to human influences and natural vari-
    ability. *WIREs Climate Change* 2, 925–930.
Tribe, D. (1994) *Feeding and Greening the World.* CAB International in association with the Crawford Fund
    for International Agricultural Research, Wallingford, UK.
Williams, A.P. and Funk, C.A. (2011) Westward extension of the warm pool leads to a westward extension
    of the Walker circulation, drying eastern Africa. *Climate Dynamics* 37(11–12), 2417–2435.
World Bank (2013) *Turn Down the Heat: Why a 4 Degree Celsius Warmer World Must be Avoided.* World
    Bank, Washington, DC.
Ziska, L.H., Epstein, P.R. and Schlesinger, W.H. (2009) Rising CO_2, climate change, and public health:
    exploring the links to plant biology. *Environmental Health Perspectives* 117, 155–158.

14 Moving to a Better Life? Climate, Migration and Population Health

Devin C. Bowles,[1] Rafael Reuveny[2] and Colin D. Butler,[1,3]

[1]*National Centre for Epidemiology and Population Health, The Australian National University, Canberra, Australia;* [2]*School of Public and Environmental Affairs (SPEA), Indiana University, Bloomington, Indiana, USA;* [3]*Faculty of Health, The University of Canberra, Australia, and Benevolent Organisation for Development, Health & Insight (BODHI)*

14.1 Introduction: Migration, Evolution and the Human Condition

Humans, like all other creatures, inhabit ecological niches that provide resources essential for life and reproduction. Numerous species modify environments in ways to improve their suitability, but the human capacity to do this far exceeds that of any competitor. Moreover, humans and their hominin ancestors seem never to have been content to stay only in one place, even if verdant. For hundreds of millennia, *Homo sapiens* has explored the next valley and, technology permitting, crossed the next river and even the next ocean.

Sometimes, the motivation for human migration may appear pure curiosity, but the struggle to gain, control and consume resources, in order to improve comfort and security, is central to the human story. The decision to stay at home or to migrate can be conceptualized as the product of four interacting forces; *glue, push, pull* and *fend*. Once humans have a comfortable niche, a territory they call home, they will seek to defend and enjoy it. They may, at times, willingly share it, but only if they do not expect any significant or prolonged decrease in their access to its resources.

At times, this attachment to place is strengthened not only by personal memories but also by knowledge of and reverence for ancestors. Departure from resource-rich and memory-laden habitats entails effort and risk; '*glue*' or attachment factors reduce this prospect.

However, both physical and social environments can change, often through co-evolution. Increasing human numbers and exploitative practices can degrade formerly attractive environments easily, creating *push* factors that propel migration, or at least the wish to migrate. Social factors can degrade faster than physical conditions, creating circumstances in which dominant groups challenge, discriminate against and, in some cases, actively endanger and expel vulnerable groups, who must choose between migration and survival or, rarely, assimilation.

Today, most fertile parts of the world are densely inhabited. Deserts, oceans, icefields and outer space have low population density, but although human life is now possible in all of these places, the permanent survival of large numbers of people is not; there are still limits to technology and fortitude. The obvious hostility of such locations acts as a *natural fend* or deterrent factor. However, not all inhabited lands are hostile to newcomers. Numerous *pull* factors lure prospective migrants and temporary workers from homes of relative (or absolute) hardship to more desirable locales, such as from the Malabar Coast of India to the Middle East as 'guest' workers, or from gloomy England to sunny Queensland. In some cases, these additional people are welcome, for example if bringing traits considered as desirable, such as a wanted trade or a large bank account.

However, in our increasingly crowded world, *socially created fend* factors dominate. Examples are numerous. All nations are demarcated by controlled borders; some countries (such as China and India) also maintain internal travel controls, restricting access and sometimes greater rights to more favoured citizens. Democratic India has almost completed construction of a fortified fence along most of the perimeter of Bangladesh, forming what some have called the 'largest human cage in the world'. The US–Mexico border is also substantially fenced and patrolled.

This chapter explores the implications and consequences for health due to migration associated with climate change. Climate change is increasingly redistributing resources, including food, water and shipping lanes. It is also changing the distribution of desirable human habitat. These trends are accelerating. Low-latitude regions have long been predicted to see crop production decline, while some countries nearer to the pole are already experiencing improved agricultural conditions. Global food and energy prices are rising, and so are many other costs of living, as global limits to growth become increasingly obvious and close. In many impoverished nations (largely tropical), birth rates remain high, making escape from poverty traps almost impossible (Campbell *et al.*, 2007). Rural poverty intensifies the push for rural to urban migration, expanding the size of slums, unplanned settlements and deepening urban poverty. Per capita economic growth, however defined, is slowing in many parts of the world. In many places, pre-existing disparities and inequalities are deepening. The factors that determine migration are continually changing; overall, however, increased demand and more attempts to migrate appear inevitable.

14.2 Migration and Past Climate

Resource scarcity and migration have numerous causes, one of which is an unfavourable climate (McMichael *et al.*, 2012). The onset of the Little Ice Age, for example, was an important cause in the decline of harvests in the 14th century, which may then have contributed to the calamitous Black Death, via large-scale undernutrition. Between 1500 and 1800, numerous poor harvests, some of them climate related, occurred in Europe, contributing to famine, warfare and regional localized population decline. However, the invasion and colonization of the 'New World', together with new technologies, allowed cross-continental migration and other buffers, such as the import of food and other resources into Europe, enabling larger populations in Europe from the late 18th century (Zhang *et al.*, 2011). While *in situ* adaptation, such as new technologies, especially agricultural innovation, have been important (Hugo, 2011), migration has always been a mechanism to cope with climatic and other forms of adverse environmental change (see reviews in Diamond, 2005; Lilleør and Van den Broeck, 2011).

14.3 Migration and Future Climate Change

The scale of migration associated with climate change in the near future (e.g. to 2050) has been contested bitterly, but high predictions exceed 100 million by 2050 (Gemenne, 2011). Similar to famine and conflict, the causation of migration is multifactorial. On balance, human-driven climate change is likely to be an increasingly significant driver of resettlement. It could also be a useful method of adaptation (Black *et al.*, 2011c) – provided migration policies focus on justice, rationally allocate people and resources and recognize that countries have different levels of responsibility for climate change.

Black and colleagues propose five drivers of migration: economic, political, social, demographic and environmental (Black *et al.*, 2011a). However, these categories are, in practice, rarely, if ever, pure; the perception of causation is mixed, influenced by the 'focal depth' or pre-existing bias of the observer (McMichael, 1999). Climate change and its effects are also viewed through a cultural lens, including by those who contemplate relocation or escape (Piguet *et al.*, 2011).

Over time, climate change is raising sea levels, and this is already increasing the salinity of some coastal agricultural areas. Higher oceans and declining river sediment (due to dams) is already decreasing the habitability of many densely populated tropical river deltas (Syvitski *et al.*, 2009). Changes to rainfall associated with climate change may already be reducing the viability of pastoralism in north-east Africa (Funk, 2011), contributing to out-migration.

Societies tend to be reactive, not proactive, including to climate change. Sudden, large-scale changes are more likely to trigger an effective response; slow, progressive change rarely receives sufficient attention (Piguet *et al.*, 2011). For example, Hurricane Katrina led not only to out-migration but also to the repair of levees, a chore previously long neglected.

Less developed countries (LDCs) are especially vulnerable to climate change. Pathways include decreased economic opportunities, which may, in turn (depending on cultural and historical factors), create harsher political realities, including ethnic tension and diminishing prospects for security and health. In turn, as climate change makes life harder for such populations, it will increase the push factors described above, driving both internal and external migration, both to nearby cities and, if possible, to developed countries.

Decisions to migrate are multicausal; climate change is but one factor. Declaring specific individuals or groups as 'climate migrants' may have legal and political implications, as acceptance of this category may impose an obligation on emitter countries to provide compensation, including resettlement (Piguet *et al.*, 2011). To date, such precise identification is elusive (Locke, 2009; Black *et al.*, 2011b). Regardless, climate migrants can be defined, without identifying individuals, as those who emigrate in a world with climate change compared to the 'counterfactual' – the same world without climate change.

Migration due to rising sea levels in some areas is likely to be permanent, eventually causing the complete emptying of some currently inhabited areas (such as low-lying islands). However, some coastal areas may adapt to sea level rise by the construction of dykes, following the example of the Netherlands. Piguet *et al.*

identify 146 million people living within 1 m of sea level to be especially at risk (Piguet *et al.*, 2011). Three-quarters of these will be in South or East Asia, and many will be farmers in river deltas (Piguet *et al.*, 2011). Even if sea walls can prevent direct inundation, increased salinization of farmland and aquifers could also prompt migration as livelihoods become untenable. Some African megacities are also vulnerable to sea level rise (Parnell and Walawege, 2011).

Other impacts of climate change are likely to induce more variable migration, both spatially and temporally. To date, rapid-onset disasters have led mostly to temporary, internal migration, with oustees often returning to their original site, even if precarious, in part as it is familiar and in part because they lack the resources to move further away (Piguet *et al.*, 2011). People are also vulnerable to a sequence of disasters, such as repeated typhoons. Even faced with multiple disasters, many people cannot move permanently. Repeated catastrophes (like multiple illnesses) can deepen poverty, creating socio-economic traps from which escape, including by migration, is almost impossible. Those unable to migrate may be most at risk (Black *et al.*, 2011c).

14.4   Climate, Migration and Fertility

Climate migration is influenced by demographic factors. Developed countries have already made the transition from high fertility, high mortality to low fertility and low mortality. China is also well advanced in this switch, especially with regard to fertility. In contrast, many LDCs, especially in Africa, South Asia and the Pacific, are at earlier stages in this transition. Some may be trapped, destined for 'irruptive' population trajectories, as occurred in Rwanda in 1994 (Butler, 2000). LDCs have long been primary drivers of world population growth and the recent source of much migration to more developed areas, both intra- and internationally (Hugo, 2011). The high proportion of children and youth, combined with ongoing agricultural mechanization in LDCs, generates labour oversupply at the same time that many developed countries (especially in Europe) are experiencing declining working-age populations (Hugo, 2011). LDCs' *in situ* adaptations are impeded by limited access to technology and capital (Hugo, 2011). This suggests an opportunity for rational human relocation, but supply of climate migrants is likely to far exceed the future demand and absorptive capacity for them. Already, millions of people languish in refugee camps, awaiting resettlement (UNHCR, 2013) (see Fig. 14.1).

Climate change is likely to accelerate the already high degree of rural to urban migration. Many Africans migrate temporarily, keeping connections to urban and rural areas, alternating as required (Parnell and Walawege, 2011). This illustrates a form of insurance to diversify family income and minimize risk (Lilleør and Van den Broeck, 2011). The harmful effects of climate change may also impair the capacity of cities to provide sufficient economic opportunities to allow permanent urban resettlement (Parnell and Walawege, 2011).

Acceleration of rural to urban migration, in part due to climate change and high fertility, is also manifest in some small Pacific Island countries (Locke, 2009). Salinization of drinking water and soil, coastal erosion, degraded reef systems and increasingly frequent king tides and flooding have decreased the habitability of many islands. Smaller islands and those just a few metres above sea level are most adversely affected, including because limited funds available for adaptation are focused on a few areas (Locke, 2009).

14.5 Analytic Framework

Health in three populations

We now discuss the health status of three groups likely to be affected by climate migration: migrants, donor populations and receiver populations (Table 14.1). Climate migrant populations will likely be small compared to the world population, even if they approach 100 million by 2050. In general, their wealth and resource security will be lower than that of receiving populations, thus providing a potential economic cofactor for emigration. However, the resource access of migrants must be sufficient to provide them with means to emigrate, potentially distinguishing them from others in their sending population (McLeman and Smit, 2006). Donor populations are likely to be relatively large and poor. Receiving populations will also be large, but tend to be wealthier, with better infrastructure and access to resources, unless in LDCs.

Migration bias

Perceived self-capacity, health, wealth and social capital are important determinants of an individual's decision to migrate, including to international and overseas destinations (McLeman and Smit, 2006; Black *et al.*, 2011c). Higher

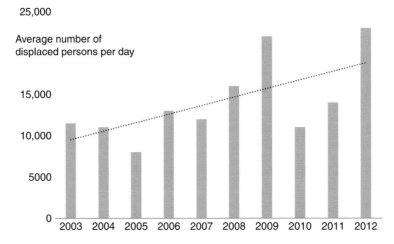

Fig. 14.1. Average number of displaced persons per day, both internally and externally (adapted from Figure 1, UNHCR, 2013).

Table 14.1. Populations impacted by climate migration.

Population	Comparative:			Change in health due to:	
	Size	Access to resources	Health pre-migration	Climate change	Climate change migration
Migrants	Small	Medium	Medium	Decrease	Increase
Donor	Large	Low	Low	Variable, generally decrease	Decline more likely
Recipient	Large	High	High	Variable	Variable, decrease likely, especially in already poor areas

migration capacity increases the likelihood of departure and also expands the range of possible destinations. Migrants, especially if travelling long distances and to different cultures, are unlikely to be a random sample of the donor population. Instead, they are likely to be healthier, wealthier and possess greater skills than their sending population. This 'migration bias' is as likely to be true for climate migrants as for any other migrant category.

Health in climate migrants

Climate migrants leave affected regions where they would probably experience diminished health conditions and prospects, as members of comparatively poor, vulnerable communities whose collective health is likely to be worsened by the adverse effects of climate change. Some may migrate successfully to richer countries; for example, by possessing a wanted qualification, such as a Filipino-trained nurse emigrating to New Zealand. However, relocation may be denied or obstructed by years in detention (as has recently occurred to non-climate migrants seeking asylum in Australia). If so, the health of even the most robust climate migrant may be damaged. Climate migrants, if travelling as asylum seekers or migrating illegally, also risk harm to their health if they are forced to work in the underground economy of a receiving nation.

Potential migrants, on the whole, are rational. They will emigrate when they judge that their life chances (whether socio-economic or related to health or freedom) are likely to be improved by moving. While their assessments are fallible, on balance, migrants are likely to judge correctly that migration is of net benefit, including for their health. It is unsurprising that immigrants thus often have lower mortality than donor populations, and the health of some migrants can even exceed that of recipient populations (Razum and Twardella, 2002; Razum, 2006). However, these conclusions may not hold in future, such as for the *average* climate migrant seeking to flee a coastal city or country in which a series of events related to climate change has made life very difficult.

The positive effects of migration are also likely to be offset partially by increased psychological distress. Acculturation can be stressful, especially if accompanied by loss of social supports. Nevertheless, these factors have been historically insufficient to cancel the 'healthy migrant effect' in developed countries.

However, the migratory journey may decrease health, especially among the poor. Living conditions in transit may be crude, crowded, insecure, stressful and unsanitary, with increased risk of exposure to new communicable diseases (Afolayan and Adelekan, 1999). Numerous African migrants have already drowned while trying to cross the Mediterranean, for example. Climate migration may also increase violence (Box 14.1).

Health in donor populations

By definition, populations generating climate migrants are vulnerable to harm from climate

---

Box 14.1   Climate Migration and Violence

Several studies observe that environmental immigration is sometimes associated with violence (Martin, 2005; Reuveny, 2007, 2008; Reuveny and Allen, 2007). Migration can alter ethnic and other forms of composition, increasing rivalry and tension within the receiving population (Salehyan and Gleditsch, 2006; Fearon and Laitin, 2011). Those fleeing strife in the donor population may also carry ideologies and resentment that foster violence in the recipient population (Salehyan and Gleditsch, 2006). However, while it is possible to predict that conflict in general will increase with climate change, no formula enabling precise forecasts exists.

Causality is bidirectional; conflict is a potent trigger for migration. Climate change is likely to increase the risk of conflict in many areas, for several reasons, of which migration is only one (see Chapter 15, this volume). Self-reinforcing cycles of violence and migration are possible.

---

change. Population reduction (through emigration) may improve the ratio of resources to people. However, the disproportionate egress of more vigorous migrants will, in many cases, reduce the human capital of donor populations and could lower resource availability, even per capita. On the other hand, remittances sent home by successful emigrants could eventually increase the living standards of donor populations.

In some scenarios, however, the health of populations affected by climate change could be damaged severely, irrespective of the fate of its emigrants. This is plausible, for example, if ethnically driven conflict engulfs such locations. Minorities are especially at risk.

This kind of climate migration may alter the healthy migrant effect. Drought, in particular, has been linked with reduced population movement (Piguet *et al.*, 2011), perhaps because populations living in drought-prone areas are familiar with it and are especially stoic. However, if drought duration, intensity or frequency increases with climate change, then such fortitude could become maladaptive. Families sapped of resources and health (both physical and mental) risk reducing their migration capacity and fitness.

Overall, the departure of comparatively able climate migrants is likely to intensify the negative effects of climate change experienced by donor populations. This harm could be relieved partially by remittances, assuming immigrants are not held in detention centres and can get jobs. However, the effects of climate change are likely to continue for many decades, if not centuries.

Health in recipient populations

The effects of climate change on recipient population health will vary greatly. Climate change to some high-latitude areas may be manageable, even welcomed, but life in some developing African and Asian megacities seems set to become increasingly difficult. The effects of climate migration on recipient population health will similarly vary, determined by their tolerance and migrant-absorptive capacity. As stated above, most climate migrants are predicted to be intranational (Lilleør and Van den Broeck, 2011), confined within LDCs. If climate migrants ever move in extremely large numbers (e.g. in the tens of millions), most seem likely to remain within LDCs, even if they move internationally. Today, the great majority of refugees are confined within camps in LDCs (UNHCR, 2013).

Climate migrants could strain infrastructure capacity, including of health services, within recipient populations. This could lead to a downward population health escalator, as those unable to receive treatment become less healthy and increase health system demands. However, a very large scale of uncontrolled climate in-migration is unlikely to be tolerated, irrespective of the demand for relocation, and is likely to be managed through the use of camps. If countries become poorer, their capacity to regulate international entry effectively may decline (Kaplan, 1994), leaving enforcement to citizens. The trend toward vigilantism is already seen in Arizona, where armed civilians patrol the border for illegal crossings from Mexico into the USA. Some populations with low fertility may accept more migrants. Improved prospects for agriculture in some places (e.g. Siberia) may create a demand for additional migrants (e.g. from China). But if economic conditions in receiving countries decline, xenophobia is likely to impede significant climate in-migration.

The key issue is of thresholds. The humanitarian impulse may permit the emergency care and financial relief of a few hundred people displaced from a small island but blanch at scaling this to provision for millions of refugees displaced, for example, by a series of climate catastrophes in India.

14.6 Discussion

Climate change is likely to affect the health of migrant, donor and recipient populations adversely. However, if well managed, climate migration has considerable capacity to lessen the ill effects of climate change on health, analogous to how other forms of migration have proved effective safety valves for many. Increased future climate migration appears certain, but its effects, for good or ill, are amenable to policy intervention. If governments cooperate in ways to foster global and regional public goods, then migration will be an important form of adaptation, at best helping to match people with resources and labour demand (Hugo, 2011). Health is a function of numerous determinants, including social, economic, environmental and health system related. Developed countries can improve the health of climate migrants not only by admitting them, and allowing them to compete for employment, but also by strengthening targeted forms of health infrastructure.

Climate change mitigation is most important. But climate change adaptation is also essential, and in LDCs it should be supported by wealthy countries, which bear disproportionate responsibility for greenhouse gas emissions. Rapid population growth can intensify the risk of climate out-migration (Kniveton et al., 2012). Female education, other forms of women's empowerment and voluntary family planning initiatives targeted at 'hot spot' areas of high population growth and low economic development would enhance resilience (Hugo, 2011) and reduce vulnerability (Bryant et al., 2009).

Effective adaptation needs also to include climate-smart development, sustainable urbanization and emergency preparedness such as flood control (Black et al., 2011c).

Conversely, hostility toward climate migrants could interact with and worsen other causal pathways by which climate change will harm health, increasing inequality, conflict and population-resource mismatch. One possible reaction to climate migration is that recipient populations will become increasingly xenophobic, creating and exaggerating perceptions that waves of poor migrants are taking scarce resources from receiving populations, such as income and jobs (Piguet et al., 2011).

Already, many wealthy governments are strengthening border security, partly in response to the perceived threat of climate migration (Welzer, 2012). Increased covert means are also being used, such as by nations entering into agreements with poorer transit countries to stop migrants from ever reaching their borders (Welzer, 2012). Such trends reduce migration's capacity to act as a safety valve and adaptation.

Another plausible adverse response is indefinite refusal by high-emitting countries to recognize the moral and putative legal claims of climate migrants, thus obstructing international solutions, including the provision of compensation or refuge. Wealthy countries may even revoke or disregard existing treaties that entail obligations to provide refuge for asylum seekers for whom climate is not the primary push factor. Tendencies in this direction are already evident in Australia.

In the worst case, high-income populations may watch large-scale tragedies passively, such as when the international community failed to intervene during the Rwandan genocide of 1994. Inadequate funding and policing could also create squalid and lawless climate migrant camps, as occurred after the Rwandan genocide.

14.7 Conclusion

The scale of future climate change grows grimmer. Refugee numbers increased substantially in 2012, in part due to conflict in several countries (including Syria and Mali), to which climate change probably contributed (UNHCR, 2013). Although the concept of climate and other forms of environmental migrants has for many years been contested, there is today growing acceptance that it is a valid conceptualization. The decision to migrate has a complex causation, and classifications of migration as purely 'economic', 'social' or 'climatic' lack intellectual rigour. But, to argue that this negates the validity of climate migration is untenable, and will increasingly be seen as such.

The fate of climate migrants, who will potentially number in the tens or even, eventually, hundreds of millions, and who will be disproportionately vulnerable and also disproportionately low emitters of greenhouse gas, is an important test for civilization. The health of donor and recipient populations could also be worsened, unless protective policies can be implemented. Indications to date suggest that rich countries will fail this test, for migrants and themselves, though some nations with low birth rates or new agricultural areas opened by climate change may be more receptive. Recognition of the dimensions of this impending failure could yet galvanize action both to mitigate climate change and to put in place the more enlightened policies and practices that are keenly required.

References

Afolayan, A.A. and Adelekan, I.O. (1999) The role of climatic variations on migration and human health in Africa. *Environmentalist* 18, 213–218.

Black, R., Adger, W.N., Arnell, N.W., Dercon, S., Geddes, A. and Thomas, D. (2011a) The effect of environmental change on human migration. *Global Environmental Change* 21, Supplement 1, S3–S11.

Black, R., Adger, W.N., Arnell, N.W., Dercon, S., Geddes, A. and Thomas, D. (2011b) Migration and global environmental change. *Global Environmental Change* 21, Supplement 1, S1–S2.

Black, R., Bennett, S.R.G., Thomas, S.M. and Beddington, J.R. (2011c) Climate change: migration as adaptation. *Nature* 478, 447–449.

Bryant, L., Carver, L., Butler, C.D. and Anage, A. (2009) Climate change and family planning: least developed countries define the agenda. *Bulletin of the World Health Organisation* 87, 852–857.

Butler, C.D. (2000) Entrapment: global ecological and/or local demographic? Reflections upon reading the BMJ's six billion day special issue. *Ecosystem Health* 6, 171–180.

Campbell, M., Cleland, J., Ezeh, A. and Prata, N. (2007) Return of the population growth factor. *Science* 315, 1501–1502.

Diamond, J. (2005) *Collapse: How Societies Choose to Fail or Succeed.* Allen Lane, London.

Fearon, J.D. and Laitin, D.D. (2011) Sons of the soil, migrants, and civil war. *World Development* 39, 199–211.

Funk, C. (2011) We thought trouble was coming. *Nature* 476, 7.

Gemenne, F. (2011) Why the numbers don't add up: a review of estimates and predictions of people displaced by environmental changes. *Global Environmental Change* 21, Supplement 1, S41–S49.

Hugo, G. (2011) Future demographic change and its interactions with migration and climate change. *Global Environmental Change* 21, Supplement 1, S21–S33.

Kaplan, R.D. (1994) The coming anarchy. *Atlantic Monthly* 273(2), 44–76.

Kniveton, D.R., Smith, C.D. and Black, R. (2012) Emerging migration flows in a changing climate in dryland Africa. *Nature Climate Change* 2, 444–447.

Lilleør, H.B. and Van den Broeck, K. (2011) Economic drivers of migration and climate change in LDCs. *Global Environmental Change* 21, Supplement 1, S70–S81.

Locke, J.T. (2009) Climate change-induced migration in the Pacific Region: sudden crisis and long-term developments. *The Geographical Journal* 175, 171–180.

McLeman, R. and Smit, B. (2006) Migration as an adaptation to climate change. *Climatic Change* 76, 31–53.

McMichael, A.J. (1999) Prisoners of the proximate: loosening the constraints on epidemiology in an age of change. *American Journal of Epidemiology* 149, 887–897.

McMichael, C., Barnett, J. and McMichael, A.J. (2012) An ill wind? Climate change, migration, and health. *Environmental Health Perspectives* 120, 646–654.

Martin, A. (2005) Environmental conflict between refugee and host communities. *Journal of Peace Research* 42, 329–346.

Parnell, S. and Walawege, R. (2011) Sub-Saharan African urbanisation and global environmental change. *Global Environmental Change* 21, Supplement 1, S12–S20.

Piguet, É., Pécoud, A. and de Guchterneire, P. (2011) Introduction: migration and climate change. In: Piguet, É., Pécoud, A. and de Guchterneire, P. (eds) *Migration and Climate Change*. Cambridge University Press and UNESCO Publishing, Cambridge, UK, pp. 1–34.

Razum, O. (2006) Commentary: Of salmon and time travellers – musing on the mystery of migrant mortality. *International Journal of Epidemiology* 35, 919–921.

Razum, O. and Twardella, D. (2002) Time travel with Oliver Twist – towards an explanation for a paradoxically low mortality among recent immigrants. *Tropical Medicine and International Health* 7, 4–10.

Reuveny, R. (2007) Climate change-induced migration and violent conflict. *Political Geography* 26, 656–673.

Reuveny, R. (2008) Ecomigration and violent conflict: case studies and public policy implications. *Human Ecology* 36, 1–13.

Reuveny, R. and Allen, A.P. (2007) On environmental refugees and implications for the future. *Ecología Política* 33, 21–36.

Salehyan, I. and Gleditsch, K.S. (2006) Refugees and the spread of civil war. *International Organization* 60, 335–366.

Syvitski, J.P.M., Kettner, A.J., Overeem, I., Hutton, E.W.H., Hannon, M.T., Brakenridge, G.R., *et al.* (2009) Sinking deltas due to human activities. *Nature Geoscience* 2, 681–686.

United Nations High Commissioner for Refugees (UNHCR) (2013) *Displacement. The New 21st Century Challenge. Global Trends 2012.* UNHCR, Geneva, Switzerland.

Welzer, H. (2012) *Climate Wars: Why People Will Be Killed in the Twenty-first Century.* Polity Press, Cambridge, UK.

Zhang, D.D., Lee, H.F., Wang, C., Li, B., Pei, Q., Zhang, J., *et al.* (2011) The causality analysis of climate change and large-scale human crisis. *Proceedings of the National Academy of Sciences* 108, 17296–17301.

15 Unholy Trinity: Climate Change, Conflict and Ill Health

Devin C. Bowles,[1] Mark Braidwood[2]
and Colin D. Butler[1,3]
[1]National Centre for Epidemiology and Population Health, The Australian
National University, Canberra, Australia; [2]School of Public
Health & Community Medicine, The University of NSW, Sydney
New South Wales, Australia; [3]Faculty of Health, The University of Canberra,
Australia, and Benevolent Organisation for Development,
Health & Insight (BODHI)

We depend upon the Nile 100 percent in our life. So, if anyone, at any moment, thinks to deprive us of our life, we shall never hesitate [to go to war] because it is a matter of life or death (Egyptian President Anwar Sadat, 1978).

(Gleick, 1991)

15.1 Introduction

Ascending through primary, secondary and tertiary effects of climate change, causal complexity increases. The more the effects of climate change interact with confounding factors, the more these factors influence eventual outcomes in pathways that are ever more entangled. This makes the prediction of tertiary (systemic) effects of climate change, including conflict, difficult at timescales such as years or even decades. Yet tertiary effects have the greatest potential impact on human health. A contributing element is that multiple interactions between climate change and other factors impede adaptation. The potential for conflict, associated with and fed partially by climatic factors, illustrates these points.

15.2 Causes of Conflict

The dominant view at the heart of both the military and science is that conflict is linked to resource scarcity (Sullivan et al., 2007; Mazo, 2010; Zhang et al., 2011; Morisetti, 2012). In contrast, the cornucopian school of thought argues against a positive causal connection between climate change and conflict (Slettebak, 2012). Often, the proposed causal pathways involved are oversimplified by this school. Christopher Butler and Scott Gates (2012) write:

> Advocates of the proposition that 'climate change causes conflict' fail to offer a convincing explanation for this relationship...they rely on a crude causal explanation – resource scarcity leads to conflict. It's not that simple (p. 32).

Indeed, it is not. Virtually all the academic proponents of a link between anthropogenic climate change or resource scarcity and conflict outline pathways far more complex than this 'straw man' portrayal (Homer-Dixon and Blitt, 1998; Kahl, 2006; Zhang et al., 2007, 2011; Mazo, 2010). Climate change and resource scarcity are predicted to exacerbate and interact with social, economic and political factors. Scholars are careful to note that characterization of any single factor as the conflict 'trigger' is problematic (Kahl, 2006).

Researchers argue that a mosaic of climate change consequences increases the risk of conflict (Mazo, 2010). Climate change combined with population growth is likely to make renewable resources scarcer in many regions, especially food and water. Foreseeable consequences include economic stress, trade imbalances, deepening local and global inequality and legions of unemployed youth. Such factors could reduce tax revenue, weaken governance and help generate terrorism and extremism; for example, from states such as Yemen, which are already showing signs of deep distress and possible failure. All of these factors can reinforce each other mutually, further increasing the risk of conflict.

Much military thinking is similarly nuanced, conceiving of anthropogenic climate change as a 'threat multiplier' (Sullivan et al., 2007) that interacts with existing challenges to peace, making them more difficult to solve. A significant fraction of the world's militaries are engaged with planning for such contingencies (Schwartz and Randall, 2003; Sullivan et al., 2007; Paskal, 2010; Morisetti, 2012). Countering these trends will require more frequent aid and peacekeeping missions from countries which remain stable, adding to the task load of militaries (Sullivan et al., 2007).

The world's militaries are unlikely only to appeal only to civil society to seek to engage with them in ways to promote peace. Some nations, including those claiming Arctic resources, such as minerals, energy and shipping routes (which are becoming more accessible due to ice melt and shorter winters), are strengthening territorial claims, creating strategic alliances based on resources and altering their military positioning based on opportunities and threats presented by future climate change. This will likely escalate with progressive climate change (Paskal, 2010).

However, in an explicit challenge to the purpose of this chapter, some writers have argued that even to discuss the possibility of a link between climate change and conflict could generate a 'self-fulfilling prophecy' (Gleditsch, 2012). Our response is that if academics and the general public fail to engage in this debate, then the vacuum of ideas will be filled solely by the military. Furthermore, apprehension of the tertiary dimensions of climate change, including its influence on conflict, and hence several important aspects of global health, may motivate community support for climate change mitigation, perhaps even playing a role akin to that of the peace movement during the Cold War (see Chapter 29, this volume).

15.3 Mechanisms and Conflict Thresholds

The 'threat multiplier' conceptualization of climate change can be likened to that of an epileptic seizure threshold. The brain is a highly complex system, so precise prediction of seizures is difficult. However, many factors (e.g. medications, alcohol and lack of sleep) lower the threshold at which a seizure may occur. Conversely, reversal of these factors (and some drugs) can make a seizure less likely. Table 15.1 lists ways climate change can lower the threshold at which conflict is likely to occur, roughly divided into ecological and social causes, though strict divisions are impossible as social context always frames ecological change.

The complex, interdependent nature of conflict makes exact temporal predictions similarly difficult. However, we can say that many of the effects of climate change will lower 'conflict thresholds', making violence more likely, but in ways influenced greatly by regional factors, both physical and social.

Mechanisms by which climate change is likely to harm human security include interruption in access to water, food, land and infrastructure, such as from natural disasters. However, human insecurity does not automatically mean increased conflict, because security or its lack is embedded within a socio-political and economic context that also influences the likelihood of conflict. Clearly, in some cases, people choose to cooperate rather than fight over limited resources. Others may lack the means to engage in conflict, and simply accept more limited resource access. In sum, the likelihood of conflict due to environmentally induced reduction in security is influenced by social and perhaps

Table 15.1. Climate change effects that may lower the threshold for violence.

Ecological	Rate or scale of climate change that overwhelms coping mechanisms
	Interruptions to food supply
	Decreased fresh water availability
	High dependence on resources affected by climate change
Social	History of conflict
	Weakened states/institutions
	Economic stress
	Trade imbalances
	Persistent local and global inequity
	Numerous unemployed youth
	Dysfunctional legal regimes and reduced capacity for their enforcement
	Ethnic tensions

other factors. But if the integrity of social systems is also weakened by climate change, then conflict-promoting synergies may occur.

States and institutions have a major influence on the conflict threshold, due to their role as a buffer against reduction in human security and providing non-violent forms of redress. A history of cooperation is important, because the political knowledge and capital of previous episodes can be applied to new challenges.

15.4 Natural Climate Change and Conflict in History

Although the scale, speed, direction and duration of modern global climate change have no precise historical analogue, previous periods of climate change are instructive (McMichael, 2012). Large-scale conflict in the archaeological record is often associated with natural changes in climate and resultant resource scarcity, particularly of food and water.

For instance, in the area of the south-west USA, no large-scale conflict was evident for millennia after humans first inhabited the area. Around a millennium ago, this region experienced climatic conditions favourable to agriculture, allowing rapid population expansion. Beginning in approximately 1150 CE, a series of droughts and altered precipitation patterns lowered agricultural productivity. This led to substantial hardship, as demonstrated in the skeletal remains of adults and by increased child and infant mortality. The first

evidence of intergroup conflict emerges at this time, with people building defensible dwellings, including the famous cliff houses, and settling in larger groups for protection. Houses and villages were burned and destroyed. There was an increase in the number of skeletons with signs of violent deaths, including embedded arrows, cut marks and missing skulls. Substantial depopulation followed, lowering resource pressure and reducing violence (Haas, 2009). Environmental scarcity has contributed to conflict in many other areas, including on Easter Island (Diamond, 2005) and the 'Wild Nile', during the early Holocene (Close, 1996).

These examples are corroborated by statistical studies. In the last millennium, several periods of natural climate change decreased agricultural productivity in Europe and China, which, in turn, was associated with economic downturn, war, famine and population reduction (Zhang *et al.*, 2007, 2011). Some analysts even suggest that between 1400 and 1900, climate was the main driver of worldwide cycles of war and peace (Zhang *et al.*, 2007). This analysis also supports the proposition that state-level organization and technological advancement cannot always eliminate the risk of conflict posed by climate change, demonstrating a perennial human dependence on climate. A recent meta-analysis of quantitative studies shows a strong positive correlation between climate change and conflict, with each standard deviation change toward a warmer climate or more extreme rainfall increasing intergroup conflict by 14% (Hsiang *et al.*, 2013).

15.5   Modern Climate Change and Conflict

Anthropogenic climate change will continue to alter resource availability and access. There will be regions, mostly in the far north, where warming temperatures and melting ice will increase the accessibility of important resources, including petroleum and minerals. Many of these are under the sea, in areas contested by major powers such as Russia, Canada and the USA (Mazo, 2010; Paskal, 2010). Warming will also open up additional agricultural land and trade routes, the control of which may rival the strategic importance of Panama and the Suez (Lee, 2009; Paskal, 2010). However, open conflict in the Arctic appears less likely than nearer the equator, as none of the powers involved is desperately poor or has any recent history of mutual conflict, and all are likely to be constrained by international opinion. However, several have substantial military strength and any conflict that does occur could have severe health consequences.

Food-related conflict

While climate change will probably enhance access to non-renewable resources in some northern countries, and may even enhance their agricultural productivity in the medium term, per capita net food resources of the planet are likely to decrease further. Ocean acidification associated with increased carbon dioxide concentrations interferes with the development of a wide range of aquatic species. Sea level rise will likely displace millions, and endanger some of the most fertile agricultural land (Sullivan et al., 2007).

Countries that employ a high proportion of their population directly in agriculture have fewer resources with which to adapt and fewer alternative industrial and post-industrial sources of income than developed countries. In this chapter, we stress that the pathways that link an increased risk of conflict with resource scarcity are complex, multiple and often interlinked (Homer-Dixon and Blitt, 1998). Countries rich in natural resources are sometimes poorly governed and often home to significant ethnic,

religious and linguistic diversity, which can interact easily with these other factors, intensifying social segmentation, opposition and tension (Kaplan, 1994).

Resource scarcity, inequality and institutions

Resource scarcity can weaken institutions or inhibit their development. For example, it is hard for nations with a limited tax base to invest in schools and health-care systems. Inequality, stagnant economic growth and poor governance can increase grievances and the potential for conflict. Weak institutions decrease opportunities for peaceful redress of grievances and increase the perception of potential challengers that violence will be successful (Homer-Dixon and Blitt, 1998).

15.6   Regression Studies

The strength of theory and case studies linking climate change with conflict have led to papers which aim to test statistically this connection in recent history. Statistical regression studies of the past must reduce the causal complexity of their models compared to those postulated by qualitative theorists (Homer-Dixon and Blitt, 1998; Kahl, 2006). For the purpose of these studies, climate change is represented by only one or two proxy variables in a single country or area, such as random annual deviations in average temperature or precipitation, and the causal link with recent conflicts tested statistically. They are thus greatly simplified and necessarily underestimate the impact of climate change on conflict. Anthropogenic climate change predicted in the near to mid-future (i.e. to 2100) will generate greater and more rapid warming and precipitation change than has occurred since the start of the Holocene 11,000 years ago. Models suggest climate change will also last longer and impact the entire globe, with global agricultural production suffering in a way that recent climate history does not parallel and which has not been captured in regression studies. Further, many climate change effects may not be related linearly to temperature, instead emerging only beyond critical thresholds (see Box 15.1). Future climate change is also expected to alter patterns in temperature extremes, the timing and amount of precipitation and the frequency of extreme weather events throughout the world. It will make the oceans more acidic, and is likely to intensify storm surges such as from Hurricane Sandy in 2012. Moreover, it is very likely that these effects will reinforce each other in ways poorly captured by existing regression models.

In addition, such studies examine the impacts of chosen proxy measures over short periods. They miss the cumulative impacts of climate change over several years or decades, and evidence from historic studies of natural climate change suggests that changes over this temporal scale, or greater, are most closely associated with conflict (Haas, 2009). Finally, statistical studies tend to use individual countries or smaller areas as the units of analysis, a design that misses cascading regional economic and security impacts of climate change. Analysis of the relatively smaller changes of the recent past can provide a guide to the expected larger climate change of the future, but is unlikely to capture causal complexity. These studies are valuable, but are an attempt to understand the past rather than directly predict the future. They must be interpreted with the knowledge that they likely underestimate the links between future climate change and conflict.

Approximately half of the published statistical studies find a connection between the chosen proxy variable for predicted climate change and conflict in recent history (Hendrix and Salehyan, 2012; Raleigh and Kniveton, 2012). Perhaps the study that best captures the regional effect of climate change compared wetter, cooler La Niña years with drier, hotter El Niño years between 1950 and 2004. It found that civil conflicts in equatorial countries were twice as likely to start in the latter (Hsiang *et al.*, 2011). This suggests that El Niño–Southern Oscillation (ENSO) was a causal factor in 21% of relevant civil conflicts (Hsiang *et al.*, 2011). The studies which find no significant connection often acknowledge that this absence is not a definitive indication that conflict is uncorrelated with climate change (Koubi *et al.*, 2012; Theisen, 2012). To our knowledge, only one statistical study found evidence that a proxy climate indicator decreased the risk of conflict significantly (Slettebak, 2012). Almost all regression studies find that climate change proxies are associated with conflict, or have no significant relationship. However, because statistical studies are biased conservatively, the results of these statistical studies are, overall, consistent with our thesis; future anthropogenic climate change is likely to increase the risk of conflict.

15.7   Case Study: Sudan

The recent history of newly partitioned Sudan (both Sudan and South Sudan) is characterized by extensive conflict. Small-scale tribal violence

---

Box 15.1   Ecological and Climatic Potential and Actual Tipping Points

Release of greenhouse gases from warming Arctic and tundra
Amazon dieback and fires, creating a net carbon source
Weakening of oceanic capacity to absorb carbon
Fisheries collapse (e.g. bluefin tuna, cod) from overfishing

Social (adverse)

Decisions to migrate, launch an attack or declare a famine
Riots, public disorder, gang attacks
Acts of terrorism, evolution of a narco-state or other forms of large-scale criminal activity, e.g. Indian Ocean piracy, Mafia

Social (beneficial)

Decision to accept an apology or reparation, restoring or preserving peace
Creation of organized groups to agitate for justice or peace
Potential emergence of the noösphere (planetary consciousness). Incipient triggers for this may include the sight of Earth from space and the global scale of communication, especially the Internet

is mostly over resources such as land or cattle. In recent decades, larger-scale conflicts between the Sudanese government and antigovernment forces have caused an estimated 2–3 million deaths (United Nations Environment Programme, 2007). In this example, factors already discussed are at play; environmental stressors interact with social, political and economic stressors to lower the conflict threshold and multiply the impacts of conflict.

A host of non-climatic factors, including continuing high population growth, corruption and weak governance, have contributed to the conflicts in this region, including in Darfur (Sudan). There is residual doubt that the observed changes in the frequency of droughts during the 'long rains' season, which have lowered the viability of agriculture, represent anthropogenic climate change or natural variation (Williams and Funk, 2011). However, it is clear that persistent harm to agriculture in this region robs many people of their livelihoods, increasing destitution and migration; this contributes to conflict. Additionally, some argue the desert has spread an estimated 50–200 km south since the 1930s, decreasing the land available for agriculture or pasture (United Nations Environment Programme, 2007).

Ethnic conflict in Darfur between herding and farming groups was fought largely over access to arable land and water, the availability of which has decreased as described above, and occurred in the context of a small portion of the population in Sudan benefitting from oil income (Lippman, 2007). Militias destroyed villages and forests. Both climate change and conflict eroded the natural resource base, creating a positive feedback loop. In an effort to retain disproportionate power in an ethnically divided country, the Sudanese government used oil revenues to pay militias and provided them with modern weaponry and communications technology (Lippman, 2007).

Mazo (2010) considers the conflict in Darfur to be the first example of large-scale violence caused by anthropogenic climate change. The harm to health caused by this civil war is immense, albeit poorly measured. The conflict is widely considered to be genocidal, with civilians, including women and children, deliberately targeted (Lippman, 2007) and reports of well-water being poisoned (United Nations

Environment Programme, 2007). Over 2 million displaced civilians were forced into camps for safety and access to basic resources (Lippman, 2007; United Nations Environment Programme, 2007). The Sudanese government slowed the flow of food and supplies, and Sudan became the most dangerous location for humanitarian workers. In June 2007, it was estimated that between 300,000 and 400,000 people had died out of a population of 6 million in Darfur (Lippman, 2007). If conflict is over access to limited resources, it becomes 'rational' to extend violence to civilians, as a smaller enemy population frees assets.

The Darfur conflict illustrates how climate can contribute to and combine with state weakness or failure (Homer-Dixon and Blitt, 1998) and can become international (Kaplan, 1994). In Darfur, fighters from other countries participated (Lippman, 2007), and at times the fighting extended into Chad and the Central African Republic (United Nations Environment Programme, 2007).

15.8   Future Scenarios

Scenarios are useful ways to explore possible futures. If climate change is confirmed as substantially limiting precipitation in East Africa, then its future as a viable habitat for stable prosperous civilization appears at high risk.[1] Imagine it is the year 2035; decades of increasing drought are punctuated by feeble 'long rains' seasons. Somalia remains a failed state beset by violence, which has spread into parts of Ethiopia. This violence hampers international relief in these countries, and raising funds is exceedingly difficult due to competition for relief dollars.

Eritrea has built military alliances with Western powers seeking a base from which to control anti-piracy campaigns and to counterbalance China's presence in oil-rich Sudan and South Sudan. The Eritrean government is oppressive and corrupt, having channelled funds toward military equipment at the expense of civil society. Its iron-fisted rule does not permit dissent, but affords a veneer of stability, so relief efforts for the region are centred here. Thousands of Somalis and Ethiopians cross the borders in search of food. To distract its hungry citizenry dissatisfied from widespread poverty, the Eritrean government uses border crossing by Ethiopian immigrants as a pretext for invading its traditional enemy, and seals its borders. Militias, covertly backed by the Eritrean government, seize the grain shipments headed into Ethiopia, creating a famine exceeding that of the 1980s.

Meanwhile, Sudan is building dams, which combine with climate change to reduce downstream flows into the Nile, including of sediment, needed to slow the encroachment of the Mediterranean into the densely populated Nile delta. Its confidence is bolstered by regular Chinese arms shipments (Klare, 2008). Downstream, the population is still growing in Egypt, America's ally, and the flooding of the delta is worsened by sea level rise (Bohannon, 2010). Egypt launches air strikes against the Sudanese dams and airfields, with the war escalating to ground battles which damage infrastructure and hinder economic growth. Cool heads in Beijing and Washington keep these superpowers out of the war. With Sudanese forces engaged in the north, South Sudan ends its 6-year ceasefire with Sudan and captures South Kordofan to support the Nubian revolutionaries in their long-standing bid for independence.

Emergent properties of conflict and health induced by climate change

The above scenario highlights three categories of effects from conflict induced by climate change (see Table 15.2). The direct effects of conflict on health, such as from trauma, are obvious. The indirect health effects of individual conflicts (e.g. resurgent infectious diseases) are well described elsewhere (Connolly *et al.*, 2004). However, in addition, a third category of health effect is evident if conflict becomes sufficiently widespread. This is largely undescribed in the literature.

A world with a higher level of conflict than today is likely to see increased military spending, with the risk of new arms races, even in countries not directly involved in conflict. Since governmental and societal resource allocation is only partially discretionary, increased defence expenditure necessitates an opportunity cost, including lower spending on health care and other health determinants such as education. A parallel shift from humanitarian to military foreign aid is also likely (Klare, 2008).

Increased xenophobia and other forms of intolerance are likely in situations of increased scarcity (Keeley, 1996; Gelfand *et al.*, 2011).

Table 15.2. Effects on health of conflict induced by climate change. (The first two sections of the table draw heavily from Connolly *et al.*, 2004.)

Direct effects	Battle deaths and injuries of combatants
	Civilian deaths and injuries from weapons of war, potentially including weapons of mass destruction
	Civilian deaths, injuries and undernutrition due to targeting of civilian food supplies and infrastructure
Indirect effects of single or repeated conflict	Reduced monitoring for new and re-emerging disease
	Reduced vector control programmes (e.g. resurgence of African trypanosomiasis)
	Increases in crowded, unsanitary conditions in refugee camps, leading to increased possibility for disease transmission
	Reduced agricultural productivity and economic output, leading to impaired nutritional status
	Increased rates of sexual violence, leading to psychological trauma and spread of sexually transmitted illnesses
	Flight or killing of medical personnel
	Shift to deliberate planting of less nutritious root crops, for example cassava, to protect from theft or destruction
Indirect effects of generalized increase in conflict due to climate change	Shift in domestic spending from public health measures to defence
	Decreased humanitarian aid, with potential for concomitant increase in military aid
	Increased xenophobia and protectionism – especially of food and other critical resources
	Reduced economic growth in conflict areas, with global effects
	Increasing regional and global instability, leading to cascading conflict

This could intensify protectionism, lower trade and deepen global economic inefficiencies and inequality. Equatorial countries, poor in human and financial resources, appear the most vulnerable.

Finally, conflict induced by climate change is itself likely to breed additional violence. Modern conflicts in Africa readily cross state boundaries. Sometimes, apparently minor wars can involve more powerful countries, as occurred during World War I. Conflict is more likely to occur when countries feel insecure and rivals increase military spending. Conflict can also give rise to militaristic governments more likely to use force than diplomacy when confronted with foreign and domestic problems. Conflict erodes the norms and institutions that promote peace.

15.9 Minimizing Negative Health Impacts

Understanding the influence of climate change on conflict as multiplying threats and lowering conflict thresholds can help decision makers design policies and programmes to minimize the potential impact on human health. Historical studies demonstrate that interruptions to food supply correlate strongly with conflict, so enhancing resilience in the agricultural sector is important. Strong states and institutions are also vital for promoting peaceful outcomes. Finally, if we can improve understanding and prediction of conflict, we can be better prepared, such as with emergency relief or pre-emptive diplomacy.

The challenges are grave. One concern is that the rate and scale of climate change will exceed the coping capacity of our institutions, undermining many of our assumptions and previous studies regarding the relationship between environmental scarcity and conflict. The Intergovernmental Panel on Climate Change (IPCC) has forecast a global increase in temperature of up to 4°C by 2100. However, the IPCC is increasingly perceived as conservative concerning temperature, rainfall change, polar ice melt and sea level rise. The IPCC also tends to understate the risk of ecological tipping points and, even more, of social tipping points, such as the decision or need to migrate.

15.10 Conclusion

Conflict driven by climate change is a major health risk and grossly underappreciated, including in the health literature (Morisetti, 2012). We have stressed that climate change does not, in and of itself, cause conflict; we must be wary of simple associations. Such complexity and redundancy mean that adaptation is likely to be only partially successful, including for preventing conflict. Many actors may react to increased conflict by increasing military spending, further withdrawing resources from public health. The already locked-in climate change appears sufficient to increase the likelihood of conflict, as Sudan demonstrates. Improved mechanisms for maintaining peace and resolving conflict are needed. Adaptation is essential, but so too are strong and urgent efforts to mitigate climate change, if we are to promote peace and protect human and planetary health.

Note

[1] We apologise for selecting a specific region; there were many candidates. It is our intention that negative scenarios such as this help motivate preventive to avoid such bleak futures.

References

Bohannon, J. (2010) The Nile Delta's sinking future. *Science* 327, 1444–1447.

Butler, C.K. and Gates, S. (2012) African range wars: climate, conflict, and property rights. *Journal of Peace Research* 49, 23–34.

Close, A.E. (1996) Plus ça change. The Pleistocene–Holocene transition in Northeast Africa. In: Strauss, L.G., Eriksen, B.V., Erlandson, J.M. and Yesner, D.R. (eds) *Humans at the End of the Ice Age*. Plenum Press, New York and London, pp. 43–60.

Connolly, M.A., Gayer, M., Ryan, M.J., Salama, P., Spiegel, P. and Heymann, D.L. (2004) Communicable diseases in complex emergencies: impact and challenges. *The Lancet* 364, 1974–1983.

Diamond, J. (2005) *Collapse: How Societies Choose to Fail or Succeed.* Viking, New York.

Gelfand, M.J., Raver, J.L., Nishii, L., Leslie, L.M., Lun, J., Lim, B.C., *et al.* (2011) Differences between tight and loose cultures: a 33-nation study. *Science* 332, 1100–1104.

Gleditsch, N.P. (2012) Whither the weather? Climate change and conflict. *Journal of Peace Studies* 49, 3–9.

Gleick, P.H. (1991) Environment and security: the clear connections. *Bulletin of the Atomic Scientists* 47, 16–22.

Haas, J. (2009) The origins of war and ethnic violence. In: Carman, J. and Harding, A. (eds) *Ancient Warfare: Archaeological Perspectives.* The History Press, Gloucestershire, UK, pp. 11–24.

Hendrix, C.S. and Salehyan, I. (2012) Climate change, rainfall, and social conflict in Africa. *Journal of Peace Research* 49, 35–50.

Homer-Dixon, T. and Blitt, J. (1998) Introduction: a theoretical overview. In: Homer-Dixon, T. and Blitt, J. (eds) *Ecoviolence: Links Among Environment, Population, and Security.* Rowman and Littlefield Publishers, Inc, Lanham, Maryland, pp. 1–18.

Hsiang, S.M., Meng, K.C. and Cane, M.A. (2011) Civil conflicts are associated with the global climate. *Nature* 476, 438–441.

Hsiang, S.M., Burke, M. and Miguel, E. (2013) Quantifying the influence of climate on human conflict. *Science* 341, doi:10.1126/science.1235367.

Kahl, C.H. (2006) *States, Scarcity and Civil Strife in the Developing World.* Princeton University Press, Princeton, New Jersey.

Kaplan, R.D. (1994) The coming anarchy. *Atlantic Monthly*, 273(2), 44–76.

Keeley, L.H. (1996) *War Before Civilization.* Oxford University Press, New York.

Klare, M.T. (2008) *Rising Powers, Shrinking Planet: The New Geopolitics of Energy.* Henry Holt and Company, LLC, New York.

Koubi, V., Bernauer, T., Kalbhenn, A. and Spilker, G. (2012) Climate variability, economic growth, and civil conflict. *Journal of Peace Research* 49, 113–127.

Lee, J.R. (2009) *Climate Change and Armed Conflict: Hot and Cold Wars.* Routledge, London.

Lippman, M. (2007) Darfur: the politics of genocide denial syndrome. *Journal of Genocide Research* 9, 193–213.

McMichael, A.J. (2012) Insights from past millennia into climatic impacts on human health and survival. *Proceedings of the National Academy of Sciences* 109, 4730–4737.

Mazo, J. (2010) *Climate Conflict: How Global Warming Threatens Security and What To Do About It.* The International Institute for Strategic Studies, Routledge, London.

Morisetti, N. (2012) Climate change and resource security. *BMJ* 344, e1352, doi:10.1136/bmj.e1352.

Paskal, C. (2010) *Global Warring: How Environmental, Economic and Political Crises Will Redraw the World Map.* Palgrave Macmillan, New York.

Raleigh, C. and Kniveton, D. (2012) Come rain or shine: an analysis of conflict and climate variability in East Africa. *Journal of Peace Research* 49, 51–64.

Schwartz, P. and Randall, D. (2003) *An Abrupt Climate Change Scenario and Its Implications for United States National Policy.* Jet Propulsion Laboratory, Pasadena, California.

Slettebak, R.T. (2012) Don't blame the weather! Climate-related natural disasters and civil conflict. *Journal of Peace Research* 49, 163–176.

Sullivan, G.R., Bowman, F., Ferrell, L.P.J., Gaffney, P.G.I., Kern, P.J., Lopez, T.J., *et al.* (2007) *National Security and the Threat of Climate Change.* The CNA Corporation, Alexandria, Virginia.

Theisen, O.M. (2012) Climate clashes? Weather variability, land pressure, and organized violence in Kenya, 1989–2004. *Journal of Peace Research* 49, 81–96.

United Nations Environment Programme (2007) *Sudan: Post-conflict Environmental Assessment.* United Nations Environment Programme, Nairobi.

Williams, A. and Funk, C. (2011) A westward extension of the warm pool leads to a westward extension of the Walker circulation, drying eastern Africa. *Climate Dynamics* 37, 2417–2435.

Zhang, D.D., Brecke, P., Lee, H.F., He, Y.-Q. and Zhang, J. (2007) Global climate change, war, and population decline in recent human history. *Proceedings of the National Academy of Sciences* 104, 19214–19219.

Zhang, D.D., Lee, H.F., Wang, C., Li, B., Pei, Q., Zhang, J., *et al.* (2011) The causality analysis of climate change and large-scale human crisis. *Proceedings of the National Academy of Sciences* 108, 17296–17301.

16 Climate Change and Health in East Asia: A Food in Health Security Perspective

Mark L. Wahlqvist

Monash University, Australia; Institute of Population Health Science, National Health Research Institute, Taiwan, and Fuli Institute of Food Science, Zhejiang University, China

There is 'a battle within us between the forces that should have caused us to protect the Earth, like our capacity to reason and our connection to nature, and our greed, materialism and alienation from nature, which, in the end, have won out.'

(Hamilton, 2010)

16.1 Introduction

Inevitably, common climate and health threads run through all world regions. But the economic dynamism of East Asia (reflected in conventional economic criteria such as gross domestic product (GDP), but also in other indicators, such as the enthusiasm of its young people for social media) brings great opportunities and perils. The time frame for preventive and corrective action is shorter.

This chapter presents a 'Food in Health Security' perspective to climate change in East Asia (Wahlqvist *et al.*, 2009). This approach embraces food and nutritional security, agriculture, ecology and health, human rights, ethics and equity. Above all, an integrated approach to issues of development and health is needed. Good population health cannot arise purely from the provision of health care, hospitals and

health systems. It also requires a sound and durable 'planetary support' system, able to provide adequate food, fuel fibre and other ecosystem 'services' for humanity to flourish (Corvalán *et al.*, 2005). Overlying and interacting with these ecological foundations is a social superstructure, which can serve to either strengthen or erode other dimensions of well-being and security, including for children, women, the elderly and other vulnerable and disadvantaged groups. The chapter also discusses several scenarios for human well-being, from an East Asian perspective.

East Asia

East Asia refers to the greater China area, the Korean Peninsula, Japan and the ten countries that form the Association of South-east Asian nations (ASEAN).[1] It therefore embraces Northeast and South-east Asia. With the exception of the Democratic People's Republic of Korea, the linkage between these countries and their sense of common destiny is growing. But before discussing climate change and health in this dynamic region, five features which contribute to both the problems and any solutions that might be envisaged need discussion.

16.2 Five Cross-cutting Features of East Asia Relevant to Climate Change and Health

Cognate and transitional cultures

The region has been influenced historically and strongly by China, whether by cultural transfer to Japan in the 700s, allegiance with the Ryuku kingdom based on Okinawa from the 1200s or by migration in Indochina and the Indonesian archipelago from at least a millennium ago. Confucian thinking and health constructs like 'hot and cold foods' are evident far from China, albeit with progressive dilution. The strength of these beliefs may override conservation of precincts or their fauna. While many aspects of Chinese culture call for respect for nature, the overriding need to feed its people and to maintain social harmony has led to the transformation of vast landscapes, most obviously in the creation of terraced rice paddies. Often, nature is honoured symbolically, rather than in actuality (Elvin, 2004). There has, however, been substantial cultural interaction between people of Chinese origin and the indigenous peoples of the region, who in general have a greater sense of place and its value. But this interaction is not without tension.

Commentators often speak of transition in East Asia as though it were toward the West, but considerable cultural shifts are taking place *within* East Asia. Their momentum will increase. This is particularly evident with the promotion of South Korean art forms and food culture through North-east Asia. While Islam is a significant way of thought and life in South-east Asia, it is generally culturally interactive, drawing on the local roots of its followers, who sometimes have a high regard for local ecology (Mangunjaya and McKay, 2012).

Climate and natural disasters

Anthropogenic climate change will add to a region that is already climatically diverse. Conditions range from monsoonal to dry: with typhoons and a hot and humid climate in coastal regions; snow in the north-east and mountainous regions; and dusty winds from western China and suffocating forest fires in Indonesia affecting respiratory health in South-east Asia. Huge topographical differences, from coastal plains to mountainous peaks, jungle to wilderness and vast estuaries, river systems and gorges, provide varying microclimates. People accustomed to living in particular localities are undergoing considerable change. Villages that may have changed little for hundreds of years are now connected to a wider world of transport, information and education, financial systems, food and health care. These changes may improve the quality of life, but paradoxically can add to ecosystem decline and climate change. Additionally, increasing population size and density make the disaster potentially worse; albeit dependent on the resource utilization and practices of the populace.

Some disasters are slow in the making, typified perhaps by those emerging in megacities in East Asia. Many, like Beijing, have outgrown their water supply. China's north is now more prone to drought, including because its aquifers are increasingly dry (Zheng *et al.*, 2010). The Chinese engineering response is to build three massive pipeline systems, all currently under construction, to bring additional water to the north, each with its own environmental controversy. Where can all this end? Hopefully, as China's population management takes effect, demand will decrease and the problem slowly dissipate. But it may not; even now, millions of people in China's south-west inhabit a parched landscape and are unable to grow enough food to feed themselves; they are net food importers. How anthropogenic is this 'natural' disaster?

Numerous environmental changes induced by megacities underscore the risk, including the paving of arable land, foliage replaced with concrete and hotter dwellings, phosphorus loss via urine into waterways and out to sea (Zhu, 2009) and food waste (which can be fully recycled; Lin *et al.*, 2009; Huang, 2010). The Chinese Academy of Sciences has documented phosphorus flows in Beijing and Tianjin, perhaps giving hope that the Chinese state will implement ways to recycle this essential element (Qiao *et al.*, 2011).

Anthropogenic ecology

It is possible for ecological systems to be changed radically by people and to take on a new kind of sustainability. The many small farmers of traditional China ran highly integrated farming systems with extensive recycling, including of human waste (Paull, 2011). These practices were not without health risk (including schistosomiasis) (Wang *et al.*, 2008a), but health literacy had the capacity to overcome much of that risk. Much of this is now challenged by urbanization and agribusiness, whose medium-to long-term net effects on climate and health are pending.

Environmental trends

The global environmental trends of warming, pollution and loss of ecosystems are well rehearsed in this volume. For East Asia, these trends are just as important as elsewhere, but the 'hot spots' are worth documenting.

These include polluted waterways in much of China; drought in northern China; drought, persistent agricultural and political inability for the Democratic People's Republic of Korea to feed itself; and earthquakes and typhoons in the Pacific Rim. These carry multiple risks, including active underwater volcanoes off the east coast of Taiwan, where four nuclear reactors are sited. Furthermore, the world's largest inland edible fish resource in the Mekong delta is threatened by numerous upstream dams (Ziv *et al.*, 2012). Issues especially relevant to Indonesia include deforestation, vast palm oil plantations and, again, volcanoes and earthquakes. The megacities of Bangkok and Jakarta have been sinking and are increasingly prone to flooding. Shanghai is also vulnerable (Wang *et al.*, 2012). These problems can only worsen as the sea level rises.

Throughout East Asia, increasing numbers of precincts will be less habitable, creating pressures for migration and resettlement in ecologically sensitive areas (Butler, 2011).

Health and ill-health pattern peculiarities

Patterns of infective illness or transmissible disease are also altering with climate change and other forms of global environmental change. Within East Asia, dengue is becoming more intense and moving north, while the distribution of malaria vectors is enlarging; though, for the moment, insecticides and antimalarials are lowering its burden of disease.

Deforestation and changes in animal husbandry can also cause infective agents to jump species, as happened when Nipah virus leapt from bats to pigs to humans in Malaysia. Severe Acute Respiratory Syndrome (SARS) in China may have emerged in part from the farming of civet cats, infected from bats (Hu and Shi, 2008). Resistance to antimicrobial agents is becoming increasingly serious, including with malaria in Indochina and tuberculosis in Indonesia. Changing strains of influenza virus present major logistic and budgetary difficulties in the region; what is less widely recognized is that viral transformation and pathogenicity are more likely to occur in nutritionally compromised intermediate host animals, especially poultry, with selenium deficiency a documented example (Beck, 2007).

Maternal (and even paternal) nutrition and personal behaviours are now recognized as precursors of intergenerational health problems, of which diabetes and its complications are the most evident. Lower birth weights are an indicator of these risks; and a much greater emphasis will be required to provide more favourable environments for women during the reproductive years.

A looming and related problem is maternal and early life exposure to environmental contaminants, especially those with 'endocrine disruptive' capacity. These include phthalates, dioxins and PCBs (polychlorinated biphenyls), which can lead to a wide range of health problems (Wang *et al.*, 2008b). The immune system, the reproductive system, with changes which alter gender orientation, and the central nervous system, with effects on mental health, may all be affected. What at first may look like a familiar feature of the food supply (such as its glycaemic index) as the basis, for example, of abdominal obesity diabetes or cardiovascular disease may, in reality, be attributable partly to a contaminant in the same food or food system. Dioxins in rice as a cause of diabetes is a well-documented example (Wang *et al.*, 2008b). The problem applies in the nutritionally and metabolically vulnerable older age groups as well, as shown in a Swedish study of diabetes in the elderly. In the wake of recent, present and, sadly, future industrialization and environmental degradation, we can expect a growing burden of this kind of disease in East Asia.

16.3   Ecology, Ecosystems and Health

The term 'ecohealth' has gained currency recently and rapidly across many disciplines, from the environmental to the health and social sciences. In mid-2010, Google had 9.9 million entries for 'eco-health' and 0.43 million for 'ecohealth'. The drivers for this reconceptualization of health include changing patterns of disease (Wahlqvist, 2002a,b), changing demography, with more long-term health needs, and climate change (Butler, 2009), along with rising food prices.

Conceptually, ecohealth recognizes that optimal health and well-being, provided through prevention and care, requires that households and communities share favourable health outcomes (Wahlqvist, 2009). It is increasingly understood, through epigenetic mechanisms, that the effects of a child's environment (including food deprivation) can be represented in the genome and transmitted to subsequent generations (Bygren, 2010). This also reinforces the importance of a more ecological approach to health than has heretofore prevailed. In turn, these are highly dependent on sustainable living and livelihood conditions, and supported by physical characteristics and biodiversity (see Fig. 16.1).

Ecosystem loss means a loss of ourselves. It is unclear how much more environmental degradation humanity can withstand. Much may be gained by cultural resilience and adaptability, but at the heart of the growing ecohealth dilemma is the viability of the ecosystems of which humans are an intrinsic part. Although ecosystems may undergo reductions in their constituents, and for some time retain functionality, they are likely to collapse at a critical point, analogous to a wave breaking on the shore, which follows the fractal or the mathematical patterns of 'chaos theory'.

The human species is an integral part of ecosystems; as these collapse, the health and survival of the species dependent on them is threatened by ecosystem loss. Human migration often represents an escape from one failing ecosystem to another. But as population expands beyond the 7 billion people now alive, can sufficient fresh ecological niches be located? Urbanization and megacities temporize this increasing mismatch, as do intensive food cultivation and technological innovations. Some of these may allow us to keep using ecologically depleted localities, such as by plant breeding for salt or flood resistance, or by increasing the nutritional value of nutrient-poor staple crops by biofortification. Food exports to locations where food cannot be grown locally also occur. However, eventually, both fertilizer and the non-renewable energy sources necessary to support these measures will be exhausted.

Access to potable and reliable sources of agriculturally safe water is of increasing concern in much of East Asia (Wahlqvist *et al.*, 2009).

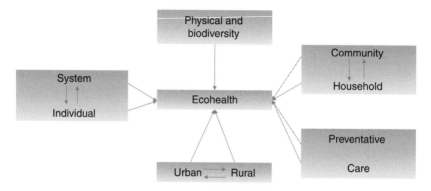

Fig. 16.1. The operations and dimensions of ecohealth.

Norman Borlaug, the father of the Green Revolution, said that he did not expect another such revolution to solve the world's food problems, but that improvements in infrastructure were the key, especially for transport by road and rail.[2] But, even here, the planned super-highways and railroads across East Asia, from China to Singapore, although enlivening economies and bringing supplies to cope with disasters, will also deliver environmental harm, such as deforestation.

The concept of ecosystems and health also reflects various values about nature that relate to the broader aspect of human development (see Table 16.1).

16.4   Climate and Health

Much has been written about how climate change may affect health and is available elsewhere in this book. In East Asia, the following are of particular relevance.

An ageing population

East Asia has four of the ten longest life expectancies (Macau, Japan, Singapore and Hong Kong) and among the best disability-adjusted life years (DALYs) globally. Despite this, health costs as a percentage of GDP are uniformly less than 10%, in contrast to about 18% in the USA. This apparent paradox reflects both diet and vigorous public health programmes and universal health insurance (accompanied by greater equity) in the better-performing East Asian countries, illustrated best by Japan and Taiwan (Beck, 2007; Maffly, 2013). In the March 2011 Fukushima tsunami, the aged were affected disproportionately and, in some cases, abandoned: a testimony to the precarious balance between extreme weather conditions and health. Aged people are especially vulnerable to extreme heat.

Maternal and child nutrition

The long-standing precarious state of maternal and child nutrition has been improving in East Asia, with a major supporting analysis from South-east Asia (Acuin et al., 2011). There have been rapid reductions in morbidity and mortality in these groups in Brunei, Singapore, Malaysia, Thailand and Vietnam, but efforts are faltering in the Philippines and Indonesia. High rates of undernutrition are now being overcome in Laos, Cambodia and Myanmar. A nexus exists between economic development and health system performance. But these trajectories are fragile and are vulnerable to the loss of healthy environments or ecosystems. The reported gains will be undermined by the rapid emergence in South-east Asia of chronic non-communicable diseases, especially on a background of stunted children.

Table 16.1. Values of nature related to human development (Wahlqvist and Specht, 1998), as proposed by Kellert (1996).

Value	Definition	Function
Utilitarian	Practical and material exploitation of nature	Physical sustenance/security
Naturalistic	Direct experience and exploration of nature	Curiosity, discovery, recreation
Ecologistic–scientific	Systematic study of structure, function and relationship in nature	Knowledge, understanding, observational skills
Aesthetic	Physical appeal and beauty of nature	Inspiration, harmony, security
Symbolic	Use of nature for language and thought	Communication, mental development
Doministic	Mastery, physical control, dominance of nature	Mechanical skills, physical prowess, ability to subdue
Humanistic	Strong emotional attachment and 'love' for aspects of nature	Bonding, sharing, cooperation, companionship
Moralistic	Spiritual reverence and ethical concern for nature	Order, meaning, kinship, altruism
Negativistic	Fear, aversion, alienation	Security, protection, safety, awe from nature

Proneness to natural disasters

The Natural Disasters Risk Index, released by the global risk advisory firm Maplecroft, is calculated by measuring the human impact of natural disasters, in terms of deaths per annum per million of population, and frequency of events over the past 30 years.[3] It reflects event likelihood and includes earthquakes, volcanic eruptions, tsunamis, storms, flooding, drought, landslides, extreme temperatures and epidemics. Using this method, combined with that of the World Bank, East Asia includes five of the 15 most natural disaster-prone countries (Taiwan, Indonesia, the Philippines, China and Myanmar). Risk management and preventive technologies (such as earthquake engineering, which is advanced in Japan and Taiwan), however, vary considerably, so that the consequences of similar disasters may be quite different within and across these countries. No country has endless resources to reconstruct repeatedly after successive and more severe disasters.

Risk planning rarely considers multiple disasters. The Morakot typhoon of August 2009, which commenced in the Philippines and extended to coastal China (with evacuations of 1 million people), was particularly disastrous for Taiwan.[4] Downpours of 3 m in 3 days followed 2–3 weeks of frequent earthquakes that had destabilized mountain rocks and soils. Mud and landslides quickly submerged whole villages in metres of mud and rock, with much loss of life. About one-third of Taiwan's agriculture, horticulture, aquaculture and agroforestry was destroyed. The displaced peoples, especially indigenous peoples and farmers, have had difficulty finding safe alternative land away from the central mountain range, because rising sea levels in the Taiwan Straits make coastal areas increasingly less habitable. A consequence was that duck farmers inadvertently moved to contaminated land, which, over several weeks, furnished contaminated duck meat to the Taiwanese market. The conjoint natural phenomenon of severe typhoons and frequent earthquakes may be made more likely by Pacific Ocean warming, which changes atmospheric pressure sufficient to increase the tectonic plate movement in Taiwan's East Rift Valley. The Morakot experience signals a set of complex climate and environmental changes which are affecting wide areas of East Asia, with varying levels of preparedness and potential for resilience. The health consequences are immediate, medium and long term.

16.5   Three Scenarios

The chapter concludes with three possible scenarios which perhaps could unfold in East Asia. Of course, none are likely to unfold exactly as discussed, but they are included in order to provide insights.

Retreat and survive

This is a scenario of winners and losers. Those with the greatest resources, such as safe and sufficient water, energy and food, and with appropriate health care, may survive. But the lessons of history are that such a scenario may see the demise of the socio-economically and technologically disadvantaged.

The originator of the Gaia hypothesis, James Lovelock, has warned, apocalyptically, that runaway climate change might lead to the collapse of civilization on a massive scale, with only tiny pockets of humanity surviving in cooler niches (Lovelock, 2009). These groups may lose contact with each other, as it is difficult to see how our current energy- and technology-dependent connectedness, relying on orbital satellites and cyber space, may long be sustained in such a breakdown world.

Scenarios such as a badly faltering trajectory for civilization have been dismissed for several decades by most mainstream experts, including eminent economists, futurists and development specialists. This is changing. The leading German climate scientist, Hans Schellnhuber, has warned that the difference between two degrees and four degrees of warming could be civilization itself, a warning echoed in the preface to his co-edited book, *Global Sustainability: A Nobel Cause* (Schellnhuber *et al.*, 2010). The German Chancellor, Angela Merkel (a physicist, like Schellnhuber), wrote the preface to this book. One of the world's most respected economists, Nicholas Stern, is a co-editor. The former President of the Royal Society, Martin Rees, is also deeply

concerned about civilization's short-term prospects, including because of climate change
(Rees, 2013).

The projected increase in population from
7 to 9 billion by 2050 will be largely of the poor
in compromised environments, some in East Asia;
they will accrue the greatest risk. In this scenario, humanity survives, but civilization enters
another dark age.

Technological rescue

In this world, there is enough technological progress to continue to match resources with additional people. Catastrophic conflict is avoided,
so too is a 'four-degree world. A 'blue revolution'
in water usage occurs (Wahlqvist *et al.*, 2009).
The people and the ecosystems of the planet pass
through a bottleneck, emerging to a future in
which population is declining voluntarily and
consciously, as is already the case in parts of
Western Europe and Japan, and is predicted to
unfold also in China by 2050.

A new ecological stability emerges. Parts
of Kalimantan and Sumatra may be repopu-
lated with orangutans and other charismatic
megafauna, kept alive for several generation
in zoos, as the economic imperative for palm
oil plantations fades, allowing reforestation.
Already, in China, great effort is being made to
sustain a small population of wild pandas,
including some that have been reintroduced to
the wild.

The interregnum, the transition, would
likely require large populations adapting to
living conditions, new energy sources and new
foodstuffs to which we are now unaccustomed.
Other new technologies might emerge, such as
personalized and renewable energy sources, safe
water generation systems and climate controlled
microshelters.

But even if such resources become widely
accessible and affordable, will they be optimal
for health? Optimistically, this order of change
might be optimized by advanced bioinformatics:
but on what would that ultimately depend and
how environmentally (or cyber war) robust will
that be? Could it be that this scenario would represent a rapid, techno-based evolution of our
species, which we might anticipate with fear or
excitement?

All-out war

A third and most horrible scenario is all-out war
for resources. In 2012, Ira Helfand, of the
International Physicians for the Prevention of
Nuclear War, reported to the 12th World Summit
of Nobel Peace Laureates in Chicago that a
nuclear war between India and Pakistan would
precipitate global famine, partly through aero-
sol-forced cooling and dimming, persisting for
several years (Helfand, 2012). In recognition of
the likely conflict-prone nature of resource
acquisition, food, health, human and planetary
security need to be embraced by the traditional
security system (McKay, 2009).

16.6   Conclusion

What are the prospects for the people of East
Asia, given these multiple interacting determinants and possibilities? Despite its development
and slowing rate of population growth, a vast,
still-growing population lives in vulnerable
areas, highly prone to natural disasters. On the
one hand, the widespread acceptance of Con-
fucianism and the importance of group cohesion may mean the region is better placed than
many to tolerate and accept a new, but less
attractive, form of ecohealth, one in which hundreds of millions experience ecological scarcity
in order to maintain a coherent central state.

In recent decades, East Asia has done well,
both for cultural and economic reasons, including the 'demographic dividend' (Eastwood and
Lipton, 2012). Generally, high human capital is
another factor, the low levels of education in
backwaters such as Cambodia and Laos being
exceptions that support the general principle.
The region's labour-intensive workforces in
small-scale local agriculture and small business
ventures provide an important resource in lieu of
financial capital. But this is altering, in many
places, due to the low birth rate. The characteristics of communities will alter as migration
quickens. This may limit the expansion of a sufficiently sized health workforce to meet the needs
of the aged. Communities will also be affected by
urbanization associated with climate change.
There is also the ever-present risk of conflict and
its displacement of people, with climate change
itself an increasing risk for conflict.

Some of what happens in East Asia will be favourable to the health and climate change equation as the region extends its small-scale farming and business enterprise around the world, enhancing the localization movement. Experience indicates that the Asian diaspora can influence the whole food system (Wahlqvist, 2012). The Chinese and Middle Eastern 'land grabs' are well under way, which will extend China's influence on global agriculture enormously, with uncertain outcomes (McKay, 2009). Countering this, food waste, an enormous contributor to food insecurity, is well managed in some Chinese societies like Taiwan (Lin *et al.*, 2009; Huang, 2010). Several East Asian countries are leading manufacturers and users of renewable energy technology. Accelerating this technological transition is vital, but East Asia's social contribution is also of great importance. East Asia cannot insulate itself fully against the global trends in climate and health, but there are signs, including a concerted effort in China to build urban environmental science (Zhu, 2009), that it may outperform other regions.

Notes

[1] Brunei Darussalam, Cambodia, Indonesia, Laos, Malaysia, Myanmar, Philippines, Singapore, Thailand and Vietnam.
[2] Personal communication to MW.
[3] Natural Disasters Risk Index (http://www.preventionweb.net/files/14169_NaturalDisasters2010.pdf, accessed 30 April 2012).
[4] See Typhoon Morakot spawns mudslide in Taiwan, burying hundreds in debris (SLIDESHOW) (http://www.huffingtonpost.com/2009/08/10/typhoon-morakot-spawns-mu_n_255327.html, accessed 27 April 2012).

References

Acuin, C.S., Khor, G.L., Liabsuetrakul, T., Achadi, E.L., Htay, T.T., Firestone, R., *et al.* (2011) Maternal, neonatal, and child health in southeast Asia: towards greater regional collaboration. *The Lancet* 377, 516–525.
Beck, M.A. (2007) Selenium and vitamin E status: impact on viral pathogenicity. *Journal of Nutrition* 137, 1338–1340.
Butler, C.D. (2009) Food security in the Asia-Pacific: climate change, phosphorus, ozone and other environmental challenges. *Asia Pacific Journal of Clinical Nutrition* 18, 590–597.
Butler, C.D. (2011) A stormy future for population health in southeast Asia? *The Lancet* 377, 885–886.
Bygren, L.O. (2010) Epigenetic epidemiology and food availability. *SCN News* 38, 54–55.
Corvalán, C., Hales, S., McMichael, A.J., Butler C.D., Campbell-Lendrum D., Confalonieri U., *et al.* (2005) *Ecosystems and Human Well-Being. Health Synthesis.* World Health Organization, Geneva, Switzerland.
Eastwood, R. and Lipton, M. (2012) The demographic dividend: retrospect and prospect. *Economic Affairs* 32, 26–30.
Elvin, M. (2004) *The Retreat of the Elephants: An Environmental History of China.* Yale University Press, New Haven, Connecticut.
Hamilton, C. (2010) *Requiem for a Species. Why We Resist the Truth About Climate Change.* Allen and Unwin, Sydney, Australia.
Helfand, I. (2012) Nuclear famine: a billion people at risk (http://www.un.org/disarmament/education/docs/nuclear-famine-report.pdf, accessed 8 July 2013). International Physicians for the Prevention of Nuclear War, Physicians for Social Responsibility.
Hu, Z. and Shi, Z. (2008) Investigation of animal reservoir(s) of SARS-CoV. In: Lu, Y., Essex, M. and Roberts, B. (eds) *Emerging Infections in Asia.* Springer, New York, pp. 57–74.
Huang, S. (2010) A recycling index for food and health security: urban Taipei. *Asia Pacific Journal of Clinical Nutrition* 19, 402–411.
Kellert, S.R. (1996) *The Value of Life.* Island Press, Washington, DC.
Lin, A.Y.C., Huang, S.T.Y. and Wahlqvist, M.L. (2009) Waste management to improve food safety and security for health advancement. *Asia Pacific Journal of Clinical Nutrition* 18, 538–545.
Lovelock, J. (2009) *The Vanishing Face of Gaia, A Final Warning.* Allen Lane, London.

McKay, J. (2009) Food and health considerations in Asia-Pacific regional security. *Asia Pacific Journal of Clinical Nutrition* 18, 654–663.

Maffly, B. (2013) Uinta Basin gas leakage far worse than most believe. Gas production, New study says up to 12 percent of basin's methane escapes (http://www.sltrib.com/entertainment/nightlife/sltrib/news/56692751-78/basin-carbon-emissions-gas.html.csp, accessed 18 November 2013).

Mangunjaya, F.M. and McKay, J.E. (2012) Reviving an Islamic approach for environmental conservation in Indonesia. *Worldviews: Global Religions, Culture, and Ecology* 16, 286–305.

Paull, J. (2011) The making of an agricultural classic: farmers of forty centuries or permanent agriculture in China, Korea and Japan, 1911–2011. *Agricultural Science* 2, 175–180.

Qiao, M., Zheng, Y.M. and Zhu, Y.G. (2011) Material flow analysis of phosphorus through food consumption in two megacities in northern China. *Chemosphere* 84, 773–778.

Rees, M. (2013) Denial of catastrophic risks. *Science* 339, 1123.

Schellnhuber, H.J., Molina, M., Stern, N., Huber, V. and Kadner, S. (eds) (2010) *Global Sustainability: A Nobel Cause*. Cambridge University Press, Cambridge, UK.

Wahlqvist, M.L. (2002a) Eco-nutritional disease or nutrition and chronic disease. *Asia Pacific Journal of Clinical Nutrition* 11, S753–S754.

Wahlqvist, M.L. (2002b) Chronic disease prevention: a life-cycle approach which takes account of the environmental impact and opportunities of food nutrition and public health policies – the rationale for an eco-nutritional disease nomenclature. *Asia Pacific Journal of Clinical Nutrition* 11, S759–S762.

Wahlqvist, M.L. (2009) Connected community and household food-based strategy (CCH-FBS): its importance for health, food safety, sustainability and security in diverse localities. *Ecology Food Nutrition* 48, 457–481.

Wahlqvist, M.L. (2012) Asian migration to Australia: food and health consequences. *Asia Pacific Journal of Clinical Nutrition* 11, S562–S568.

Wahlqvist, M.L. and Specht, R.L. (1998) Food variety and biodiversity: econutrition. *Asia Pacific Journal of Clinical Nutrition* 7, 314–319.

Wahlqvist, M.L., Keatinge, J.D.H., Butler, C.D., Friel, S., McKay, J., Easdown, W., *et al.* (2009) A Food in Health Security (FIHS) platform in the Asia-Pacific region: the way forward. *Asia Pacific Journal of Clinical Nutrition* 18, 688–702.

Wang, J., Gao, W., Xu, S. and Yu, L. (2012) Evaluation of the combined risk of sea level rise, land subsidence, and storm surges on the coastal areas of Shanghai, China. *Climatic Change* 115, 537–558.

Wang, L., Utzinger, J. and Zhou, X.-N. (2008a) Schistosomiasis control: experiences and lessons from China. *The Lancet* 372, 1793–1795.

Wang, S.-L., Tsai, P.-C., Yang, C.-Y. and Guo, Y.L. (2008b) Increased risk of diabetes and polychlorinated biphenyls and dioxins: a 24-year follow-up study of the Yucheng cohort. *Diabetes Care* 31, 1574–1579.

Zheng, C., Liu, J., Cao, G., Kendy, E., Wang, H. and Jia, Y. (2010) Can China cope with its water crisis? – perspectives from the North China plain. *Ground Water* 48, 350–354.

Zhu, Y.G. (2009) Soil science in the understanding of the security of food systems for health. *Asia Pacific Journal of Clinical Nutrition* 18, 516–519.

Ziv, G., Baran, E., Nam, S., Rodríguez-Iturbe, I. and Levin, S.A. (2012) Trading-off fish biodiversity, food security, and hydropower in the Mekong River Basin. *Proceedings of the National Academy of Sciences* 109, 5609–5614.

17 Climate Change and Health in South Asian Countries

Manpreet Singh[1] and Mala Rao[2]
[1]*Dalberg Global Development Advisors, Nairobi, Kenya;* [2]*Institute for Health and Human Development, University of East London, London, UK*

17.1 Introduction

This chapter focuses on geographically defined South Asia, with brief discussion of the Southeast Asian archipelago of Indonesia, part of the World Health Organization (WHO) region associated mostly with South Asia. It describes the major potential impacts of climate change on health in this region.

South Asia is home to over 1.7 billion people, almost a quarter of the world's population. By 2050, it is estimated that the population may have grown another 40%, to about 2.4 billion (International Institute for Applied Systems Analysis, 2012). As the impact of anthropogenic climate change increases, the effects on health in South Asia are likely to be profound, and may have global repercussions.

South Asia is particularly vulnerable to the impacts of climate change, for several reasons. High population densities exist in vulnerable coastal megacities. Endemic, widespread poverty and inequality increases ill health and limits individual and communal adaptive capacity. This landscape is worsened by inadequate institutional capacity and governance, compounded by political systems ranging from governments with scant respect for human rights to democracies that are far more preoccupied with economic growth (as conventionally defined) than with climate change. These problems are intensified by high population growth, especially in urban areas (Rao *et al.*, 2011).

Models (mathematical simulations) that project the likely environmental effects of climate change in South Asia suggest that average temperatures and sea levels will rise, Himalayan glacial melt will increase and extreme weather events will be more frequent (IPCC, 2007a). The cumulative effects on agriculture may be very severe.

Data on the specific links between climate change and health in South Asia are limited (Dogra and Srivastava, 2012). In addition, research has been focused on the relatively medium term and the likely primary and secondary health effects (Butler and Harley, 2010). Even at these levels, there is a limited availability of high-quality quantitative predictions, due to the uncertainty of modelling a range of different climate and population scenarios. The tertiary effects, operant at the intersection of complex systems and large-scale societal changes, do not fit easily into mathematical models and are harder to quantify.

An attempt to project the tertiary effects of climate change may therefore seem speculative, but there is a growing consensus that if climate change continues unabated, its impacts are likely to be massive and potentially catastrophic. They are also likely to be global in scope, challenging the assumption in many developed countries that a combination of economic wealth, technology and better governance will insulate them from the worst effects.

Adapting to climate change and minimizing its potential health impacts requires urgent global effort on an unprecedented scale. It is hoped that by defining and drawing attention to the potentially disastrous tertiary effects of climate change on health in South Asia, governments, non-governmental organizations and other health stakeholders will integrate action to address climate change with their immediate and long-term planning.

17.2   Environmental Change

The impact of climate change in South Asia has been characterized through models developed by the Intergovernmental Panel on Climate Change (IPCC, 2007a). Its findings of physical climate change, soon to be updated, include:

- Mean surface air temperature rises of 2.7–3.6°C by the end of the 21st century compared to the pre-industrial era (IPCC, 2007a).
- Increase in annual rainfall of around 11% across South Asia (IPCC, 2007a).
- Increase in intensity of heavy rainfall events over northern Pakistan, north-west and north-east India, Bangladesh and Myanmar (IPCC, 2007a).
- Increased frequency and intensity of tropical cyclones in the Bay of Bengal (IPCC, 2007a).
- Increasing Himalayan glacial melt (IPCC, 2007a).
- Sea level rise between 0.23 m and 0.47 m globally (IPCC, 2007a).

Changes in water flow

Meltwater from the Hindu Kush Himalayan glaciers is a major supply for the Indus, Ganges and Brahmaputra river basins, supplying water to over a billion people in Asia, including India, Pakistan, Bangladesh and China (Immerzeel et al., 2010). This water is used for domestic purposes, agriculture and industry. As average temperatures rise across South Asia, a likely impact is increased glacial melt (IPCC, 2007a). In the short-term, this is likely to cause glacial lake outburst floods, a situation already occurring in Bhutan and Nepal

(Behrman, 2010). In the long term, the resulting reduction in glacier mass may lead to an 8–20% decrease in annual river flows in the Indus and Brahmaputra rivers[1] and an associated decrease in crop yields for people throughout South Asia, by 2050 (Immerzeel et al., 2010).

Rainfall patterns are likely to alter due to climate change, although the changes projected are highly variable. In general, it is suggested that India will experience increased mean annual rainfall, but the annual monsoon will be more variable in magnitude and duration. Although it seems counterintuitive, this increased variation may increase the risk of periodic severe drought and flooding (Behrman, 2010). When considering population growth alone, there is a projected decrease in per capita water availability from 1830 m³/year in 2001 to 1140 m³/year in 2050 in India (IPCC, 2007a). Given increased uncertainty and variation in water availability as a result of climate change, the capacity of South Asian countries to meet increased population demands for water may decline.

Sea level rise

Climate change causes sea level rises through melting of polar ice sheets, reductions in mountain glaciers and thermal expansion of oceans. The sea level is estimated to have risen already by 17 cm over the past century (Foresight International Dimensions of Climate Change, 2011). Its possible rise by 2100 is thought by some analysts to surpass the IPCC upper boundary of 0.59 m, but is unlikely to exceed 2 m globally (Pfeffer et al., 2008).

Approximately 60% of the population of Indonesia live in low-lying coastal areas (Measey, 2010). In the Maldives, over 90% of the population live in low-elevation coastal zones and may be displaced due to sea level rise (McGranahan et al., 2007). In South Asia, a rise of 1 m could displace up to 6.4 million people across South Asia, without significant coastline protection (Dasgupta et al., 2008).

17.3   Natural Disasters

While extreme weather events and associated natural disasters have always occurred, evidence suggests that their frequency and intensity has increased in the past 50 years (IPCC, 2007a). The former position that no extreme weather event can be attributed entirely to climate change is shifting to the view that all extreme weather events occur in an anthropogenically changed climate, interacting with climate variability and human vulnerability (Trenberth, 2011). Climate change is associated with increased heatwaves and heavy precipitation events and a predicted increase in the severity of intense tropical cyclones (see Chapter 5, this volume; IPCC, 2012). In addition, the impact of such storms is worsening due to sea level rise, an issue of critical importance to South Asia, especially to the populations who live near the Bay of Bengal.

Cyclone Nargis, which killed over 138,000 people in Myanmar in 2008, is an example of the devastating impact natural disasters can have on human life (IPCC, 2012). Even if its wind speeds and precipitation were not contributed by a warmer atmosphere, it was worsened by sea level rise, which is particularly problematic in the Bay of Bengal (Han *et al.*, 2010). Storm surges in shallow seas are made worse if the coastal geometry can funnel water, as is the case in the northern Bay of Bengal (Mitchell *et al.*, 2006).

17.4   Primary Health Impacts

Temperature

In South Asia, climate change is associated with higher mean temperatures and an increase in the frequency and magnitude of heatwaves (IPCC, 2012). The effects of temperature changes on health are variable. Rising temperatures can reduce mortality associated with cold weather, but high temperatures are associated with increased respiratory and cardiovascular mortality, as well as the direct effects of heatstroke (Costello *et al.*, 2009). Modelling the relationship between temperature and health in Delhi, India, demonstrated that every 1°C increase in maximum temperature above 29°C was associated with a 3.9% increase in mortality (McMichael *et al.*, 2008). Temperatures are higher in cities than in rural areas, due to the urban heat island effect. As urbanization continues, greater proportions

of the population will be exposed to high temperatures. Extreme temperatures are already having an impact on human health in South Asia. Heatwaves in Andhra Pradesh in 2002 are reported to have killed over 600 people (IPCC, 2012). People most at risk of death during heatwaves are manual labourers and people who work outdoors – usually the poorest (Kovats and Akhtar, 2008). These estimates are likely to be extremely conservative and, furthermore, do not account for any exacerbation in chronic illness.

Flooding

Flooding can result from sea level rises, glacial melt, changes in monsoon patterns and increased precipitation. These effects are often worsened by inadequate drainage. Flooding has direct impacts on health through trauma and drowning. Damage to water and sanitation infrastructure can also occur due to flooding, thus increasing the chance of diarrhoeal disease, as well as vector-borne diseases such as malaria and dengue (Costello *et al.*, 2009). Flooding may also result in forced human migration, as well as damage to roads, schools and health facilities. South Asian megacities situated along Himalayan-sourced rivers are particularly at risk of flooding – these include Delhi, Dhaka, Kolkata and Karachi (Steinbruner *et al.*, 2012). Large coastal cities, such as Mumbai, Jakarta and Chennai, are also at high risk (Kovats and Akhtar, 2008). The pressures of rapid urbanization, poverty and population growth throughout South Asia have led to the proliferation of low-quality housing and water and sanitation infrastructure, exacerbating vulnerability to the health impacts of flooding.

A particularly vulnerable megacity is Mumbai, which in 2011 was home to 18 million people. Over 40% of this population lives in squatter communities (Census of India, 2011), with poor water and sanitation infrastructure. Monsoon-related flooding in 2005, worsened by blocked drains, killed more than 1000 people and was associated with the increased transmission of infectious diseases, including leptospirosis, the incidence of which rose eightfold (Kovats and Akhtar, 2008).

17.5 Secondary Health Impacts

As mentioned, warmer temperatures and flooding are likely to alter the epidemiology of vector-borne disease; changes in sea temperature are associated with an increased risk of diarrhoeal disease. In coastal Bangladesh, sea level rise has already increased salinity in soil and drinking water, with likely adverse health implications for large populations, particularly by elevating blood pressure (Vineis and Khan, 2012).

Vector-borne diseases

There are six major vector-borne diseases endemic in South Asia: malaria, dengue, chikungunya, filariasis, Japanese encephalitis and visceral leishmaniasis. Of these, malaria causes the greatest morbidity and mortality, with an estimated toll of over 45,000 deaths/year in India alone in 2010 (Murray *et al.*, 2012). Higher temperatures may prolong the malaria transmission window and reduce the incubation period required for replication of the parasite in an infected mosquito. As regional temperatures change across India, the transmission window for malaria is likely to increase by 2–3 months in the northern states of Punjab, Haryana and Jammu and Kashmir, but to decrease in more southerly Odisha, Andhra Pradesh and Tamil Nadu as temperatures exceed 40°C (Dhiman *et al.*, 2010). Malaria and other vector-borne diseases are also projected to spread to new parts of previously low-risk countries such as Bhutan and parts of Indonesia (National Environment Commission, 2009).

Dengue fever, transmitted by the *Aedes* mosquito, is also highly sensitive to changes in temperature and humidity; climate change is predicted to increase the global population at risk of dengue from 1.5 billion people to 4.1 billion people by 2055 (Hales *et al.*, 2002). Current dengue transmission in India is greatest in southern India. With 2–4°C rises in temperature, dengue transmission windows are predicted to increase in north India (Dhiman *et al.*, 2010).

Other vector-borne diseases in South Asia may also rise, secondary to climate change. Chikungunya, a mosquito-borne viral illness, once thought to have been eliminated from the subcontinent, has re-emerged in southern India. Again, as temperatures rise, transmission windows are likely increase in north India. Visceral leishmaniasis, a climate-sensitive disease, is re-emerging in parts of Nepal and north India (Dhiman *et al.*, 2010), though this is probably more due to a rebound following reduced insecticide use rather than from climate change.

Diarrhoeal disease

Climate change will probably increase the incidence of infective diarrhoea, due to the causes discussed, including rising temperatures, reduced access to clean drinking water due to drought and damage to water and sanitation infrastructure due to flooding (Costello *et al.*, 2009). There is evidence from Bangladesh that warmer coastal sea temperatures are associated with phytoplankton blooms, which support the spread of cholera and other bacteria causing diarrhoeal disease (IPCC, 2007b). In the WHO South-east Asian Region-D (SEAR-D: Bangladesh, Bhutan, Democratic People's Republic of Korea, India, Maldives, Myanmar and Nepal), there may be an excess incidence of over 40 million cases of diarrhoea/year by 2030,[2] as a result of climate change (Ebi, 2008).

17.6 Tertiary Health Impacts

These impacts arise at the intersection between people, politics and ecology, and are likely to be profound. They need consideration in the context of broader structural and societal shifts, brought about by economic growth, globalization and urbanization.

Food security

Temperature increases (in already hot places), drought, increased glacial melting and extreme weather events are all likely to reduce crop yields. Under the A2 scenario, an IPCC model representing a world of less integrated global economic growth, crop yields will decrease by

5–30% in Asia by 2050, leading to 132 million additional people at risk of chronic hunger (IPCC, 2007b). These estimations are, however, inexact, as crop yields are very sensitive to changes in agricultural technology, including the use of fertilizer and other intensive farming techniques. Current agricultural yields in South Asia are relatively low, and it is possible that technological improvements may mitigate some of the effect of climate change. On the other hand, wheat yield growth in India is flattening, at a comparatively low level (Cassman *et al.*, 2010).

The health consequences of undernutrition are well established. Acute undernutrition causes ill health through starvation; chronic undernutrition increases susceptibility to other illnesses, including infectious diseases, such as pneumonia (Rao and Beckingham, 2013). The current burden of undernutrition on health is enormous in south-central Asia, causing almost a million deaths of children under a year old (Black *et al.*, 2008). Its broader development effects are greater still – including poorer cognitive development, lower educational achievement and reduced economic status in adulthood (Victora *et al.*, 2008). This effect is likely to be multiplied as poverty is worsened by rises in food prices, due to several factors.

As competition for global resources intensifies, a breakdown of civilization is even possible (McMichael, 2012). By 2050, positive feedbacks could release more greenhouse gases, driving accelerated climate change (Liggins, 2009); desertification of tropical regions alone represents a serious threat to human survival.

Migration

Several effects of climate change affect the determinants of migration and population displacement. Flooding, droughts and land degradation against a background of poverty and limited opportunity have already driven the migration of millions of Bangladeshis since the 1950s, both internally and to India (Reuveny, 2007). More recently, economic growth in Bangladesh is driving internal migration to urban centres. But Dhaka is particularly vulnerable to climate change, and this pattern of internal migration may reverse, with future displacement back toward inland areas (Podesta and Ogden, 2008).

The Maldives is particularly vulnerable to migration as a result of sea level rises. The worst-case scenario is for almost its whole landmass to be submerged. The Maldivian government is preparing for internal migration – preparing five 'safer islands' with coastal protection against sea level rise, to where the entire population of 300,000 could retreat (Republic of Maldives, 2007).

The impacts of migration on health include trauma, undernutrition and transmission of infectious disease during the travel phase, and worse outcomes once at the destination, across a number of domains. Mental health, reproductive health and childhood health are also likely to be affected (Zimmerman *et al.*, 2011).

Conflict

The empirical evidence that climate change can directly cause conflict is limited and contested. The literature suggests that climate change is linked to conflict through migration, competition over scarce natural resources and a breakdown of international and national governance institutions (Foresight International Dimensions of Climate Change, 2011).

Migration induced by climate change has the potential to cause conflict by increasing competition for food, potable water and land, and by exacerbating existing ethnic and religious tensions (Reuveny, 2007). In South Asia, extreme rainfall has been suggested as contributing to Hindu–Muslim riots. Reduced agricultural income and forceful reallocation of property have been cited as possible cofactors (Hsiang *et al.*, 2013).

The multilateral global governance structures set up post World War II have played a significant role in stabilizing international relations. However, the international system has, to date, failed to address the complexities and compromise demanded by climate change, as demonstrated by the inability to agree to international carbon reduction targets (Haldén, 2007). This failure challenges the legitimacy of the international system, and leaves it less able to cope with political instability and potential conflict (Foresight International Dimensions of Climate Change, 2011). If multilateral institutions are perceived as ineffective and unable to cope with the climate change challenge, a risk arises that countries will take unilateral action to protect their resources (Haldén, 2007), or even become predatory in a quest to secure essential supplies as their own become scarce. South Asia is particularly vulnerable to this, with high population densities, two nuclear powers with a history of conflict and water shortages likely to be worsened by climate change.

At a national level, natural disasters, migration and social disruption induced by climate change may be associated with a breakdown of governance structures. This has been seen not only in the developing world but also in the developed world; for example, following Hurricane Katrina in New Orleans. The inability of national government institutions to plan and implement adequate responses to protect populations from disasters may fuel public unrest and create the conditions for radical political groups to enhance their presence and influence. This is a particular concern in South Asia, where there is some evidence that such changes are already taking place. During the 2010 floods in Pakistan, militant groups were among the first organizations to respond in the worst-hit areas, providing aid and assistance (Foresight International Dimensions of Climate Change, 2011). There is also evidence of rising extremism in Bangladesh and Myanmar. Nepal and parts of north-eastern India are vulnerable to such influences, with increased conflict as a possible consequence (Podesta and Ogden, 2008).

Linking climate change to conflict may seem speculative, but the British, American and Swedish military communities (among others) are making climate change part of their long-term economic and military contingency planning. The American intelligence community is planning periodic stress tests to 'inform national security decision makers about places that are at risk of becoming security concerns as a result of climate change' (Steinbruner et al., 2012).

Economic instability

Increasing globalization has generated complex, tortuous and vulnerable international supply chains (Steinbruner et al., 2012). Though not associated with climate change, flight bans caused by the eruption of an Icelandic volcano in 2010 placed the UK at risk of shortage of fresh fruit and vegetables (Rao and Beckingham, 2013), highlighting the modern vulnerability even of strong economies to disruptions of supply. Sea level rises and extreme weather events may disrupt ports and airports, affecting freight transportation (Foresight International Dimensions of Climate Change, 2011). This may affect health directly, as the international distribution of medicines and vaccinations is dependent on these supply chains. Floods in Pakistan in 2010 damaged roads and the transportation of health supplies, and undermined access to health facilities, worsening the humanitarian impact of the disaster (WHO, 2011).

More downstream, disruption of supply chains, for even a short period, can lead to economic instability, and even threaten large-scale economic collapse. Natural disasters are also associated with disrupted banking and financial services. The 2005 floods in Mumbai caused temporary closure of the Stock Exchanges (IPCC, 2012). This does not include the unquantifiable economic impact natural disasters have on the daily livelihoods of millions of people. Any weakening of international financial systems and insurance markets will affect health across biological, psychological and social domains.

17.7   Mental Health

Climate change is associated with mental ill health through exposure to natural disasters, increased physical illness and impacts on community well-being (IPCC, 2012). People with existing poor mental health, whether diagnosed or not, are at particular risk of adverse health outcomes during significant societal disruption.

Survivors of natural disasters are at risk of developing post-traumatic stress disorder. Displaced populations are also at increased risk of mental health problems, due to social exclusion, socio-economic disadvantage and exposure to racial or cultural discrimination in receiving communities (Fritze et al., 2008).

South Asia is at high risk of natural disasters, conflict and involuntary migration related with climate change. As a result, the mental health impacts are likely to be substantial across the region. Given that mental illness is already significantly underdiagnosed in South Asia, and that hospital and health resources are likely to be even more stretched, there is a strong likelihood that mental ill health will represent a large, unmet need as a result of climate change.

17.8   Economic Cost of Climate Change

Climate change represents a significant economic burden to South Asian governments when taking into account the cost of mitigation, adaptation and of treating the health impacts, in the absence of suitable alternative solutions. Mitigation represents a significant opportunity cost – Indonesia and India are currently two of the ten highest carbon emitters in the world (World Resources Institute, 2012). While this does not take into account population size or historical emissions, there remains an expectation that South Asian countries will have to mitigate carbon emissions, or at least limit their rate of growth, alongside reductions from developed countries. Given that, historically, economic development has depended substantially on fossil fuel combustion, restricting carbon emissions in the developing world has the potential to slow economic growth and compromise gains made in alleviating poverty, illiteracy, hunger and disease.

Adaptation is also expensive: India has been reported as spending 2.6% of its gross domestic product (GDP) on responding and adapting to environmental threats such as flooding (Padukone, 2010). Internationally, adaptation by coastal hubs to sea level rise has been estimated to cost between US$28 billion and US$90 billion/year until 2050 (Foresight International Dimensions of Climate Change, 2011).

The additional global cost of diarrhoea, malaria and malnutrition, as a result of climate change, is predicted to be $10 billion by 2030. At present, over 40% of all disability adjusted life years attributable to climate change are experienced in South Asia (Ebi, 2008). Given the patterns of predicted population growth, it can be expected that this proportion will increase.

Similarly, the counterfactual expenditure of money spent on adaptation cannot be predicted, but given current spending priorities, it seems likely that a substantial proportion would be spent on health.[3]

17.9   Impacts on Women

Women are especially vulnerable to the health impacts of climate change, due to skewed power relations, financial poverty and exclusion from policy- and decision-making institutions (Nellemann et al., 2011). Women are also more at risk of morbidity at all stages of a natural disaster. Household and childcare responsibilities can make it harder for women to escape to safety. This was documented in floods in Bangladesh in 1991, when the mortality of flood-affected females aged 20–44 was four times higher than flood-affected males (Nellemann et al., 2011). Post-disaster, inequality in distribution of food means that women and girls are at higher risk of undernutrition. Rape and other forms of gender-based violence is also a significant concern in association with natural disasters (Neumayer and Plümper, 2007).

17.10   Conclusion

Climate change is already increasing mortality and morbidity in South Asian people; impacts are likely to increase as climate change intensifies. The tertiary effects of climate change on health are the hardest to quantify, but in the long-term, are likely to generate the greatest health impact via interlinked, synergistic disruption of the supporting natural, economic and social systems that underpin health.

Overall, the poorest people are the most vulnerable to the health impact of climate change. Poverty is associated with low-quality shelter, an increased risk of trauma from natural disasters, poor sanitation infrastructure and an increased risk of communicable disease. Furthermore, the poorest are most likely to migrate from areas made uninhabitable by climate change. Climate change is likely to exacerbate existing inequalities. Farmers will be affected by declining crop yields, and the poor by rising food prices. Models suggest that, as a result of climate change, 50 million fewer people in India will escape from poverty by 2040 (Jacoby et al., 2011).

1

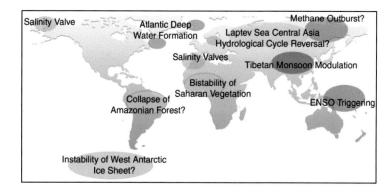

Plate 1. Potential tipping elements in the Earth System. These are critical regions where small changes in driving variables can trigger abrupt changes in critical processes that have significant implications for the functioning of the Earth System as a whole (Schellnhuber, 2002). (Reprinted with kind permission of Springer Science+Business Media.)

2

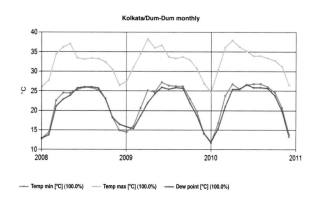

Plate 2. A graphical representation of the nine planetary boundaries. The inner green circle represents the safe operating space for humanity in which all boundaries are respected. The red wedges represent the current values of the control variables. Three boundaries (rate of biodiversity loss, climate change and interference with the nitrogen cycle) have already been crossed (Rockström *et al.*, 2009b). (Reprinted with kind permission of *Nature*.)

3

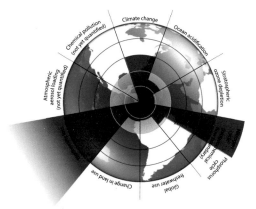

Plate 3. Variation of monthly averages of maximum and minimum temperature and dew point at Kolkata's Dum-Dum airport (data from NOAA/GSOD; output from Hothaps-Soft software, www.climatechip.org; Otto, M., *et al.*, unpublished). The 100% note means that data for every day are available.

4a

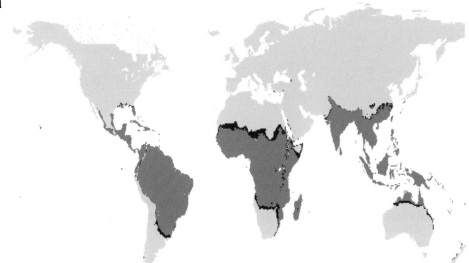

- ■ Contraction
- ■ Expansion
- ■ Stable presence
- ▫ Stable absence

4b

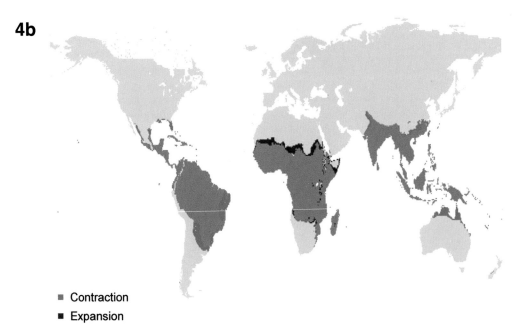

- ■ Contraction
- ■ Expansion
- ■ Stable presence
- ▫ Stable absence

Plate 4. Projected changes to the distribution of dengue under climate change in the 2050s, compared to modelled 'current' distribution. The effect of climate change alone is shown at the top (map a) and the effect of both climate change and the projected changes in per capita GDP is shown at the bottom (map b). (Figure from Åström *et al.* (2012). Reprinted with kind permission of Springer Science+Business Media.)

5

Pre-peak fires: 11–21 July 2010

Peak fires: 22–29 July 2010

Peak fires: 1–10 Aug. 2010

Peak fires: 11–18 Aug. 2010

Post-peak: 19–26 Aug. 2010

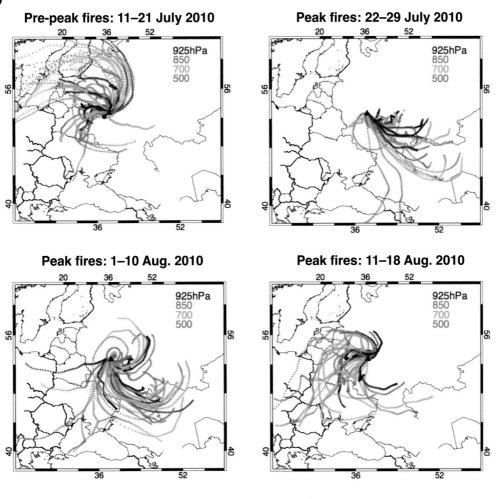
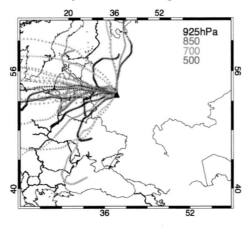

Plate 5. Back trajectories from Moscow (black triangle) with pressure levels plotted at 925 hPa, 850 hPa, 700 hPa and 500 hPa for five periods between July and August 2010. These show the path by which smoke was transported to and then contained above Moscow by the anticyclone circulation. (Adapted from Witte *et al.*, 2011; published under the creative commons attribution conditions.)

6

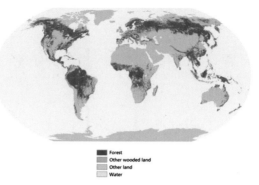

Forest
Other wooded land
Other land
Water

Plate 6. Global forest distribution (Amiro *et al.*, 2009). (Reprinted with permission of FAO and © 2008 Canadian Science Publishing or its licensors.)

7

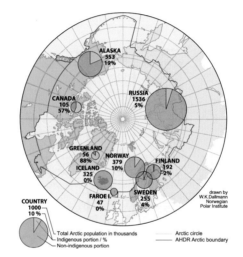

Plate 7. Indigenous and non-indigenous populations in the Arctic portions of the Arctic countries. (Adapted from figure in Young and Einarsson (2004), with updated data for Russia (estimate, based on 2002 census numbers) and Iceland (from http://www.statice.is/Statistics/Population). Published with permission of the Arctic Institute.)

8

Plate 8. Typical hearth in Benin (© Fotolia.com).

Climate change will harm every aspect of South Asian development over the next century and beyond. Few organizations appear to be appropriately prepared for the potential impact. Bangladesh, Bhutan and the Maldives have prepared National Adaptation Programmes of Action for climate change – an effort to be applauded, but similar plans need to be developed by other South Asian nations. The UK Department for International Development has undertaken a review entitled ORCHID (Tanner *et al.*, 2007), to ensure that aspects of climate change are included in all its programmes in India. Discussion of the potentially disastrous tertiary consequences may accelerate international efforts to mitigate carbon emissions. Appropriate early planning will help to reduce the health impact of climate change in South Asia; it also represents an opportunity for stakeholders in this region to develop areas of global expertise in innovative approaches to research, mitigation and adaptation.

Notes

[1] These rivers are thought more vulnerable than the other great rivers arising on the Tibetan plateau.
[2] Compared to about 1 billion cases per annum at present in these countries.
[3] Additional health expenses are unlikely to address the full needs for adaptation, even if such funds are available.

References

Behrman, N. (ed.) (2010) *The Waters of The Third Pole: Sources of Threat, Sources of Survival.* Aon Benfield UCL Hazard Research Centre, University College London, China Dialogue, Humanitarian Futures Programme, King's College London, London.

Black, R.E., Allen, L.H., Bhutta, Z.A., Caulfield, L.E., de Onis, M., Ezzati, M., *et al.* (2008) Maternal and child undernutrition: global and regional exposures and health consequences. *The Lancet* 371, 243–260.

Butler, C.D. and Harley, D. (2010) Primary, secondary and tertiary effects of the eco-climate crisis: the medical response. *Postgraduate Medical Journal* 86, 230–234.

Cassman, K.G., Grassini, P. and van Wart, J. (2010) Crop yield potential, yield trends, and global food security in a changing climate. In: Rosenzweig, C. and Hillel, D. (eds) *Handbook of Climate Change and Agroecosystems: Impacts, Adaptation, and Mitigation.* Imperial College Press, London, pp. 37–51.

Census of India (2011) *Provisional Population Totals.* Press Information Bureau, Government of India, New Delhi.

Costello, A., Abbas, M., Allen, A., Ball, S., Bell, S., Bellamy, R., *et al.* (2009) Managing the health effects of climate change. *The Lancet* 373, 1693–1733.

Dasgupta, S., Laplante, B., Meisner, C., Wheeler, D. and Yan, J. (2008) The impact of sea level rise on developing countries: a comparative analysis. *Climatic Change* 93, 378–388.

Dhiman, R.C., Pahwa, S., Dhillon, G.P.S. and Dash, A.P. (2010) Climate change and threat of vector-borne diseases in India: are we prepared? *Parasitology Research* 106, 763–773.

Dogra, N. and Srivastava, S. (eds) (2012) *Climate Change and Disease Dynamics in India.* The Energy and Resources Institute, New Delhi.

Ebi, K.L. (2008) Adaptation costs for climate change-related cases of diarrhoeal disease, malnutrition, and malaria in 2030. *Globalization and Health* 4, 9, doi:10.1186/1744-8603-4-9.

Foresight International Dimensions of Climate Change (2011) Final Project Report. The Government Office for Science, London.

Fritze, J.G., Blashki, G.A., Burke, S. and Wiseman, J. (2008) Hope, despair and transformation: climate change and the promotion of mental health and wellbeing. *International Journal of Mental Health Systems* 2, 13, doi:10.1186/1752-4458-2-13.

Haldén, P. (2007) *The Geopolitics of Climate Change. Challenges to the International System.* FOI, Stockholm.

Hales, S., de Wet, N., Maindonald, J. and Woodward, A. (2002) Potential effect of population and climate changes on global distribution of dengue fever: an empirical model. *Lancet* 360, 830–834.

Han, W., Meehl, G.A., Rajagopalan, B., Fasullo, J.T., Hu, A., Lin, J., *et al.* (2010) Patterns of Indian Ocean sea-level change in a warming climate. *Nature Geoscience* 3, 546–550.

Hsiang, S.M., Burke, M. and Miguel, E. (2013) Quantifying the influence of climate on human conflict. *Science* 341, doi:10.1126/science.1235367.

Immerzeel, W.W., van Beek, L.P.H. and Bierkens, M.F.P. (2010) Climate change will affect the Asian water towers. *Science* 328, 1382–1385.

International Institute for Applied Systems Analysis (2012) *Asian Demographic and Human Capital Data Sheet* (http://www.iiasa.ac.at/web/home/research/researchPrograms/WorldPopulation/Publications MediaCoverage/ModelsData/models_and_data.html, accessed 31 March 2013).

IPCC (Intergovernmental Panel on Climate Change) (2007a) *The Physical Science Basis. Contribution of Working Group I to the Fourth Assessment Report of the Intergovernmental Panel on Climate Change.* Cambridge University Press, Cambridge, UK, and New York.

IPCC (2007b) *Climate Change 2007: Impacts, Adaptation and Vulnerability. Contribution of Working Group II to the Fourth Assessment Report of the Intergovernmental Panel on Climate Change.* Cambridge University Press, Cambridge, UK, and New York.

IPCC (2012) *Intergovernmental Panel on Climate Change Special Report on Managing the Risks of Extreme Events and Disasters to Advance Climate Change Adaptation.* (Field, C.B., Barros, V., Stocker, T.F., Qin, D., Dokken, D.J., Ebi, K.L., *et al.* (eds)). Cambridge University Press, Cambridge, UK, and New York.

Jacoby, H., Rabassa, M. and Skouas, E. (2011) Distributional implications of climate change in India. World Bank Policy Research Working Paper Series.

Kovats, S. and Akhtar, R. (2008) Climate, climate change and human health in Asian cities. *Environment and Urbanization* 20, 165–175.

Liggins, F. (2009) Greenhouse gas emissions: the hard facts. In: Griffiths, J., Rao, M., Adshead, F. and Thorpe, A. (eds) *The Health Practitioner's Guide to Climate Change: Diagnosis, Cure.* Routledge, London, pp. 4–32.

McGranahan, G., Balk, D. and Anderson, B. (2007) The rising tide: assessing the risks of climate change and human settlements in low elevation coastal zones. *Environment and Urbanization* 19, 17–37.

McMichael, A.J. (2012) Insights from past millennia into climatic impacts on human health and survival. *Proceedings of the National Academy of Sciences* doi:10.1073/pnas.1120177109.

McMichael, A.J., Wilkinson, P., Kovats, R.S., Pattenden, S., Hajat, S., Armstrong, B., *et al.* (2008) International study of temperature, heat and urban mortality: the 'ISOTHURM' project. *International Journal of Epidemiology* 37, 1121–1131.

Measey, M. (2010) Indonesia: a vulnerable country in the face of climate change. *Global Majority E-Journal* 1, 31–45.

Mitchell, J.F.B., Lowe, J., Wood, R.A. and Vellinga, M. (2006) Extreme events due to human-induced climate change. *Philosophical Transactions of the Royal Society A – Mathematical Physical and Engineering Sciences* 364, 2117–2133.

Murray, C.J., Rosenfeld, L.C., Lim, S.S., Andrews, K.G., Foreman, K.J., Haring, D., *et al.* (2012) Global malaria mortality between 1980 and 2010: a systematic analysis. *The Lancet* 379, 413–431.

National Environment Commission (2009) *Strategizing Climate Change for Bhutan.* Government of Bhutan, Thimphu.

Nellemann, C., Verma, R. and Hislop, L. (2011) *Women at the Frontline of Climate Change: Gender Risks and Hopes: A Rapid Response Assessment.* United Nations Environment Programme, Birkeland, Norway.

Neumayer, E. and Plümper, T. (2007) The gendered nature of natural disasters: the impact of catastrophic events on the gender gap in life expectancy, 1981–2002. *Annals of the Association of American Geographers* 97, 551–566.

Padukone, N. (2010) Climate change in India: forgotten threats, forgotten opportunities. *Economic and Political Weekly* XLV, 47–54.

Pfeffer, W.T., Harper, J.T. and O'Neel, S. (2008) Kinematic constraints on glacier contributions to 21st-century sea-level rise. *Science* 321, 1340–1343.

Podesta, J. and Ogden, P. (2008) The security implications of climate change. *Washington Quarterly* 31, 115–138.

Rao, M. and Beckingham, A. (2013) Food, nutrition and public health. In: Pielke, R.A. Snr (ed.) *Climate Vulnerability: Understanding and Addressing Threats to Essential Resources.* Elsevier, Amsterdam, pp. 87–94.

Rao, M., Lakshmi, J. and Samarth, A. (2011) *Addressing Health of the Urban Poor in South-East Asia Region: Challenges and Opportunities*. WHO South East Asia Regional Office, New Delhi.

Republic of Maldives (2007) *National Adaptation Programme of Action*. Ministry of Environment, Energy and Water, Malé, Republic of Maldives.

Reuveny, R. (2007) Climate change-induced migration and violent conflict. *Political Geography* 26, 656–673.

Steinbruner, J.D., Stern, P.C. and Husbands, J.L. (eds) (2012) *Climate and Social Stress: Implications for Security Analysis*. The National Academies Press, Washington, DC.

Tanner, T.M., Nair, S., Bhattacharjya, S., Srivastava, S.K., Sarthi, P.P. and Sehgal, M. (2007) *ORCHID: Climate Risk Screening in DFID India Synthesis Report*. Institute of Development Studies, London.

Trenberth, K.E. (2011) Attribution of climate variations and trends to human influences and natural variability. *WIREs Climate Change* 2, 925–930.

Victora, C.G., Adair, L., Fall, C., Hallal, P.C., Martorell, R., Richter, L., *et al.* (2008) Maternal and child undernutrition 2: maternal and child undernutrition: consequences for adult health and human capital. *The Lancet* 371, 340–357.

Vineis, P. and Khan, A. (2012) Climate change-induced salinity threatens health. *Science* 338, 1028–1029.

WHO (World Health Organization) (2011) *Pakistan Floods 2010. Early Recovery Plan for the Health Sector*. World Health Organization, Geneva.

World Resources Institute (2012) Climate Analysis Indicators Tool 2012. CAIT version 9.0. World Resources Institute, Washington, DC (http://cait.wri.org).

Zimmerman, C., Kiss, L. and Hossain, M. (2011) Migration and health: a framework for 21st century policy-making. *PLoS Medicine* 8, e1001034.

18 Climate Change and Global Health: A Latin American Perspective

Ulisses Confalonieri,[1] Germán Poveda,[2] Horacio Riojas,[3]
Marilyn Aparicio Effen[4] and Ana Flávia Quintão[5]

[1]CPqRR – Fundação Oswaldo Cruz (FIOCRUZ), LAESA,
Minas Gerais, Brazil; [2]Department of Geosciences and Environment,
Universidad Nacional de Colombia, Medellín, Colombia; [3]Instituto
Nacional de Salud Pública, Cuernavaca Morelos, México; [4]Climate
Change, Environmental and Health Unit, Instituto Boliviano de Biologia
de Altura (IBBA), Universidad Mayor de San Andrés, Bolivia;
[5]Escola de Saúde Pública do Estado de Minas Gerais (ESP-MG),
Minas Gerais, Brazil

18.1 Introduction

Latin America (LA) is a very heterogeneous region, with 41 countries covering 20,393,600 km². The total population in 2011 was about 600 million; about 93% of whom had access to safe water supplies and 79% were provided with some type of waste disposal facilities (Pan American Health Organization (PAHO), 2011). It is considered the most urbanized region in the world; almost 80% of its population lives in cities; 14% of whom live in megacities (UN Habitat, 2012).

The epidemiological profile of LA is marked by diseases of poverty and underdevelopment. As a consequence of the strong historical and current socio-economic inequalities found in Latin American countries, conditions such as diarrhoea and respiratory infections in children still constitute a serious problem in many regions. Among the socio-environmental diseases are those arising from water shortage and pollution, indoor and outdoor air pollution, exposure to chemical substances and, recently, those attributable to climate variability, climate change and natural disasters (Magrin et al, 2014). As yet, there is no precise estimate of the morbidity and mortality attributable to this last factor. Recent studies have tried to demonstrate the link between the deterioration of ecosystems and public health problems in LA (Riojas-Rodriguez et al., 2010). Chronic illnesses such as diabetes and cardiovascular diseases also contribute significantly to this epidemiological profile.

Globalization has introduced, accelerated or triggered health problems in LA. Examples span from influenza epidemics to transboundary chemical pollution throughout the continent. The climate crisis has contributed to the generation of problems in the region, such as food insecurity (in many Central American countries); mobilization of vectors toward sites where they were not found previously (as in Bolivia); increases in exposure to air pollutants such as ozone resulting from changing temperature patterns (as in Mexico City); and increases in morbidity and mortality as a consequence of extreme hydrometeorological events across the region.

In 2010, about 3% of the regional population was exposed to the risk of malaria transmission (637,801 cases of malaria were reported, with 1.7 million cases of dengue fever; PAHO, 2011).

Global health solutions require highly inter and multidisciplinary approaches, focussing on issues that can transcend national boundaries. This also implies solutions based on a global cooperation aimed at health equity among all nations and for all people (Koplan *et al.*, 2009).

Recently, a Latin American Alliance for Global Health was formed as a collaborative network between regional public health institutions. Its major aim is to address global health issues considered as priorities for the LA region, to be approached as a public good informed by social justice and universal rights. Workshops and meetings are being organized periodically to tackle broad themes such as the impacts of the economic crisis, population displacements, universal coverage of health services, disaster management and global climate change (www. saludglobal.uchile.cl).

18.2 Natural Disasters and Extreme Weather Events

Diverse climate change impacts are already being observed in the region. These include changes in hurricane intensity and frequency in the Caribbean, alterations in precipitation patterns and in temperature levels and increased droughts. Additionally, rising sea levels affect coastal LA; glaciers in Patagonia and the Andes are receding; and ice sheets in West Antarctica are shrinking (Programa de las Naciones Unidas para el Medio Ambiente (PNUMA) *et al.*, 2010). In Andean countries, glacial retreat is reducing the quality and availability of drinking water, and also threatens hydroelectricity supply. In Bolivia, 80% of glaciers are retreating, for example the Chacaltaya glacier shrank from 0.22 km^2 of surface ice in 1940 to only 0.01 km^2 in 2005.

One of the most important impacts of climate change is to drinking water availability in major cities, caused by altered rainfall patterns. For La Paz, future climate scenarios indicate drier dry seasons and rainy seasons with small increases in water availability. Water systems in two neighbourhoods in La Paz are more sensitive to an imbalance between supply and demand from demographic processes, and three are more sensitive to an imbalance of supply and demand

owing to the potential short-term disappearance of glaciers (Aparicio *et al.*, 2013a).

In Bolivia, owing to its high vulnerability and topography, extreme climatic events are harming economic development and investment. The 1997/98 El Niño generated economic disasters in Bolivia, with losses totalling US$530 million, equivalent to 7% of gross domestic product (GDP) (CEPAL, 2007).

Climate change is also reinforcing the intensity and frequency of climate variability, with more intense events, associated with either pole of the El Niño–Southern Oscillation (ENSO). Floods and droughts have become more extreme in Bolivia and nearby countries such as Peru and Ecuador. Although more recent ENSO cycles have been moderate, their impacts were still significant in Bolivia (Arana *et al.*, 2007a). Extreme weather events recurrent in urban and rural areas manifested as flooding, landslides, droughts, snowstorms, windstorms and hailstorms, as in La Paz, and seem to be increasing (see Fig. 18.1).

On 19 February 2002, a rainfall of 70 mm in 45 min, with hail, overwhelmed all drainage systems in La Paz, leading to flash floods, with 70 deaths, 19 missing and many persons injured (Arana *et al.*, 2007b). This event generated the creation of an early warning system (EWS) by the La Paz municipal government.

On 26 February 2011, La Paz, at 3640 m above sea level, a city in a mountainous ecosystem, suffered a huge landslide, affecting nine neighbourhoods and causing significant economic damage. On 24 and 25 February, street cracks preceded the landslide, triggering the EWS and neighbourhood evacuation. It was the rainiest February since 1919, with 25 wet days. Also contributing to the landslide were saturated soil and possible uncontrolled sanitary discharges (Aparicio *et al.*, 2013b).

Extreme climate events can also have important other public health consequences, including crop destruction, population relocation and damage to health systems infrastructure.

Better civil protection alert systems have decreased mortality from hurricanes; however, the number of people affected and displaced has increased, especially since the 1990s, probably due mainly to population increase. The effects of extreme weather events (EWEs), however, in the intermediate and long term have not been quantified adequately. Some Central American studies

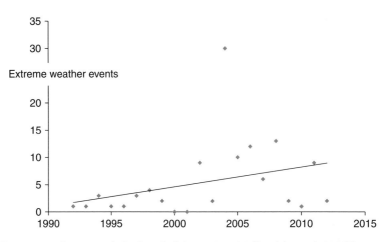

Fig. 18.1. Extreme weather events in La Paz, Bolivia, 1992–2012 (Aparicio *et al.*, 2013b).

show the impact of EWEs on food safety, especially in the degree of undernutrition found in children after these events (Barrios *et al.*, 2000).

The number of storms between 2000 and 2009 has multiplied by 12, and the number of floods has quadrupled since the 1970s. The number of people affected by extreme temperatures, storms, droughts and forest fires has increased from 5 million during the 1970s to 40 million during the past decade (PNUMA *et al.*, 2010).

Statistics on the health impacts of disasters (Centre for Research on the Epidemiology of Disasters (CRED), 2011) indicate that during 2000–2010 around 630 weather and climate extreme events occurred in LA, resulting in about 16,000 deaths among 46.6 million people affected, with losses of US$208 million.

The effects of climate change and climate variability are also being felt strongly in Colombia. Massive floods during 2010–2012, caused by climate change, La Niña and excessive deforestation, resulted in hundreds of deaths and thousands of displaced people, with infrastructure and agricultural losses (Poveda *et al.*, 2011a,b).

A 1991–2010 Civil Defence Report (Brazil, 2012) for Brazil found that 73% of all accidents related to EWEs since 1991 occurred between 2000 and 2010. Most deaths (*n* = 3404) related to these events were associated with landslides (41%) – usually resulting from heavy rains – and floods (42.9%). Casualties were concentrated in the densely populated south-eastern region,

with 20 deaths/million: 69% of all disaster-related deaths in the country.

18.3   Infectious Diseases

Climatic variables and EWEs provide the ecological preconditions for an augmented risk of infectious disease transmission via vectors and hosts (Team and Manderson, 2011). Climate variability associated with both phases of ENSO have affected the incidence of diverse tropical and infectious diseases significantly in LA, including malaria, dengue, leishmaniasis (cutaneous and visceral), yellow fever, cholera, diarrhoea and probably Chagas disease. Climate change is now triggering the increase, persistence and re-emergence of some diseases in non-endemic areas, including some once considered eliminated, eradicated or controlled (Rodriguez-Morales *et al.*, 2010).

The recession of Andean tropical glaciers is modifying the geographic range of different species, with consequences to infectious diseases. Natural habitats are receding and disappearing, and imbalances between species and their predators are altering food chains. The modified plant and animal biodiversity has created suitable habitats for vector development, resulting in an increasing incidence of vector-borne diseases at higher altitudes, including malaria in non-traditional endemic areas (Aparicio *et al.*, 2010). Microorganisms and Chagas disease vectors including *Tripanosoma cruzi* and *Triatoma infestans* are adapting rapidly to the higher altitudes of mountain ecosystems, with new cases reported at 2800–3000 m in La Paz (Aparicio *et al.*, 2010).

Heavier rainfalls and drier seasons are altering human behaviour, increasing home water storage, further enhancing mosquito habitat and risking biological water contamination. Since 1998, the dengue seasonality for Santa Cruz, Bolivia, has extended from January to May, peaking in March or April (Aparicio *et al.*, 2010).

Dengue fever serotype 1 re-emerged in Bolivia between 1987 and 1998, coincident with strong ENSO events. Serotype 2 produces many dengue cases, considering that actual climate is warmer than the baseline, predominantly in the summer months. The relationship between climate, weather and dengue is being analysed to develop EWSs. In Santa Cruz, Bolivia, average and minimum temperature thresholds combined with weekly rainfall thresholds for the same period were found to be good predictors for dengue outbreaks, if fitted with a 5-week lag period (Arana *et al.*, 2007a).

Carbajo *et al.* have evaluated the role played by climate in the epidemiology of dengue in Argentina (1998–2011) (Carbajo *et al.*, 2012). By modelling annual risk using a temperature-based mechanistic model, they found that temperature, although useful to estimate annual transmission risk, did not describe dengue occurrence adequately at the country level. Climatic variables separately performed worse than geographic or demographic ones.

Using historical, climatic and epidemiological data sets, diverse studies have investigated the effects of El Niño events on the distribution of malaria and dengue in countries such as México, Costa Rica, Argentina and Brazil. In Mexico, the dengue epidemiological curve during cold weather has changed in such a way that incidence is higher than before. Outbreaks of malaria have occurred in unusually high latitudes, such as Bolivia (Rutar *et al.*, 2004). There are also preliminary studies suggesting that climate change will alter the distribution of leishmaniasis (Salomón *et al.*, 2012). In other cases, due to the absence of field study evidence, projections have been made based on possible ecosystem changes associated with climate change and their probable effects on vector distribution, as with Ecuador (Pinault and Hunter, 2012). Other contributions include the use of research findings to develop EWSs for malaria in Colombia (Poveda *et al.*, 2011) and to generate a vulnerability atlas to anticipate better the effects of climate change. For example, in Mexico, infectious diseases (acute diarrhoea) and vector-borne diseases (dengue) are included in this atlas (Riojas-Rodríguez *et al.*, 2012).

The incidence of malaria in Colombia has increased steadily in the past 50 years (see Fig. 18.2). This may be explained by historically increasing trends in average air temperatures, due to local and regional climate change, as well as massive deforestation (Poveda *et al.*, 2011). Additionally, Colombia experiences strong outbreaks of malaria during the warm phase of ENSO events that occur over the central-eastern topical Pacific (Ruiz *et al.*, 2006; Poveda *et al.*, 2011). Malaria in Colombia, as elsewhere, is also influenced by human migration in endemic areas (Osorio *et al.*, 2007; Rodriguez-Morales *et al.*, 2010).

Recent studies have indicated the role of climate-driven river flow variations in the Amazon as contributors to increases in the incidence of malaria in particular localities. Assuming that the precipitation–malaria relationships depend on the surface water conditions, researchers have observed that, while in the uplands, the malaria risk in relation to rainfall is variable, it is negative in areas dominated by wetlands and large rivers, where more precipitation indicates less malaria (Olson *et al.*, 2009). In some municipalities in Amazonas, Brazil, malaria increases have been observed to occur 1–2 months after peaks of temperature, and also a few days after sudden oscillations in river water levels, a hydrological phenomenon known locally as 'repiquete' (Wolfhart, 2011).

In 1998, a malaria outbreak near Titicaca Lake in the inter-Andean valleys near La Paz (elevation 2615–3592 m) led to the first local documentation of *Anopheles pseudopunctipennis* associated with new ecosystem conditions that emerged in the context of a changing climate. The current microclimate is 0.85°C warmer than at the baseline (1960–1990) (Aparicio *et al.*, 2013b).

The malaria assessment for Pando, Beni and La Paz, in northern Bolivia, showed an

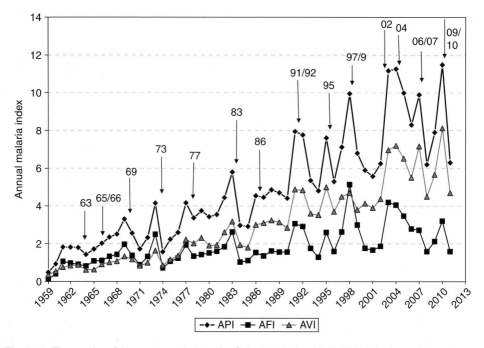

Fig. 18.2. Time series of three malarial indices for Colombia during 1959–2011. The Annual Parasitic Incidence (API) index aggregates malaria caused by *Plasmodium falciparum* (AFI) and *Plasmodium vivax* (AVI). Dates and arrows denote historical El Niño events (http://www.cpc.ncep.noaa.gov/products/analysis_monitoring/ensostuff/ensoyears.shtml). AFI = Annual Falciparum Index; AVI = Annual Vivax Index. (Data: Colombian National Institute of Health (http://www.ins.gov.co/Paginas/inicio.aspx.)

incremental increase since March 1993, caused by *Plasmodium falciparum* and, since April 1994, for cases caused by *Plasmodium vivax*. It was also observed that the climatic factors were found to explain 27.4% of malaria cases (*P. vivax* 11.3% and 43.6% for *P. falciparum*) (Aparicio and Ortiz, 2000).

The tropical Americas have witnessed a marked increase in both classic dengue fever (DF) and dengue haemorrhagic fever (DHF) incidence in the past two decades, imposing an economic burden of around US$2.1 billion/year (2010 US dollars) (Shephard *et al.*, 2011). Environmental and climatic variability affects dengue incidence by increasing vector (*Aedes aegypti*) densities, as has been documented in recent outbreaks in Honduras and Nicaragua (Rodriguez-Morales *et al.*, 2010) and Costa Rica (Mena *et al.*, 2011). Developed and tested predictive spatial ecological models of DF occurrence based on environmental and topographical variables for the metropolitan area of Medellin,

Colombia, have good predictive capacity, including for non-sampled areas (Arboleda *et al.*, 2009).

Diarrhoeal diseases are of concern for different reasons: there continues to be a high frequency in poor regions; the bacteria and viruses are sensitive to temperature changes; and, in some regions, food-cooling chains are affected during extreme events. Epidemiological studies have shown an association of increased incidence of 2% for each degree in average temperature rise. Floods, heavy rains and high temperatures have been associated with an incremental rise in the number of people admitted to hospitals with acute diarrhoea. This is caused by high temperatures aiding the proliferation of bacteria and parasites ingested through drinking water (Checkley *et al.*, 2000).

In the state of Pernambuco, Brazil, rainfall intensity is associated with increased rates of infant mortality (Rocha, 2012). In the semi-arid part, low precipitation was linked to higher infant mortality rates, while in the non-arid area, heavy rainfall was positively associated with infant mortality, caused mostly by diarrhoea.

Yellow fever, though vaccine preventable, is endemic in tropical South America below an elevation of 2300 m, owing to climatic and environmental conditions. In 2003–2004, Colombia witnessed the largest outbreak of yellow fever in 50 years: more than 200 cases and 50 deaths were reported in rural areas, with a high risk of the epidemic reaching urban areas.

Schistosomiasis is a prevalent neglected tropical disease (NTD) in South America, mostly in Brazil, Suriname, Venezuela and the Andes. Ambient temperatures affect its distribution and prevalence, which is likely to alter due to climate change (Mangal *et al.*, 2008). Schistosomiasis is also likely to increase its local infection and geographic range with climate change (Mas-Coma *et al.*, 2009).

Cutaneous leishmaniasis (CL) is a parasitic endemic cutaneous disease transmitted to human and other mammal hosts by the bite of phlebotomine sandflies. The disease is associated with climatic factors in LA, including ENSO. Bolivia exhibits a high incidence of CL, with 33 cases/100,000 inhabitants. In the 1990s, the disease increased markedly during La Niña, but fell by 40% during El Niño years (Gomez *et al.*, 2006). Climatic variability was considered responsible for the occurrence of 34% of new leishmaniasis cases in northern Bolivia (Aparicio and Ortiz, 2000). The impact of interannual ENSO variability on the incidence of cutaneous and visceral leishmaniasis (VL) in Colombia during 1985–2002 has also been studied. Contrary to Bolivia, researchers found that leishmaniasis increased 15% during El Niño, but decreased by 12% during La Niña (Cárdenas *et al.*, 2006). Further analysis in a broader area of Colombia has shown mixed results, with some areas showing an increase of incidence during La Niña. A study of the 2003 outbreak of CL in Colombia, the largest ever, found that land use, elevation and climatic variables (mean temperature and precipitation) were statistically significant predictors (Valderrama-Ardila *et al.*, 2010).

The correlation between positive anomalies in sea surface temperatures associated with El Niño and cholera in Peru, Ecuador, Colombia, Mexico and Venezuela is well documented (Moreno, 2006; Gavilán and Martínez-Urtaza, 2011).

Modelling studies about future changes in the geographical distribution of animal reservoirs of viral diseases have been developed for some Latin American countries. This is the case of rodent species in Argentina that harbour the hantavirus pulmonary syndrome virus, which are expected to change their distribution in response to different climate change scenarios. In Patagonia (southern Argentina), climate change may lower the transmission risk by reducing the potential distribution of the reservoir (Carbajo *et al.*, 2009).

18.4 Vulnerability

Global health is committed to promoting better health for all, but focuses on the most vulnerable populations (Feldbaum *et al.*, 2006; Fried *et al.*, 2010). In Latin American countries, issues related to the vulnerability and adaptation of the health sector comprise part of the agenda of regional inter-American and intergovernmental organizations, such as the Inter-American Development Bank (IDB) and the PAHO (PAHO, 2011). The IDB 'Integrated Strategy for Climate Change Adaptation and Mitigation, and Sustainable Renewable Energy' seeks to promote the development and use of public and private instruments to strengthen the capacity of countries to address climate change challenges. Priority sectors for support include sanitation, water resources and disaster risk management, especially through vulnerability assessments of geographical areas and improved methods for risk reduction (IDB, 2011).

During the past decade, several studies in Brazil have examined the social, environmental and health vulnerability to the impacts of climate change (Confalonieri *et al.*, 2009, 2013; Barata *et al.*, 2011). These studies have introduced and quantified composite vulnerability indices at national, regional and local levels, using diverse indicators of population health, socio-economic information and environmental data, in combination with diverse downscaled climatic scenarios. Nationally, the most vulnerable region was the semi-arid north-east, characterized by low socio-economic development and high rates of endemic (neglected) infectious diseases (Confalonieri *et al.*, 2009).

In Colombia, a pilot adaptation strategy is being implemented by the Ministry of Health and National Institute of Health in such a way that the Epidemiological Surveillance System responds to changes in the dynamics of malaria transmission and exposure brought about by climate variability and climate change (Poveda *et al.*, 2011).

Bolivia has developed the National Climate Change Mechanism, which includes strategies for health vulnerability and adaptation assessment. It is being implemented at different administrative levels to reduce climate change health impacts (Arana *et al.*, 2007a).

18.5   Conclusions

Many health outcomes – besides infectious diseases and accidents – are being impacted by climate change and climate variability in LA. These include outcomes associated with nutrition, occupational health (heat strain and heatstroke), respiratory diseases resulting from poor air quality, mental health, skin cancers and diseases associated with environmental pollutants and allergens.

The inhabitants of the Caribbean and Central America, including Colombia and Venezuela, are subject to higher vulnerability from tropical storms and to more intense hurricanes. Populations living in intra-Andean valleys are more vulnerable to intense storms, triggering landslides and large floods. Inhabitants of the low hot and humid regions of the tropical Americas are more vulnerable to climate-sensitive infectious diseases, including malaria, dengue, yellow fever and leishmaniasis. Climate change favours the transmission of those diseases at higher altitudes. In the tropical Americas, human populations are often living with temperatures that are close to intolerable thresholds.

The vulnerability of Central and South American population to the impacts of climate variability and change is well recognized (Moreno, 2006). Many factors contribute to this, including urbanization patterns, poverty, institutional and cultural aspects, poor sanitation and lack of access to clean water (Rodriguez-Morales *et al.*, 2010). Additionally, ecosystem degradation and decline of life-support systems will affect human health and well-being.

Mega- and large cities in LA are particularly vulnerable to climate and global change. Provision of potable water and energy is exacerbated by air and water pollution, and periodic inundations (Borsdorf and Coy, 2004). Increasing levels of social fragmentation and inequality in Latin American countries can exacerbate these problems.

Most Latin American countries have national plans of adaptation to the impacts of global climate change, which usually involve different sectors. Health protection measures are starting to be organized. The PAHO is committed to supporting countries in the development of adaptation plans that promote health protection strategies (PAHO, 2012). Global climate change is one among other recent global crises – economic, food and energy – that make health efforts more challenging, both worldwide and in LA (Beaglehole and Bonita, 2010).

References

Aparicio, M. and Ortiz, P. (2000) *Vulnerabilidad y adaptación de la salud humana ante los efectos del cambio climático en Bolivia.* MDSP-VMARNDF, PNCC PNUD, OPS/OMS – GEF 2000, La Paz, Bolivia.

Aparicio, M., García, J., Aparicio, J., Ríos, N. and Ocampo, M. (2010) El Dengue, el cambio climático y la variabilidad climática en Santa Cruz de la Sierra. *Revista del Colegio Médico de Bolivia* 2.

Aparicio, M., Aparicio, J., Ríos, N., Ocampo, M., Arana, I. and Vargas, J. (2013a) *Agua, cambio climático y salud en Ancoraimes.* IDH IBBA-Universidad Mayor de San Andres, La Paz, Bolivia.

Aparicio, M., Olmos, C., Aparicio, J., Ocampo, M., Pasten, M. and Giménez, A. (2013b) *Climate Health Vulnerability for Extreme Events and Water Sanitation in La Paz – Bolivia.* PAHO WHO, Kobe Center, Japan.

Arana, I., Garcia, M. and Aparicio, M. (2007a) *El cambio climático en Bolivia (análisis, síntesis de impactos y adaptación).* PNCC,VMPTA-MPD, La Paz, Bolivia.

Arana, I., Garcia, M., Aparicio, M. and Cabrera, M. (2007b) *Mecanismo nacional de adaptación al cambio climático.* Ministerio de Planificación del Desarrollo, VPTA, PNCC, La Paz, Bolivia.

Arboleda, S., Jaramillo-O., N. and Peterson, A.T. (2009) Mapping environmental dimensions of dengue fever transmission risk in the Aburrá valley, Colombia. *International Journal of Environmental Research and Public Health* 6, 3040–3055.

Barata, M., Confalonieri, U., Lima, A., de Lima, A.C.L., Marinho, D.P., Luigi, G., *et al.* (2011) Mapa de vulnerabilidade da população do estado do Rio de Janeiro aos impactos das mudanças climáticas nas áreas social, saúde e ambiente. Secretaria do Ambiente, Rio de Janeiro (http://download.rj.gov.br/documentos/10112/364217/DLFE-40943.pdf/rel_vulbilidade.pdf, accessed 19 April 2012).

Barrios, R.E., Stansbury, J.P., Palencia, R. and Medina, M.T. (2000) Nutritional status of children under 5 years of age in three hurricane-affected areas of Honduras. *Revista Panamericana de Salud Pública* 8, 380–384.

Beaglehole, R. and Bonita, R. (2010) What is global health? *Global Health Action* 3, 5142, doi:10.3402/gha.v3i0.5142.

Borsdorf, A. and Coy, M. (2004) Megacities and global change: case studies from Latin America. *Die Erde* 140, 1–20.

Brazil (2012) *Atlas Brasileiro de desastres naturais 1991–2010.* Volume Brasil. CEPED-UFSC, Florianopólis, Brazil.

Carbajo, A.E., Vera, C. and González, P.L. (2009) Hantavirus reservoir *Oligoryzomys longicaudatus* spatial distribution sensitivity to climate change scenarios in Argentine Patagonia. *International Journal of Health Geographics* 8, 44, doi:10.1186/1476-072X-8-44.

Carbajo, A.E., Cardo, M.V. and Vezzani, D. (2012) Is temperature the main cause of dengue rise in non-endemic countries? The case of Argentina. *International Journal of Health Geographics* 11, 26, doi:10.1186/1476-072X-11-26.

Cárdenas, R., Sandoval, C.M., Rodríguez-Morales, A.J. and Franco-Paredes, C. (2006) Impact of climate variability in the occurrence of leishmaniasis in northeastern Colombia. *American Journal of Tropical Medicine and Hygiene* 75, 273–277.

CEPAL (2007) *Alteraciones Climáticas en Bolivia: Impactos Observados en el Primer Trimestre de 2007* (http://www.eclac.org/publicaciones/xml/4/29494/L792-1.pdf, accessed 10 August 2012).

Checkley, W., Epstein, L.D., Gilman, R.H., Figueroa, D., Cama, R.I., Patz, J.A., *et al.* (2000) Effects of El Niño and ambient temperature on hospital admissions for diarrhoeal diseases in Peruvian children. *The Lancet* 355, 442–450.

Confalonieri, U.E.C., Marinho, D.P. and Rodriguez, R.E. (2009) Public health vulnerability, to climate change in Brazil. *Climate Research* 40, 175–186.

Confalonieri, U.E.C., Lima, A.C.L., Brito, I. and Quintão, A.F. (2013) Social, environmental and health vulnerability to climate change in the Brazilian northeastern region. *Climate Change* doi:10.1007/s10584-013-0811-7.

CRED (Centre for Research on the Epidemiology of Disasters) (2011) *Annual Disaster Statistical Review. The Numbers and Trends.* UC Louvain, Brussels.

Feldbaum, H., Patel, P., Sondorp, E. and Lee, K. (2006) Global health and national security: the need for critical engagement. *Medicine, Conflict and Survival* 22, 192–198.

Fried, L.P., Bentley, M.E., Buekens, P., Burke, D.S., Frenk, J.J., Klag, M.J., *et al.* (2010) Global health is public health. *The Lancet* 375, 535–537.

Gavilán, R.G. and Martínez-Urtaza, J. (2011) Environmental drivers of emergence and spreading of vibrio epidemics in South America. *Revista Peruana de Medicina Experimental y Salud Pública* 28, 109–115.

Gomez, C., Rodríguez-Morales, A.J. and Franco-Paredes, C. (2006) Impact of climate variability in the occurrence of leishmaniasis in Bolivia. *American Journal of Tropical Medicine and Hygiene* 75, 273–277.

IDB (2011) IDB *Integrated Strategy for Climate Change Adaptation and Migration, and Sustainable Renewable Energy.* Inter-American Development Bank, Washington, DC.

Koplan, J.P., Bond, T.C., Merson, M.H., Reddy, K.S., Rodriguez, M.H., Sewankambo, N.K., *et al.* and for the Consortium of Universities for Global Health Executive Board (2009) Towards a common definition of global health. *The Lancet* 373, 1993–1995.

Magrin, G., Marengo, J., Boulanger, J.-P., Buckeridge, M.S., Castellanos, E., Poveda, G., *et al.* (2014) Latin America. In: Barros, V., Field, C.B., Dokken, D.J., Mastrandrea, M.D., Mach, K.J., *et al.* (eds) *Climate Change 2014: Impacts, Adaptation, and Vulnerability. Volume II: Regional Aspects. Contribution of Working Group II to the Fifth Assessment Report of the Intergovernmental Panel on Climate Change.* Cambridge University Press, Cambridge, UK and New York.

Mangal, T.D., Paterson, S. and Fenton, A. (2008) Predicting the impact of long-term temperature changes on the epidemiology and control of schistosomiasis: a mechanistic model. *PLoS One* 3, 1438–1441.

Mas-Coma, S., Valero, M.A. and Bargues, M.D. (2009) Climate change effects on trematodiases, with emphasis on zoonotic fascioliasis and schistosomiasis. *Veterinary Parasitology* 163, 264–280.

Mena, N., Troyo, A., Bonilla-Carrión, R. and Calderón-Arguedas, Ó. (2011) Factors associated with incidence of dengue in Costa Rica. *Revista Panamericana de Salud Pública* 29, 234–242.

Moreno, A.R. (2006) Climate change and human health in Latin America: drivers, effects, and policies. *Regional Environmental Change* 3, 157–164.

Olson, S.H., Gangnon, R., Elguero, E., Durieux, L., Guégan, J.-F., Foley, J.A., *et al.* (2009) Links between climate, malaria, and wetlands in the Amazon Basin. *Emerging Infectious Diseases* 15, 659–662.

Osorio, L., Todd, J., Pearce, R. and Bradley, D.J. (2007) The role of imported cases in the epidemiology of urban *Plasmodium falciparum* malaria in Quibdó, Colombia. *Tropical Medicine and International Health* 12, 331–341.

PAHO (Pan American Health Organization) (2011) Health situation in the Americas – basic indicators (http://ais.paho.org/chi/brochures/2011/BI_2011_ENG.pdf, accessed 20 August 2012).

PAHO (2012) *Protecting Health from Climate Change.* PAHO, Washington, DC (http://www.who.int/global-change/publications/Final_Climate_Change.pdf, accessed 20 August 2012).

Pinault, L.L. and Hunter, F.F. (2012) Malaria in highlands of Ecuador since 1900. *Emerging Infectious Diseases* 18, 615–622.

PNUMA (Programa de las Naciones Unidas para el Medio Ambiente), CEPAL and GRID-Arendal (2010) Gráficos vitales del cambo climático para América Latina y el Caribe. PNUMA/CEPAL, Panamá city/ Santiago.

Poveda, G., Álvarez, D.M. & Rueda, Ó.A. (2011a) Hydro-climatic variability over the Andes of Colombia associated with ENSO: a review of climatic processes and their impact on one of the Earth's most important biodiversity hotspots. *Climate Dynamics* 36, 2233–2249.

Poveda, G., Estrada-Restrepo, Ó.A., Morales, J.E., Hernandez, O.O., Galeano, A. and Osorio, S. (2011b) Integrating knowledge and management regarding the climate-malaria linkages in Colombia. *Current Opinion in Environmental Sustainability* 3, 448–460.

Riojas-Rodriguez, H., Romero, M.F. and Capítulo, X. (2010) El deterioro de los ecosistemas y la biodiversidad: sus implicaciones en la salud humana. In: Galvão, L.A.C., Finkelman, J. and Henao, S. (eds) *Determinantes Sociales y Ambientales de la Salud Humana.* Organizaciòn Panamericana de la Salud. McGraw-Hill, Washington, DC, pp. 233–258.

Riojas-Rodríguez, H., Hurtado-Diaz, M., Moreno-Banda, G.L., Castañeda-Martínez, A., Saavedra-Lara, N. and Moreno-López, G. (2012) Atlas de Cambio Climático y Salud Humana. México. Instituto Nacional de Salud Pública (http://atlas.insp.mx, accessed 27 August 2012).

Rocha, M.A.G. (2012) Associação espacial entre mortalidade infantil e precipitação pluviométrica no Estado de Pernambuco. Tese de Doutorado. Faculdade de Saúde Pública da Universidade de São Paulo, São Paulo, Brazil.

Rodriguez-Morales, A.J., Risquez, A. and Echezuria, Z. (2010) Impact of climate change on health and disease in Latin America. In: Simar, S. and Austin, M.E. (eds) *Climate Change and Variability.* SCIYO, Rijeka, Croatia, pp. 463–486.

Ruiz, D., Poveda, G. and Vélez, I.D. (2006) Modelling entomological–climatic interactions of *Plasmodium falciparum* malaria transmission in two Colombian endemic-regions: contributions to a national malaria early warning system. *Malaria Journal* 1, 66, doi: 10.1186/1475-2875-5-66.

Rutar, T., Salgueiro, E. and Maguire, J. (2004) Introduced *Plasmodium vivax* malaria in a Bolivian community at an elevation of 2,300 meters. *American Journal of Tropical Medicine and Hygiene* 70, 15–19.

Salomón, O.D., Quintana, M.G., Mastrángel, A.V. and Fernández, M.S. (2012) Leishmaniasis and climate change-case study: Argentina. *Journal of Tropical Medicine* 601242, doi:10.1155/2012/601242.

Shephard, D.S., Coudeville, L., Halas, Y.A., Zambrano, B. and Dayan, G.H. (2011) Economic impact of dengue illness in the Americas. *American Journal of Tropical Medicine and Hygiene* 84, 200–207.

Team, V. and Manderson, L. (2011) Social and public health effects of climate change in the '40 south'. *Wiley Interdisciplinary Reviews: Climate Change* 2, 902–918.

UN Habitat (2012) *Estado de las ciudades de América Latina y el Caribe 2012. Rumbo a una nueva transición urbana.* Nairobi.

Valderrama-Ardila, C., Alexander, N., Ferro, C., Cadena, H., Marín, D., Holford, T.R., *et al.* (2010) Environmental risk factors for the incidence of American cutaneous leishmaniasis in a sub-Andean zone of Colombia (Chaparral, Tolima). *American Journal of Tropical Medicine and Hygiene* 82, 243–250.

Wolfhart, B.R. (2011) Análise epidemiológica, espacial e temporal e suas relações com as variáveis ambientais sobre a incidência da malária no período 2003 a 2009 em 4 municípios do Estado do Amazonas, Brasil. Dissertação (Mestrado), INPA, Manaus, Brazil.

19 Small Island States – Canaries in the Coal Mine of Climate Change and Health

Elizabeth G. Hanna and Lachlan McIver
National Centre for Epidemiology and Population Health,
The Australian National University, Canberra, Australia

> Islands are the barometers of international environmental policies. The entire world will first witness their success or their failure on our islands.
>
> James Michel's, the President of Seychelles, address to the Rio+20 Conference
> (D'Offay, 2012)

19.1 Introduction

Climate change will bring mixed fortunes to Earth's inhabitants. The inherent interconnectedness of Earth's ecosystems, and their dependence on a stable climate, means that anthropogenic perturbations in the climate system will inevitably disturb the ecosystem services which underpin all human existence. The rate of change in the global climate, and the difficulty in predicting with precision its future effects, thus pose a hazard to our global community's capacity to work and rest, feed and flourish. If some regions do benefit from climate change, such benefits may be comparatively limited in terms of area and population involved. At the other extreme are regions and populations that will experience, precipitously and substantively, the *negative* effects of the numerous manifestations of a warming world.

As they herald the unfolding future for the rest of the planet, highly climate-sensitive regions such as small island developing states (SIDS)[1] are, in effect, the 'canaries in the climate change coal mine'. In this chapter, we detail the drivers of the vulnerability[2] of SIDS to the impacts of climate change and outline some of the key health risks and adaptive options these nations face. Examples and case studies cited will be taken predominantly from Pacific island nations, although their plight will be shared more broadly across developing island nations of the Caribbean and Indian Ocean regions.

19.2 Characteristics of Small Island Developing States

SIDS have been variously classified and clustered according to their geographic, political and/or historical characteristics. As an example of the latter, the Indian Ocean Commission (IOC) voices the concerns of its (French-speaking) member states, namely: Comoros, Reunion (France), Madagascar, Mauritius and the Seychelles. The roughly equivalent regional organizations for island countries and territories in the South Pacific and Caribbean are, respectively, the Pacific Islands Forum and the Caribbean Commission (CARICOM). Other, non-regional groupings include the Association of Small Island States (AOSIS) – a relatively influential non-aligned group within the UN – and a conglomeration of Atlantic, Indian Ocean, Mediterranean and South China Sea (AIMS) countries.

A list of SIDS, as defined by the UN Office of the High Representative for the Least Developed Countries, Landlocked Countries and Small Island Developing States (UN-OHRLLS), is provided in Table 19.1, along with a number of their key characteristics, which are discussed in more detail below.

Geography of SIDS

Many SIDS are burdened by extreme isolation, whether due to distance or in combination with the vagaries of travel (e.g. weekly flights from New Zealand to Niue; monthly boats from Samoa to Tokelau); these factors are often compounded by their small size. The tiny Pacific island nations of Tuvalu and Nauru compete for the title of the world's smallest independent countries, with populations of approximately 10,000 each and land areas of less than 27 km².

The small size of SIDS limits vital natural resources, notably fresh water, food and land for agriculture and settlement. Particularly relevant in the context of climate change, relocation options – either internal or external – are often limited or non-existent. Many SIDS are partly, or entirely, low lying. Kiribati, the Marshall Islands, Tuvalu and the Maldives are exclusively atoll countries,[3] with an average elevation above sea level of only a few metres. Resource scarcity also limits population; many inhabitants live outside SIDS; their remittances are vital to their communities who remain behind.

To ensure access to vital life-sustaining resources, human settlements throughout history have tended to develop close to the sea and fresh water, while agriculture developed on fertile valleys and coastal lands. Urbanization, infrastructure and productive activities are therefore commonly located close to coastal and riverine areas that, while critical for food production, are disaster prone (Kreimer, 2001). This phenomenon is typical for SIDS – in the Caribbean, for example, more than half of the population live within 1.5 km of the coast (Mimura et al., 2007).

Population and demography of SIDS

Population size and trends constitute another vulnerability of SIDS, whether by nature of being precariously small, too large (relative to the land available) or growing too rapidly.

As can be seen from Table 19.1, most SIDS have small populations; the age structure of many are also skewed toward the young. Such demographic profiles, which usually correspond with high fertility rates, pose challenges with respect to the provision of sufficient infrastructure for education, employment, sanitation and health services (Ware, 2005).

Urbanization compounds these population pressures. In 1950, only 8.5% of Pacific islanders lived in urban centres. Mirroring global trends, by 2011 that proportion had risen to almost a quarter. Although total population numbers in SIDS remain small, some urban areas in the Pacific (e.g. Tarawa in Kiribati and Majuro in the Marshall Islands) have population densities in excess of 7000 people/km² (Pacific Islands Forum Secretariat, 2012). Approximately one-third of households have nine or more inhabitants (SPC Demographic and Health Surveys for Kiribati and RMI, 2007, 2009).

Economies of SIDS

As Table 19.1 shows, SIDS are generally poor, ranking low on human development indices. Typically, SIDS economies depend on few – often single – industries, such as fishing, agriculture or tourism. This economic fragility exposes SIDS to the impact of global market perturbations, conflict and natural disasters. A single, relatively localized event can devastate a small island state economy for years, as occurred in Niue following Cyclone Heta (when the country was transformed from a net food exporter to being dependent on food imports for 2 years) (Food Secure Pacific Working Group, 2008).

Low levels of monetary wealth also manifest as relative powerlessness among trading partners and in diplomatic negotiations. SIDS are dependent on foreign aid – the largesse, often programme specific and donor driven – of more powerful nations (WHO, 2011). Foreign aid is rarely without obligations, such as to accept

Table 19.1. Selected characteristics of small islands states. (Data: UN-OHRLLS, unless otherwise stated, www.unohrlls.org.)

SIDS[a]	Geography		Demography			Economy		
Country	Land area (thousand km²)	Max elevation (m)[b]	Population[c] (thousands)	Per cent of population <15[d]	Life expectancy[c]	Per cap GDP (PPP)[c] (thousands)	HDI (2011) [of 187]	Main industries[e] (per cent of GDP)
Antigua and Barbuda	0.4	402	90 (2012)	25.2	72.6	22	60	T 60%
Bahamas	14	63	347	24	74.5 (2010)	31	53	T 60%
Bahrain	0.7	122	1248 (2012)	20.2	75.1	27	42	P 11%
Barbados	0.4	336	238	18.8	76.8	24	47	S 80%
Belize[g]	23	1124	328	36.3	76	8.3	93	T
Cape Verde	4	2829	501	31.9	74.2	4	133	S
Comoros	2.2	2360	754	42.2	61.1	1.2	162	A 40%
Cuba	111	1974	11,254	17	79.1	9.9	51	S
Dominica	0.8	1447	68	22.6	77.5	13.6	81	A, T
Dominican Republic	48	3098	10,056	29	73.4	9.3	98	S, A
Fiji	18	1324	868	28.6	69.2	4.6	100	T, A
Grenada	0.3	840	105	25	76	13	67	T
Guinea-Bissau[g]	36	300	1547	40.2	?	1.1	176	A
Guyana[g]	215	2750	756 (2010)	31.1	69.9	7.5	117	A, M
Haiti	28	2680	10,124	35.3	62.1	1.2	158	R, A
Jamaica	11	2256	2751	29.5	73.1	9	79	S, T
Kiribati	0.8	3 (x), 81 (y)	101	33.1	68.1	6.2	122	Copra, F, R
Maldives	0.3	2	320	21.1	76.8	8.4	109	T 30%
Marshall Islands	0.2	10	64	37.8	72	2.5 (2008)	##	Aid (USA)
Federated States of Micronesia	0.7	791	112	33	69	2.2	116	A, F
Mauritius	2	828	1307	21.6	73.4	15	77	A, T, S
Nauru	0.02	71	10	32.7	79.9[f]	5 (2005)	##	Phosphate
Palau	0.5	242	21	21	71.8	10 (2009)	49	T A, F
Papua New Guinea	463	4509	7014	35.9	62.8	2.5	153	A, M, P
Samoa	2.9	1857	184	34.7	72.2	6	99	R, T, A, F
Sao Tome and Principe	1	2024	168	44.4	64.7	2	144	A (cocoa)

Continued

Table 19.1. Continued.

SIDS[a] Country	Geography		Demography			Economy		
	Land area (thousand km²)	Max elevation (m)[b]	Population[c] (thousands)	Per cent of population <15[d]	Life expectancy[c]	Per cap GDP (PPP)[c] (thousands)	HDI (2011) [of 187]	Main industries[e] (per cent of GDP)
Singapore[h]	0.69	164	5188	14	81.1	60	26	Diverse
St Kitts and Nevis	0.26	1156	53	22.3	73.1	16	72	T
St Lucia	0.62	950	176	22.2	74.6	13	82	T
St Vincent and Grenadines	0.39	1234	109	24	72.3	12	85	A, T
Seychelles	0.46	905	87	21.4	73.6	25	52	T
Solomon Islands	28	2335	552	37.3	67.9	3.3	142	A, F, forestry
Suriname[g]	163	1230	529	27.5	70.6	9.5	104	M
Timor-Leste	15	2963	1153	33.5	62.5	3.1	147	A, P (recent)
Tonga	0.75	1033	105	36.7	72.3	7.5	90	A, F
Trinidad and Tobago	5	940	1346	19.5	70.1	20	62	P
Tuvalu	0.03	5	10	30.2	67.2	3.4 (2010)	Not ranked	R, trusts, A
Vanuatu	12	1877	246	29.1	71	2.7 (2009)	125	A, F, T

[a]Only full members of the UN are listed, with the exception of Niue, a full UN member but not a SIDS according to UN-OHRLLS.
[b]Wikipedia.
[c]2011 estimate unless otherwise stated.
[d]Source: CIA World Factbook 2012.
[e]c.12% of SIDS GDP is foreign direct investment and development assistance; notable exceptions include Marshall Islands and Kiribati (The Little Data Book on Climate Change 2011; World Bank, 2012).
[f]Some figures provided by UN-OHRLLS differ from national census data.
[g]Majority of land area part of a continent, i.e. not technically an 'island state' in geographic terms.
[h]Singapore is the obvious outlier, raising the question of why it is still classified as a SID. A = agriculture; F = fishing; HDI = Human Development Index; M = mining; P = petroleum; PPP = purchasing power parity; R = remittances; S = services; T = tourism; x = South Tarawa; y = Banaba; ## = no data.

unfair trade treaties or to support the position of 'patron' countries in international fora. After decades of aid, most SIDS remain highly dependent on it. Arguably, pre-existing international inequalities have deepened, highlighting the importance and potential 'combined clout' of the regional alliances mentioned above.

19.3   Health Status of SIDS

The majority of SIDS are poor, tropical countries and suffer from both 'old world' infectious diseases, such as malaria, tuberculosis, pneumonia and diarrhoea, and more modern epidemics, such as Human Immunodeficiency Virus (HIV). Like other developing countries, many SIDS are also experiencing the 'epidemiological transition', whereby non-communicable diseases (NCDs) (e.g. diabetes, circulatory diseases and cancer) are overtaking infectious diseases as leading causes of illness and death.

NCDs now account for 75% of all deaths in the Pacific region (Asian Development Bank, 2011). This burden is particularly problematic for SIDS, killing at a younger age in low- and middle-income countries, with 29% of NCD deaths occurring under the age of 60, compared to 13% in high-income countries (WHO, 2011). NCDs thus act to impede economic growth and prosperity, compounding the effects of disasters and young demographic structures.

In the Pacific, this burden is expected to become even greater, due in part to the outmigration of skilled people, including those who could provide desperately needed health care, both preventive and curative (Pacific Islands Health Officers Association, 2010).

The health status of impoverished people is substantially determined by nutrition, which is often limited in range and quality (Hanna, 2013). Global food prices have remained high since 2008. Some agricultural crops in SIDS are already showing signs of stress under current climatic conditions, such as in Vanuatu (Reti, 2007). Decreasing island food yields worsens food stress imposed by rising import prices, and the greatest toll is borne by the poorest households, due to their relatively high expenditure on food (Food Secure Pacific Working Group, 2008). Urbanization potentially exacerbates

food insecurity, as urban dwellers can rarely supplement food supplies by subsistence farming or fishing, as rural poor often can (Miskelly et al., 2011).

Poor, hungry people have little capacity to absorb price spikes and variable availability, which drives shifts towards cheaper, nutrient-poor alternatives such as noodles, mutton flaps and turkey tails – all common fare in the Pacific (von Grebmer et al., 2011). Nutritional risks arise as fresh foods are, in general, the most nutrient rich. An estimated 27% of the population in the Pacific island region is undernourished, contributing to high maternal mortality (Asian Development Bank, 2011), intensifying an intergenerational health deficit which includes developmental delay, cognitive impairment and decreased economic productivity (Hanna, 2013). According to the 'thrifty phenotype hypothesis'[4] (Hales and Barker, 2001), undernutrition during pregnancy predisposes to altered metabolic conditions in later life (Eberle and Ament, 2012). Risks incurred include harmful serum lipid profiles, high glucose concentrations and elevated blood pressure, which impart adverse lifelong health consequences (Victora et al., 2008).

On the other hand, many small island states have an extraordinary dependence on fish for food security and animal protein, particularly in rural areas (Bell et al., 2011). Acidification and ocean warming, in combination with overfishing, are stressors on marine life that threaten biogeochemical cycles and the goods and services the ocean ecosystems provide (Plymouth Marine Laboratory, 2012).

19.4   SIDS and Natural Disasters

Analysis of natural hazards and vulnerability research, patterned with development, social and political factors, generates the World Risk Index, which indicates the probability that a country or region will be affected by a disaster (Alliance Development Networks, 2012). On a per capita basis, Oceania has the highest incidence of 'environmental disasters' (Mahany and Keim, 2012). Their disproportionately high vulnerability relates not only to event frequency and severity but also to national coping capacity.

19.5 Climate Change Impacts on SIDS

The Australian Bureau of Meteorology, the Commonwealth Scientific and Industrial Research Organisation (CSIRO) and meteorological services around the Pacific region have collaborated to provide downscaled climate projections for South Pacific SIDS (Australian Bureau of Meteorology and CSIRO, 2011). The main projections arising from this work to date include:

- increased ambient air and sea surface temperatures;
- more frequent extremely hot days and warm nights;
- increases in annual mean rainfall in specific areas and widespread increase in the number of heavy and extreme rain days;
- increased aridity from greater evapotranspiration;
- decreased sea surface salinity (i.e. freshening), which, in association with intensified warming, makes the surface ocean less dense and more stratified;
- sea level rise;
- ocean acidification.

These findings are generally consistent with those of the Intergovernmental Panel on Climate Change (IPCC) which, in its chapter on small islands in its Fourth and Fifth Assessment Reports, highlights the exquisite vulnerability of SIDS to warming induced by climate change, altered rainfall patterns, increased storm severity and rising seas (Mimura et al., 2007, Nurse et al., 2014). Climatic changes present significant concerns for large countries, but SIDS face the prospect of existential threats to land, livelihoods and perhaps even sovereignty due to rising seas and coastal inundation. A realistic threat exists that some may prove uninhabitable in the long term, due to loss of habitable, arable land and secure supplies of food and water. Rainwater harvesting and technical adaptations such as energy-intensive desalination plants may not fully protect SIDS from a less 'water-secure' future.

19.6 Climate Change and Health in SIDS

In 2009, the Commonwealth Health Ministers identified climate change as:

> … among the most serious challenges facing Pacific Island countries and territories, threatening every aspect of their environment, social and economic development, and political and human security. … For some of the 22 member countries and territories of the Pacific Community, it is a question of survival, and for all of them, a time for action.
>
> (Rodgers et al., 2009)

Assessing the vulnerabilities of SIDS to the health impacts of climate change

Butler and Harley's triphasic categorization of the population health consequences of climate change (Butler and Harley, 2010) applies well to SIDS. 'Primary effects' include direct injury and health problems arising from extreme weather events (EWEs) such as storms and floods. Other less direct, 'secondary' manifestations include water- and food-borne diseases; or altered distributions of arthropod vectors, intermediate hosts and pathogens, which alter the epidemiology of many infectious diseases that occur as a result of climatic variation. Of particular relevance to SIDS are the more severe, protracted future health consequences of climate change, classified as 'tertiary effects'. These include widespread and lingering economic disadvantage following EWEs or environmental degradation, such as loss of freshwater supplies or arable land.

When whole communities are exposed to catastrophic episodes which overwhelm their recovery capacity, conflict and large-scale migration can result (McMichael *et al.*, 2012). Cascading negative alterations to lifestyle attributable to climate change are likely also to impact mental health severely, arising, to various degrees, from any of those three categories of effects (Berry *et al.*, 2010). A summary of the assessments of the likely health impacts of climate change in Pacific SIDS, carried out by the World Health Organization (WHO) between 2010 and 2013 in collaboration with the health sector in each country, is presented in Table 19.2.

In addition to the priority risks to health identified in Table 19.2, other, more subtle, threats are posed by climate change to the health of island communities. A singular example of this island-specific vulnerability is the potential for sea level rise as an additional driver of overcrowding in atoll communities. This, in turn, may amplify infectious disease transmission in atoll countries such as Kiribati, Tuvalu and the Marshall Islands, all of which experience high rates of contagious diseases (such as diarrhoeal illness and tuberculosis), together with a high prevalence of diabetes and smoking.

Another plausible health risk is the exacerbation of the burden of NCDs. Though not unique to SIDS, the prospect of hotter, more humid conditions decreasing people's ability or willingness to exercise or perform outdoor work (Hanna *et al.*, 2011a), coupled with the compromised food security and the effects on NCDs of heat stress, is a potent potpourri (McIver and Hanna, in press).

19.7 Climate Change and Health Adaptation Options

Despite the many challenges inherent in the attempts to avert the most serious impacts of climate change on the health of island communities, there is cause for cautious optimism. Humans have managed to inhabit most of Earth's landmass, across vastly diverse environments. This success is directly attributable to human ingenuity, adaptability and technology. Small economies, on isolated islands, are not a new phenomenon. Similarly, island nations have long endured exposures to climatic extremes, responding with ingenious strategies for survival and recovery.

Communities develop their lifestyles, agriculture and cultural practices in response to

Table 19.2. Climate change and health vulnerabilities of Pacific island countries (source: McIver, 2012).

Country	Main climate-sensitive health issues
Cook Islands	Dengue fever, diarrhoeal disease
Federated States of Micronesia	Water- and mosquito-borne diseases, malnutrition
Fiji	Dengue fever, typhoid fever, leptospirosis, diarrhoeal disease
Kiribati	Food (safety, security, food-borne diseases), water (safety, security, water-borne diseases) and vector-borne diseases
Nauru	Air quality, food security, non-communicable diseases (NCDs)
Niue	Vector-borne diseases, ciguatera, diarrhoeal disease, respiratory disease, heat-related illness, NCDs, trauma from extreme weather events
Palau	Vector-borne diseases, zoonotic infections, gastroenteritis, respiratory disease, NCDs, trauma from extreme weather events, mental health issues
Republic of the Marshall Islands	Food-, water- and vector-borne (dengue) diseases, respiratory diseases, malnutrition
Solomon Islands	Vector-borne diseases (malaria), respiratory diseases
Tonga	Diarrhoeal diseases, vector-borne diseases (dengue), food security/nutrition, non-communicable diseases, injuries and deaths from extreme weather events
Tuvalu	Diarrhoeal disease, respiratory disease, compromised food security and impacts on NCDs
Vanuatu	Food- and water-borne diseases

their natural environment, including extreme fluctuations that occur infrequently (Hanna, 2011). While these cultural practices are enduring, they are not static. Ecosystems also change, responding to natural and human-induced interventions. Long-term population survival is determined by the capacity to develop effective resource utilization and management. SIDS have been striving toward this end; yet, with climate change, the goal posts are moving.

Essentially, there are three major forms of climate change adaptive options: (i) to boost local capacity; (ii) to share hardships and responses; or (iii) ultimately, to relocate.

19.8 Boosting Local Capacity

Education for all, food self-sufficiency, effective sanitation, safe drinking water and effective affordable primary health care are all aims of strengthened public health systems. These are key climate change adaptation strategies (McMichael et al., 2003; Ebi et al., 2005; WHO, 2010; McIver, 2012; Hanna, 2013).

Adaptation strategies should combine heightened protection of communities and vulnerable groups (e.g. improved disease surveillance and effective early warning systems for epidemics and EWEs) with strategies that render populations and habitats more climate-proof in the long run (e.g. improved water supplies and sanitation systems, sustainable food systems and heat- and storm-tolerant infrastructure) (Hanna et al., 2011b). Raising the standard of these fundamentals – particularly for poor and otherwise disadvantaged communities – unquestionably fits with building resilience and exemplifies a 'no regrets' approach. Other basic population health issues identified as critical climate change adaptation strategies for low-income countries such as SIDS include gender equality and improved reproductive health, which usually corresponds with slowing population growth[5] (Bryant et al., 2009).

19.9 Cooperative Responses

Improved disaster risk management – reducing risk and improving response, recovery and rehabilitation – will be key to reducing the substantial health impacts of disasters in SIDS attributable to climate change. The Hyogo Framework for Action (for disaster risk reduction) 2005–2015 identified the development of financial risk-sharing mechanisms, particularly insurance and reinsurance against disasters as a priority action for highly vulnerable SIDS (ISDR Scientific and Technical Committee, 2009). The concept of mutual assistance in disaster risk management and response is gaining traction. An example is the Caribbean Catastrophe Risk Insurance Scheme (CCRIS), where, on average, at least one major hurricane and numerous tropical storms cross the Caribbean each year, causing annual aggregate economic losses of US$613 million (World Bank, 2012). Under CCRIS, small island countries in the region all contribute to a disaster insurance pool, which is 'topped up' by developed countries and international financial institutions (World Bank, 2012).

A different type of cooperation can be seen in the approach taken by several Pacific Island countries (PICs), along with a range of other developing countries, in combining climate change adaptation with disaster risk reduction at the national level. The resulting 'Joint National Action Plans' provide a streamlined platform for practical implementation and sourcing of financial support, in acknowledgement of the shared risks and management strategies for the parallel threats posed by disasters and other climate change phenomena.

The sharing of information and 'best practices', including the use of information and communication technologies (ICT), is another promising area for SIDS collaboration. Knowledge management platforms like SIDSnet, the Islands Communication Network and the University Consortium of Small Island States play an important role for these purposes (UNESCO, 2012). Specific to climate change and health, a recent online discussion forum hosted by the Pacific Solution Exchange yielded fascinating views, both novel and traditional, on the link between climate change and NCDs in the Pacific, as well as on their management.

19.10 Relocation

The ambivalent attitude of many Pacific islanders toward relocation reflects a natural aversion to contemplate this drastic measure. While voluntary migration offers potential benefits to SIDS, forced migration due to climate change is likely to harm mental health and culture.

In addition, studies suggest that following relocation, migrants continue to face many problems, including landlessness, unemployment, homelessness, social marginalization, food insecurity, reduced access to common-property resources and increased morbidity (Barnett and O'Neill, 2012).

19.11   External Support Requirements

Climate change poses enormous threats to the health, and indeed survival of SIDS populations. Addressing such threats will likely exceed the internal resources of these countries. Consensus exists that the primary external assistance required by SIDS is for larger, wealthier countries to reduce, immediately and significantly, and then reverse their carbon emissions. Today, obsessed with conventionally defined economic growth and an addiction to fossil fuels, most developed and rapidly developing countries continue to reveal the low priority of mitigation.

This negligence contrasts with that of some SIDS. As one example, Tokelu is on track to achieve 100% renewable energy, via solar energy and locally sourced palm oil.

While mitigation must remain the overarching international priority, SIDS must focus on adaptation. For this to occur, wealthier countries and international institutions must rapidly scale up their support for these processes.

With external assistance, the five Member States of the Indian Ocean Commission launched the ACCLIMATE Project in 2008 to help strengthen coping capacity. The project aimed to create a regional climate profile and to define, for each country, vulnerability by 12 key sectors (including health, food security, fresh water, biodiversity and infrastructure) (D'Offay, 2012). This has been followed by a Regional Adaptation Strategy that includes health as one of four main priority areas.

In the Caribbean, the Regional Framework for Achieving Development Resilient to Climate Change found that some countries spent 30–40% of foreign exchange earnings on fossil fuels. Reducing the cost of energy, in particular

for the poorest of those countries, became a key priority. Adaptation projects can incorporate regional approaches to accelerate the transition to low-carbon energy, complementing work on adaptation and resilience (Ellis et al., 2013).

There is a clear and urgent need to develop an overarching framework for addressing the public health adaptive needs in SIDS as climate change intensifies. Mechanisms for inter-institutional coordination with the health sector are often lacking; however, the work conducted by WHO in the northern Micronesian region of the Pacific demonstrated some novel and important linkages between health and other sectors. Several PICs have included climate change adaptation in their national health policies and plans, and more than a dozen now have specific National Climate Change and Health Action Plans (NCCHAPs) (or equivalent) (McIver, 2012).

The focus must now shift toward implementation of these plans, enabled by substantial sustained assistance. An example of progress in implementing such plans is in Kiribati, where the NCCHAP has attracted health sector-specific funding via the European Union Global Climate Change Alliance for Small Island States mechanism (McIver et al., 2013). Similar opportunities may be available to other SIDS, and the potential advantages of 'clubbing together' to source funding, carry out activities and address common priorities (e.g. vector-borne diseases, malnutrition, mental health) should be made clear.

Finally, there are yawning gaps in knowledge regarding the health impacts of climate change in SIDS and how the local and global communities may best manage them. The focus to date of the majority of climate change and health research has been on infectious diseases. While these pose significant hazards to island communities in the context of climate change, it is plausible that climate change-attributable increases in the burden of NCDs and mental health disorders are of equivalent or greater concern (McIver and Hanna, in press).

19.12   Summary and Recommendations

This chapter's leading quote by James Michel, describing islands as the 'barometers of international environmental policies' (D'Offay, 2012), is a blunt reminder of the grave future that looms due to global fossil fuel addiction. The day when the first SID is totally and forever abandoned will be a tragic occasion for our species. Will developed countries then act aggressively to reduce atmospheric CO_2 emissions? Or, will they procrastinate until the next nation is forced to migrate, or the next? Island communities forced to abandon their homes and cultures will exemplify an exceedingly selfish global culture, seemingly ignorant that we have only one planet and that our collective folly is rapidly converting it into a hot and nasty place, increasingly unsuitable for human habitation. Many children born in 2015 in rich countries will have life expectancies of 85 years. Their parents will anticipate them to be alive in 2100. However, today's climate trajectories and projections suggest that their world will be 4–6°C warmer. Their survival to 2100 may be precarious.

The pathway toward and beyond 2100 will be shaped by our collective efforts. In addition to mitigation, substantial improvements can be achieved with careful planning. Climate change amplifies the imperative to invest in tackling fundamental population health challenges such as the MDGs. In addition to actions within the health sector, reducing the climate change-related health risks for SIDS requires intersectoral collaboration to increase access to clean water and sanitation; improve the management of precious water and food resources; increase resilience to extreme events; and strengthen social capital, gender equality and education. There needs also to be increased support for community organizations and investments in information, monitoring and early warning systems, including to remote areas.

Climate vulnerability assessments must be conducted appropriately and integrated into health planning and policy documents. Upskilling of the health workforce in climate risks increases the likelihood that climate-orientated health information is absorbed and that pro-health responses are adopted. This requires comprehensive, multisectoral climate risk management programmes that can implement these priorities holistically and ensure their sustainability. We again stress that SIDS cannot execute climate change adaptation options without a reliable stream of external funding. Finally, we remark that while such external funding seems like more of the same old aid, it needs to be delivered and used in genuine ways that enhance independence.

Notes

[1] Small Island Developing States (SIDS) were recognized as a distinct group of developing countries facing specific social, economic and environmental vulnerabilities at the United Nations Conference on Environment and Development (UNCED), also known as the Earth Summit, held in Rio de Janeiro, Brazil (3–14 June 1992).
[2] Vulnerability to climate change is defined by the IPCC as 'the degree to which a community, individual or the environment is susceptible to and unable to cope with adverse effects of climate change' (UNFCCC, 2011, Glossary of climate change acronyms. UN Framework Convention on Climate Change).
[3] An atoll is a coral reef that partially or completely encircles a lagoon. Atoll countries are those made up exclusively of atolls.
[4] This expands the Barker Hypothesis, published in 1995, 'Low birth weight are at greater risk of developing coronary heart disease'. It proposes that the epidemiological associations between poor fetal and infant growth, and the subsequent development of type 2 diabetes and the metabolic syndrome, result from the effects of poor nutrition in early life, which produce permanent changes in glucose–insulin metabolism.
[5] Population control measures are advocated using only human rights based means – such as improved education (especially for girls); provision and increased enforcement of women's rights; and the provision of adequate family planning services.

References

Alliance Development Networks (2012) *World Risk Report 2012 – Focus: Environmental Degradation and Disasters.* The United Nations University Institute for Environment and Human Security, Berlin.

Asian Development Bank (2011) *The Millennium Development Goals in Pacific Island Countries: Taking Stock, Emerging Issues, and the Way Forward. Workshop Report*. Asian Development Bank, Manila.

Australian Bureau of Meteorology and CSIRO (2011) *Climate Change in the Pacific: Scientific Assessment and New Research. Volume 1: Regional Overview*. Australian Bureau of Meteorology, CSIRO, Melbourne, Australia.

Barnett, J. and O'Neill, S.J. (2012) Islands, resettlement and adaptation. *Nature Climate Change* 2, 8–10.

Bell, J.D., Johnson, J.E., Ganachaud, A.S., Gehrke, P.C., Hobday, A.J., Hoegh-Guldberg, O., *et al.* (2011) *Vulnerability of Tropical Pacific Fisheries and Aquaculture to Climate Change: Summary for Pacific Island Countries and Territories*. Secretariat of the Pacific Community, Noumea, New Caledonia.

Berry, H.L., Bowen, K. and Kjellstrom, T. (2010) Climate change and mental health: a causal pathways framework. *International Journal of Public Health* 55, 123–132.

Bryant, L., Carver, L., Butler, C.D. and Anage, A. (2009) Climate change and family planning: least-developed countries define the agenda. *Bulletin of the World Health Organization* 87, 852–857.

Butler, C.D. and Harley, D. (2010) Primary, secondary and tertiary effects of eco-climatic change: the medical response. *Postgraduate Medical Journal* 86, 230–234.

D'Offay, C.D. (2012) Indian Ocean Islands Call on Rio+20 to Focus on Climate Change. Guest Article No 76. International Institute for Sustainable Development (http://climate-l.iisd.org/guest-articles/indian-ocean-islands-call-on-rio20-to-focus-on-climate-change/, accessed 23 March 2014).

Eberle, C. and Ament, C. (2012) Diabetic and metabolic programming: mechanisms altering the intra-uterine milieu. *ISRN Pediatriatics* 2012, 975685.

Ebi, K.L., Lewis, N.D. and Corvalán, C. (2005) *Climate Variability and Change and Their Potential Health Effects in Small Island States: Information for Adaptation Planning in the Health Sector*. UNEP, WHO, WMO, Geneva, Switzerland.

Ellis, K., Cambray, A. and Lemma, A. (2013) *Drivers and Challenges for Climate Compatible Development*. Climate and Development Knowledge Network (CDKN), London.

Food Secure Pacific Working Group (2008) *Climate Change and Food Security in Pacific Island Countries*. Food and Agriculture Organization of the United Nations (FAO), Rome.

Hales, C.N. and Barker, D.J.P. (2001) The thrifty phenotype hypothesis: Type 2 diabetes. *British Medical Bulletin* 60, 5–20.

Hanna, E.G. (2011) Health hazards. In: Dryzek, J.S., Norgaard, R.B. and Schlosberg, D. (eds) *The Oxford Handbook of Climate Change and Society*. Oxford University Press, Oxford, UK, pp. 271–231.

Hanna, E.G. (2013) Vulnerable countries – Pacific Island Countries. In: Pielke, R. Sr, Adegoke, J. and Wright, C. (eds) *Climate Vulnerability. Volume 1 Health*. Elsevier, Amsterdam, pp. 165–182.

Hanna, E.G., Kjellstrom, T., Bennett, C. and Dear, K. (2011a) Climate change and rising heat: population health implications for working people in Australia. *Asia Pacific Journal of Public Health* 23, suppl 14S–26S.

Hanna, E.G., McMichael, A.J. and Butler, C.D. (2011b) Climate change and global public health: impacts, research and actions. In: Parker, R. and Sommer, M. (eds) *The Routledge International Handbook on Global Public Health*. Routledge, Oxfordshire, UK, pp. 133–144.

ISDR Scientific and Technical Committee (2009) *Hyogo Framework for Action 2005–2015: Building the Resilience of Nations and Communities to Disasters*. United Nations International Strategy for Disaster Reduction, Geneva, Switzerland.

Kreimer, A. (2001) Social and economic impacts of natural disasters. *International Geology Review* 43, 401–405.

McIver, L. (2012) Climate change and health in the Pacific: cause for concern; opportunities for adaptation. *Inform'Action Bulletin* 36, 3–6.

McIver, L. and Hanna, E.G. (in press) Fragile paradise – health and climate change in the South Pacific. In: Butler, C.D., Dixon, J. and Capon, A.G. (eds) *From Healthy Workers to a Healthy Planet*. ANU E-Press, Canberra.

McIver, L., Hoy, D., Lepers, C., Naicker, J. and Souares, Y. (2013) The first Pacific regional climate change and health symposium. *Inform'Action Bulletin* special issue, 3–6.

McMichael, A.J., Woodruff, R.E., Whetton, P., Hennessy, K., Nicholls, N., Hales, S., *et al.* (2003) *Human Health and Climate Change in Oceania: A Risk Assessment*. Commonwealth of Australia, Canberra.

McMichael, C., Barnett, J. and McMichael, A.J. (2012) An ill wind? Climate change, migration, and health. *Environmental Health Perspectives* 120, 646–654.

Mahany, M.J. and Keim, M.E. (2012) Challenges and strategies for climate change adaptation among Pacific Island nations. *Disaster Medicine Public Health Preparedness* 6, 415–423.

Mimura, N.L., Nurse, R.F., McLean, J., Agard, L., Briguglio, P., Lefale, R., *et al.* (2007) Small islands. In: Parry, M.L., Canziaani, O.F., Palutikof, J.P., van der Linden, P.J. and Hanson, C.E. (eds) *Climate Change 2007: Impacts, Adaptation and Vulnerability. Contribution of Working Group II to the Fourth Assessment Report of the Intergovernmental Panel on Climate Change.* IPCC, Cambridge University Press, Cambridge, UK and New York, pp. 687–716.

Miskelly, R., Cocco-Klein, S. and Abbott, D. (2011) *Situation Monitoring: Food Price Increases in the Pacific Islands.* United Nations Children's Fund (UNICEF), Pacific, Suva, Fiji Islands.

Nurse, L., McLean, R., Agard, J., Briguglio, L.P., Duvat, V., Pelesikoti, N., *et al.* (2014) Small islands. In: Barros, V., Field, C.B., Dokken, D.J., Mastrandrea, M.D., Mach, K.J., *et al.* (eds) *Climate Change 2014: Impacts, Adaptation, and Vulnerability, Volume II: Regional Aspects. Contribution of Working Group II to the Fifth Assessment Report of the Intergovernmental Panel on Climate Change.* Cambridge University Press, Cambridge, UK and New York.

Pacific Islands Forum Secretariat (2012) Remarks by Secretary General Slade at UNICEF State of the World's Children 2012 report launch. UNICEF, Fale, Tuvalu.

Pacific Islands Health Officers Association (2010) 'Declaring a Regional State of Health Emergency Due to the Epidemic of Non-Communicable Diseases in the United States-Affiliated Pacific Islands'. Board Resolution No 48-01.

Plymouth Marine Laboratory (2012) *Hot, Sour & Breathless – Ocean Under Stress. How is the Biggest Ecosystem on Earth Faring in the Lead up to +20.* Plymouth Marine Laboratory, Scripps Institution, Plymouth, UK.

Reti, M.J. (2007) *An Assessment of the Impact of Climate Change on Agriculture and Food Security in the Pacific: A Case Study in Vanuatu.* FAO, Apia, Samoa.

Rodgers, J., Parr, B., Taylor, M., Nicol, S., Bell, J., Borg, B., *et al.* (2009) Climate change and health: a Pacific perspective. Commonwealth Health Ministers' Update 2009. Regional case study. Noumea.

SPC Demographic and Health Surveys for Kiribati and RMI (2007 and 2009) (http://www.spc.int/sdd/index.php/en/downloads/cat_view/46-dhs/67-kiribati-2009-dhs-report; http://www.spc.int/sdd/index.php/en/downloads/cat_view/46-dhs/47-marshall-islands-2007-dhs-fact-sheets, accessed 19 March 2014).

UNESCO (2012) *Climate Change Education for Sustainable Development in Small Island Developing States. Report and Recommendations.* Expert Meeting Nasau, Commonwealth of the Bahamas. September 2011.

Victora, C.G., Adair, L., Fall, C., Hallal, P.C., Martorell, R., Richter, L., *et al.* (2008) Maternal and child undernutrition: consequences for adult health and human capital. *The Lancet* 371, 340–357.

von Grebmer, K., Torero, M., Olofinbiyi, T., Fritschel, H., Wiesmann, D., *et al.* (2011) *Global Hunger Index. The Challenge of Hunger: Taming Price Spikes and Excessive Food Price Volatility.* International Food Policy Research Institute (IFPRI), Washington, DC.

Ware, H. (2005) Demography, migration and conflict in the Pacific. *Journal of Peace Research* 42, 435–454.

WHO (2010) *Asia Pacific Strategy for Emerging Diseases.* Technical Papers. World Health Organization, Western Pacific Regional Office.

WHO (2011) *Food Security – Trade, Foreign Policy, Diplomacy and Health.* World Health Organization, Geneva, Switzerland.

World Bank (2012) *Implementation, Completion and Results Report – Caribbean Catastrophe Risk Insurance.* Report No ICR00002332. World Bank (http://www.ccrif.org/sites/default/files/publications/2012WBImplementationCompletionReportCCRIF.pdf, accessed 24 March 2014).

20 Climate Change Adaptation to Infectious Diseases in Europe

Jan C. Semenza

Health Determinants Programme, Office of the Chief Scientist, European Centre for Disease Prevention and Control, Stockholm, Sweden

20.1 Introduction

Climate is changing rapidly due to anthropogenic activity such as extensive fossil fuel combustion and widespread alterations in land use. Global climate change manifests itself both through gradual changes, such as increasing air and ocean temperatures and shifting mean annual precipitation or altered seasonality, as well as through erratic events, such as heavy rainfall and other extreme weather events. Europe will experience differential impacts depending on the regions, since biogeographic regions display specific vulnerabilities to changing environmental conditions (Fig. 20.1) (EEA, 2008). The regional variability is related to Europe's location in the northern hemisphere with its seas, continental land mass and part of the Arctic. Projections of annual average temperature indicate a rise of 1.0–5.5°C (for 2080–2100 compared to the 1961–1990 average), but regional differences far outstrip these estimates (Giorgi *et al.*, 2004).

- During the summer, the Mediterranean is expected to warm by 6°C. Southern Europe will be most affected by droughts, with yearly rainfall dropping by up to 30% of current annual precipitation (Christensen *et al.*, 2007). Less precipitation and much warmer temperatures will lead to higher risks of water scarcity, droughts, lower crop yields, heatwaves, forest fires, biodiversity losses, soil and ecosystem degradation and eventually desertification.

- In Western and Atlantic Europe, extreme events such as violent storms and floods are projected to become more frequent, due to warmer temperatures and higher volumes and intensities of precipitation, particularly during the winter. This will result in increased river flow and higher risk of coastal flooding.

- In central and Eastern Europe, the annual mean temperature increase is projected to be in the order of 3–4°C, except for the more continental regions of central Europe and the Black Sea Region, like Romania, where temperatures could increase by as much as 4–4.5°C. Projections also indicate more temperature extremes. Annual mean precipitation should increase by up to 10% in most regions. Precipitation would increase mainly in winter, while there would be reductions in summer precipitation in several areas. Thus, more river floods are projected for the winter and more forest fires for the summers.

- In northern Europe (boreal region), conditions will be similar to Western Europe, but with more change in temperatures and precipitation. There will be less snow and lake and river ice cover, but increased river flows. Crop yields could increase by 10–30%, if warming were limited to 1–3°C above pre-industrial levels, but agriculture could

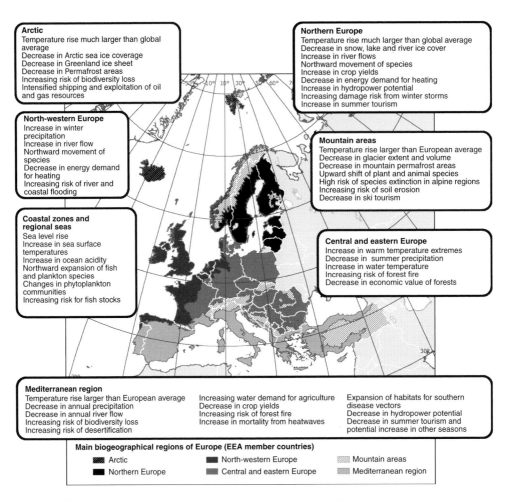

Fig. 20.1. Key observed and projected climate change and impacts for the main regions in Europe. (Source: European Environment Agency.)

suffer from new pests and diseases. With this warmer climate, the Baltic Sea could be affected increasingly by eutrophication (algal bloom) and pollution.

Differential changes in local climate and weather patterns (Fig. 20.1) will alter the infectious disease burden in Europe (Semenza and Menne, 2009). However, it is unlikely that the effect of climate change on individual microorganisms is idiosyncratic; rather, a multitude of impacts would be expected because dispersion, transport, fate and environmental exposure pathways of these pathogens are linked intricately to local climate and weather conditions (Boxall et al., 2009). Interactions of these factors and other

drivers with climate change create another layer of complexity (Suk and Semenza, 2011).

The distribution of vector-borne diseases will shift, since temperature influences: survival, reproduction rates, habitat suitability, distribution and abundance of vectors; intensity and temporal pattern of vector activity (particularly biting rates); and rates of development, survival and reproduction of pathogens within vectors (Semenza and Menne, 2009). Higher temperatures may foster the dispersal of vector-borne diseases in northern and central Europe, but projected excessive temperatures during the summer months may decrease transmission in southern Europe (Giorgi et al., 2004). Thus, drought and water scarcity might reduce the circulation of vector-borne pathogens there. The burden of food- and waterborne diseases will also shift, since ambient temperature and precipitation patterns have a bearing on their environmental exposure routes (Semenza *et al.*, 2012a). Growth rates of food-borne microorganisms increase with higher temperatures, and extended summer seasons may increase the opportunities for deficiencies in food handling. Parasites that resist chlorination can be flushed into water treatment and distribution systems during extreme and erratic rain events, which can result in waterborne outbreaks (Semenza and Nichols, 2007). The complexity of these relationships calls for a concerted public health approach, with changes in surveillance, preparedness and response (Lindgren *et al.*, 2012; Semenza *et al.*, 2012c). The European Centre for Disease Prevention and Control (ECDC) has devised a process of climate change adaptation to infectious diseases that involves the following: (i) data exploration of the peer-reviewed literature to assess the established evidence; (ii) feeding this information into risk assessments; (iii) developing forecasts and predictions of potential threats; (iv) determining potential vulnerabilities in the public health infrastructure; and (v) adapting accordingly (Fig. 20.2). The different steps are described in more detail below.

20.2 Evidence: Climate Change Knowledge Base of Food- and Waterborne Diseases

Cataloguing and mapping the association between climate change and food- and waterborne diseases can shed light on this complex relationship (Semenza *et al.*, 2012b). It can also help to disentangle climatic from other socio-economic, environmental or behavioural drivers of population health (Suk and Semenza, 2011). We used a systematic approach to retrieve all papers in the peer-reviewed literature on climate change/climate variability and food- and waterborne diseases (Semenza *et al.*, 2012b). The data were entered into an online, relational knowledge base and tagged with terms from a predefined vocabulary. This ontology of terms was composed of five thematic aspects (pathogens, food, water, climate/environment and reservoirs) and three spatial aspects (continent, region and country). The applied ontology formally represented a set of concepts and vocabulary in the climate change literature; they were used to classify related variables and model interactions between them. As such, this provided a conceptual model for analysing the relationships between climatic variables and food- and waterborne diseases. Thus, information was extracted

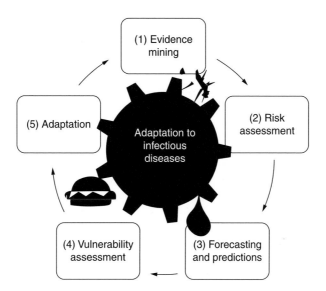

Fig. 20.2. The process of climate change adaptation to vector-, food- and waterborne infectious diseases in Europe.

from selected articles and organized and structured according to the predefined specifications of the ontology. Once entered into the ontological knowledge base, the information was searched to quantify and map some of the complex relationships between food- and waterborne pathogens and meteorological variables (Table 20.1). A summary of this analysis is listed below.

Campylobacter

Campylobacter does not replicate outside its animal host and the seasonal incidence peak does not occur during the hottest time of the year (Semenza *et al.*, 2012a). Thus, assessment of the relationship between climatic variables and campylobacter has been difficult, as reflected in the inconsistent findings in different studies. For example, the association between campylobacter incidence and average temperatures has been unreliable (Kovats *et al.*, 2005; Fleury *et al.*, 2006; Bi *et al.*, 2008). High environmental temperatures and relatively low humidity have been associated with elevated incidence (Patrick *et al.*, 2004; Kovats *et al.*, 2005). Precipitation in early spring can cause campylobacteriosis

outbreaks (World Health Organization, 2008). Private water sources in rural areas are more vulnerable to contamination during extreme weather events, which is reflected in the number of waterborne outbreaks (Pebody *et al.*, 1997; Hearnden *et al.*, 2003). The risk of groundwater and surface contamination is predicted to increase with the projected rise in heavy rainfall events in northern Europe due to climate change, as discussed above. In southern Europe, episodes of drought might call for rainwater harvesting; campylobacter in untreated water from cisterns or roof runoff might contribute to an elevated risk of both human and animal disease (Palmer *et al.*, 1983; Savill *et al.*, 2001).

Salmonella

Seasonal temperatures are suspected drivers of reported *Salmonella* cases, as an increase in weekly temperature is followed by a rise in salmonellosis in different European countries (Kovats *et al.*, 2004; Naumova *et al.*, 2007). Monthly monitoring of *Salmonella* sp. in aquatic environments during summer and autumn was associated with maximum precipitation subsequent to faecal contamination events

Table 20.1. Selected pathogens with environmental/climatic variables (and factors associated with climate). Climate change knowledge base for food- and waterborne diseases, 1998–2009.

	Campylobacter	Salmonella	Listeria	Vibrio	Cryptosporidium	Norovirus
Temperature	⇔	⇔	?	⇔	⇔	⇔
Extreme temperature	⇔	?	?	⇔	⇔	?
Temperature threshold	⇔	⇔	?	⇔	⇔	?
Precipitation	⇔	⇔	?	⇔	⇔	?
Precipitation pattern[a]	⇔	?	?	?	⇔	⇔
Extreme precipitation	⇔	?	?	⇔	⇔	⇔
Humidity	⇔	⇔	⇔	?	?	?
UV light	⇔	⇔	⇔	⇔	⇔	?
Seasonality	⇔	⇔	?	⇔	⇔	⇔
Salinity	◌	◌	◌	⇔	◌	◌
Floods	⇔	⇔	?	⇔	⇔	⇔
Drought	⇔	?	?	◌	⇔	?
Storms	?	?	?	⇔	?	?
Irrigation[b]	?	⇔	?	◌	⇔	?
Recreational activities	⇔	⇔	?	⇔	⇔	⇔
Shellfish production	⇔	⇔	⇔	⇔	?	⇔
Consumption habits	⇔	⇔	⇔	⇔	⇔	⇔

[a]For example: seasonality of rain events; [b]water.
Note: ⇔ = impact; ◌ = no impact; ? = impact unknown.

(Craig *et al.*, 2003; Martinez-Urtaza *et al.*, 2004). Such heavy precipitation events might also result in floods that may inundate water treatment and sewage systems and lead to elevated exposure to *Salmonella* sp. and other waterborne pathogens. Salmonellosis has continued to decrease in Europe over the past decade, in part due to control measures. Thus, a contribution of temperature to disease incidence might be modulated by animal husbandry and public health interventions. Hopefully, health promotion and food safety policies should be able to mitigate adverse impacts on public health.

Cryptosporidium

The concentration of *Cryptosporidium* oocysts in river water and drinking water supplies can increase significantly during heavy rainfall events and lead to outbreaks of cryptosporidiosis (Aksoy and Tuncay, 2007; Semenza and Nichols, 2007; Hoek *et al.*, 2008). Conversely, a low level of precipitation preceding a heavy rain event has also been linked with drinking water outbreaks (Nichols *et al.*, 2009). Thus, heavy precipitation can lead to infiltration of *Cryptosporidium* oocysts in drinking water reservoirs from springs and lakes, which then persist in the water treatment and distribution system due to their resistance to chlorine. The projected increase in precipitation in northern Europe due to climate change might result in more cryptosporidiosis outbreaks, but the strength of the relationship varies by climate category (Jagai *et al.*, 2009).

Norovirus

Norovirus epidemics have been associated with climate variability; for example, heavy precipitation and floods may result in wastewater overflow, which can contaminate shellfish harvesting sites. Flood waters in Austria have been linked to a norovirus outbreak (Schmid *et al.*, 2005). The degree of rainfall has also been related to viral contamination of the coastal environment and with peaks in diarrhoea incidence (Miossec *et al.*, 2000). The projected increase of heavy rainfall events in northern Europe under climate change scenarios could result in elevated norovirus infections, since floods can result in norovirus outbreaks.

Vibrio (non-cholera)

At elevated summer temperatures, *Vibrio* sp. replicates in marine waters (Pedersen *et al.*, 1997). *Vibrio vulnificus* infections have been reported during hot summer weather with elevated water temperatures in the Baltic Sea above 20°C (Andersson and Ekdahl, 2006). While such infections can be potentially fatal in immuno-compromised individuals, the current disease incidence is low. Elevated summer (water) temperatures, extended summer seasons and non-cholera *Vibrio* sp. infections could potentially become a public health concern under climate change projections.

In summary

This climate change knowledge base of food- and waterborne diseases is a useful tool to assess potential health impacts. We found climate change affected food- and waterborne diseases in Europe through elevated water and air temperatures and through more heavy rainfall events or severe events such as flooding. Yet, such impacts can be modulated or exacerbated by food safety interventions, the water treatment infrastructure, human behaviour and a range of other drivers. Attributing individual events to climate change is challenging due to data gaps for selected pathogens and climatic variables. Moreover, alterations in weather patterns might indirectly influence consumption habits, recreational activities or adaptation behaviours.

20.3 Risk Assessment: Decision-making Tool for Climate Change and Food- and Waterborne Diseases

Climate change models for Europe project higher minimum and maximum temperatures, heatwaves, excessive precipitation, storm surges and droughts, depending on the biogeographic region discussed above (Fig. 20.1) (IPCC, 2007). Since environmental exposure routes of food- and waterborne pathogens are susceptible to changing climatic conditions, future human exposures, both direct and indirect, might differ significantly from current patterns. Moreover, new types of pathogens might emerge from a new ecological niche, taking advantage of new exposure pathways.

These changes pose considerable challenges to the existing public health infrastructure. Amid competing demands, allocating scarce resources to pressing needs is the predicament of public health. In order to alleviate this difficult decision-making process, the ECDC has developed a decision support tool to estimate infection risks from climate change (Schijven *et al.*, 2013). Through quantitative microbial risk assessment of food- and waterborne diseases, this tool can assist decision makers to prioritize different adaptation options. Quantitative microbiological risk assessment has traditionally been used to estimate the health impacts from exposure to pathogens (Haas *et al.*, 1999; Vose, 2000). The interactive computational tool presented here computes food and water disease risk from climate change, under current and future climate conditions.

A total of 22 modules break down each step of the fate and behaviour of the selected microorganism during its transmission from environmental sources to humans. The exposure pathways include drinking water, bathing water, oysters, egg (products), or chicken fillets. A total of 13 pathogen–pathway combinations were created with the selected pathogens and exposure pathways. The individual quantitative microbial risk assessment modules were programmed and presented in an interactive computational tool with Mathematica 7, which is free software under Mathematica Player 7 (Wolfram Inc).

The tool is built for use by default values from the literature (see above) or under user settings with local data (Fig. 20.1). Thus, these 13 combinations can also be fed with location-specific current climate conditions, projected climate conditions and specific data depending on the selected modules, in order to estimate the direction (increase or decrease) and size of the relative infection risks for the selected pathogens caused by climate change. The final output of the decision support tool is an estimate of the relative infection risk between current and future climate conditions (−CC) (+CC), respectively, for a particular pathogen–pathway combination. For example, the tool can help quantify

the risk of norovirus infection from drinking water under climate change scenarios (Fig. 20.3). The tool models a combined sewer overflow, temperature-dependent inactivation or die-off in surface water, the effectiveness of the water treatment plan and the volume of unboiled drinking water consumed per person per day. The dose–response model computes the risk for current climatic conditions and climate change conditions such as a 2°C increase in temperature, 20% increase in precipitation and a doubling of days with heavy precipitation per year (Fig. 20.3). The relative risk is computed compared to current climatic conditions. The tool can be utilized to estimate the disease burden from food- and waterborne diseases for a number of pathogen–pathway combinations. Computing disability-adjusted life years (DALYs) can then support prioritization of different adaptation options, by focusing on exposure pathways rather than pathogens. Thus, this tool is a valuable resource to assess risk from climate change to infectious diseases for different geographic regions.

It can help both to quantify the anticipated impact, such as the possible burden of additional cases of infectious diseases, and to assess the significance or the relative importance of such an event. The tool can also estimate the degree to which adaptation options would likely reduce vulnerability to the anticipated health impact. It can be used to evaluate the effectiveness of the intervention and whether the intervention can be implemented realistically in the context of current and planned programmes and activities. For example, a number of adaptation interventions can be considered to reduce waterborne disease: upgrading water treatment and distribution systems; strengthening drinking water regulations (and compliance with these regulations) to control the introduction and spread of waterborne diseases; beach closures; boil-water notices; and surveillance programmes for waterborne disease outbreaks. Similarly for foodborne diseases, a number of adaptation options could be considered: stricter regulations (and compliance to these regulations) of food production, processing, transport and storage in order to control the introduction and spread of diseases; surveillance programmes for food-borne disease outbreaks; or education programmes on appropriate food handling.

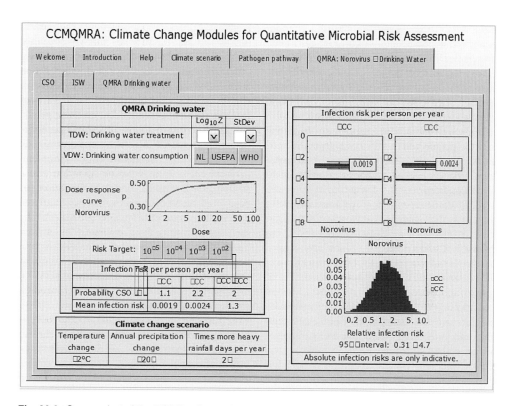

Fig. 20.3. Screen shot of the ECDC software for the quantitative microbial risk assessment tool for food- and waterborne diseases and climate change.

In summary

The decision-making tool for public health practitioners has been developed to help assess the threats of climate change-associated food- and waterborne diseases in European Union (EU) member states. It is designed to assist decision makers in risk analysis in order to prioritize different intervention options and to allocate funds accordingly. A conceptual framework was developed for diverse and complex transmission routes for food- and waterborne diseases under the influence of climate factors. Food- and waterborne diseases are caused by pathogens that are transported through food, water and soil. The characteristics, fate and behaviour of these pathogens are described for selected pathogens. These environmental, climate and pathogen factors are the building blocks of the mathematical model of the decision-making tool.

20.4 Forecasting: European Environment and Epidemiology Network (The 'E3 Network')

Environmental factors are significant drivers of infectious diseases. Land-use patterns, such as intensive agriculture, deforestation, irrigation or road construction, are important determinants of infectious disease spread (Suk and Semenza, 2011). Urbanization and urban sprawl have encroached on agricultural and semi-natural areas in Europe, and the trend is expected to continue. One consequence is that, increasingly, wildlife may need to find new habitats, sometimes in urban or abandoned environments which affect exposure to infectious pathogens. Climatic conditions are also significant drivers of infectious diseases, and climate change can shift the distribution of infectious diseases (Semenza and Menne, 2009). Many of these environmental drivers can be considered as epidemic precursors of disease, and monitoring changes in environmental conditions can help anticipate or even forecast an upsurge of disease (Mills *et al.*, 2010). On an ecological level, the utility of predictive models has been documented for a number of case studies: deer mice density and infection with hantavirus (Glass *et al.*, 2002); rainfall and temperature to identify high-risk areas of plague and hantavirus pulmonary syndrome (Eisen *et al.*, 2007); sea surface temperature, elevated rainfall and vegetation index to predict outbreaks of Rift Valley fever (Anyamba *et al.*, 2009); expansion of schistosomiasis in China based on climate change scenarios (Zhou *et al.*, 2008); effect of climate change on Lyme disease risk and forecast of the emergence of a tick-borne infectious disease (Brownstein *et al.*, 2005); and shifts in patterns of transmission of plague and tularaemia due to climate change (Nakazawa *et al.*, 2007). Merging environmental, satellite, demographic and other data sets and integrating their data requirements would permit the analysis of complex interactions and generate essential information that

increases our understanding of these systems (ECDC, 2008). Identifying long-term projections would build the evidence base for strategic public health action, while identifying short-term events linked to environmental conditions would improve and accelerate early warning and response capability.

With a vision toward 21st century surveillance, the ECDC has embarked on the construction of the European Environment and Epidemiology (E3) Network (Semenza and Menne, 2009) (Fig. 20.4). A data repository has been built and populated with environmental data including the EDEN project, a research consortium funded by the European Commission. These data are available through an ECDC geoportal and include environmental data such as climate change data sets, land cover information, meteorological data, vegetation, hydrology, soil data and wind speed (Semenza *et al.*, 2013). Also included are socio-economic data such as population, economic, education levels, health care, hospitals, transport networks and statistics, migrant populations, demographic profiles,

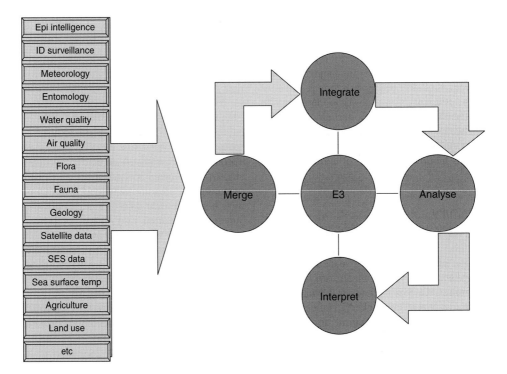

Fig. 20.4. European Environment and Epidemiology Network.

agriculture and livestock. The E3 Network serves as a data repository and provides access to climatic/environmental and infectious disease data to member states, academia, legislators and other authorized users.

These data have now been used to model the risk of tick-borne encephalitis in southern Sweden with cumulative spring temperatures, and the environmental suitability of malaria transmission in Greece (Sudre *et al.*, 2013). The long-term goal of the E3 Network is to connect environmental data with epidemic intelligence and infectious disease surveillance data (currently housed at the ECDC) for early warning, prediction and forecasts. For example, heavy precipitation may be linked to *Cryptosporidium* outbreaks or coastal water contamination with Enterococcus, or Coliform may be linked to increased risk from recreational water use. Alternatively, a particularly warm winter may sustain vector populations, which would warrant vector abatement measures in the spring.

In summary

The E3 Network will monitor the environmental precursors of epidemics to increase our understanding of these complex relationships and drive public health action. Identifying short-term environmental events can help improve early warning systems, while identifying long-term trends can inform strategic public health action. Linking such data strengthens public health capacity by enabling the coordination between environmental and public health agencies.

20.5 Vulnerability Assessment: ECDC Handbook

Reducing the current (see Sections 20.2 and 20.3 above) and projected (see Section 20.4 above) health risks attributable to climate change is an important public health challenge. As climate change is only one of many public health issues that need to be addressed, policies and measures should ensure that actions to reduce climate-related health risks support current programmes to address health burdens and explicitly consider key uncertainties. The ECDC has developed a handbook as an aid for EU member states to assess and manage changes in the risk of infectious diseases posed by climate change (ECDC, 2010). The handbook draws on current scientific knowledge, as well as experiences and best practices from previous national risk, vulnerability and adaptation assessments (Ebi *et al.*, 2013). The aim is to provide suggestions, tools and hands-on approaches on how to access data and choose organizational structure and analysis methods for a national assessment based on local conditions, competence and aims. With these data, local vulnerability can be assessed (Semenza *et al.*, 2012c; Fig. 20.1). Vulnerability is the susceptibility to harm defined in terms of a population or a location. The Intergovernmental Panel on Climate Change defines vulnerability to climate change as the degree to which a system is susceptible to, or unable to cope with, the adverse effects of climate variability and change.

The ECDC handbook can help identify the nature and extent of climate change impacts on infectious disease risk (ECDC, 2010). This risk is a function of regional vulnerability, which in turn is a function of local conditions, their resilience to change and capacity to adapt (Fig. 20.1). Thus, a baseline evaluation of the current situation needs to be performed first in order to predict what may occur in the future in different regions of each country. Variables to consider depend on the diseases or the location being assessed. Generally, the following factors are of interest (in addition to current climatic conditions): demographic and socio-economic factors; the health system (including the healthcare sector, public health infrastructure and programmes); epidemiologic factors; land cover and land use; and sectors other than the health sector. Of particular concern are vulnerable populations such as the elderly, low-income households, populations in remote areas with limited access to medical care, children, pregnant women and mentally and physically disabled individuals who are living alone, which increases the risk of not seeking medical attention when needed.

Societal vulnerability to climate change requires a transdisciplinary approach to public health. Lack of cross-sectoral collaboration is an indication of potential institutional vulnerability to climate change events due to the interconnected nature of modern society. Sectors that need to be involved include water management for water sources and infrastructure; the food industry for agriculture issues including animal keeping, veterinary medicine for domestic animals and wildlife; nature conservation and management for forestry issues; and housing and planning for transport and energy issues.

Indicators should also include the incidence and geographic range of climate-sensitive infectious diseases (see Section 20.2 above), relevant environmental variables (i.e. changes in temperature, precipitation and other weather variables) and land-use change, as well as possible confounding variables such as demographic change, status of the public health infrastructure and economic development.

In summary

The reason for assessing climate change vulnerability is to identify current susceptibilities in the public health infrastructure and evaluate the risks associated with climate-related hazards (Semenza et al., 2012c). Climate change is a risk multiplier which exacerbates existing vulnerabilities, and the ECDC handbook aims to counter this predicament by mounting a proactive response with a number of strategic operations described below.

20.6   Adaptation Assessment: ECDC Handbook

Adaptation is the term used by the climate change community to describe the process by which strategies and measures to moderate, cope with and take advantage of the consequences of climatic events are enhanced, developed, implemented and monitored. Based on the adaptation process described in Fig. 20.2, effective adaptation includes both the present and future health risks from climate change (ECDC, 2010). Constituting the basis of such an adaptation assessment are the current incidence and prevalence of infectious diseases, the sources of vulnerability for these risks and the interventions to address them. Thus, adaptation assessments build on the process described in

Section 20.5. In public health, prevention is the term analogous to adaptation. Adaptive capacity describes the general ability of individuals, communities and institutions to prepare effectively for and cope with the consequences of climate variability and change (Semenza et al., 2011).

First, different adaptation options need to be identified in collaboration with stakeholders from a range of sectors (ECDC, 2010). Often, existing programmes and activities are not equipped adequately to handle extreme weather events and adjustments are needed. Consequently, human and financial resources are required to implement changes in a timely fashion. For example, food-borne disease outbreaks might be linked to changing temperature or precipitation patterns. In such a scenario, regulations and compliance with those regulations are warranted, controlling the spread of diseases during food production, processing, transport and storage. This situation might also call for enhanced surveillance programmes for food-borne disease outbreaks or educational programmes on appropriate food handling. Conversely, waterborne disease outbreaks under a changing climate might be an issue; then regulations (and compliance with those regulations) are needed to control the introduction and spread of waterborne diseases in public drinking water sources or surveillance programmes for waterborne disease outbreaks. The use of early warning systems for beach closures and other sources of waterborne disease outbreaks is another adaptation strategy. Vector-borne disease distribution might shift under a changing climate. This would justify employing, for example, vector abatement strategies; educational campaigns to increase awareness of the risks of a particular vector-borne disease and the actions needed to protect individuals from exposure to vectors; vaccination programmes (e.g. tick-borne encephalitis); and surveillance programmes.

These are a range of adaptation options that should prioritize which issues to tackle first, based on a cost–benefit analysis (ECDC, 2010). The benefits of the interventions should exceed their cost. The metrics for measuring gain should include the benefits and effectiveness of reducing vulnerability, the human and financial resources required, feasibility of and constraints to implementation. It is also important to ensure active and continued stakeholder participation, to consider the changes in climate and vulnerability over time and to incorporate uncertainties in climate projections, development pathways and social justice. These adaptation options should be prioritized and integrated in public health strategies and policies at the national, regional and community levels to manage the risks to health from climate change.

In summary

Adaptation assessment can help to evaluate and manage changes in the risk of infectious disease transmission posed by climate change. The ECDC handbook can provide guidance, tools and hands-on approaches on how to identify data and structure and manage assessments. It is important to engage different sectors besides the health sector, since climate change rarely acts in isolation.

20.7 Conclusion

Climate change adaptation to infectious disease risks is a complex process, as illustrated in Fig. 20.2. Public health practitioners are obliged to address credible risks – even if that requires acting in the absence of conclusive evidence. By examining the existing evidence and conducting quantitative risk assessments and forecasts, vulnerability, impact and adaptation assessment can prepare society for impending changes to the shifting infectious disease burden from climate change. Climate change adaptation to infectious disease can strengthen the public health infrastructure and provide co-benefits for society at large.

Acknowledgements

I would like to acknowledge my collaborators, without whom none of this work would have been possible; in particular, Jonathan Suk, Bertrand Sudre, Tolu Oni, Massimiliano Rossi and Virginia Estevez at the ECDC. The work on the knowledge base was performed with Christoph Höser, Susanne Herbst, Andrea Rechenburg and Thomas Kistemann. The decision tool was developed with Jack Schijven, Ana Maria Roda Husman and Marjin Bouwknegt and the adaptation handbook with Elisabet Lindgren and Kris Ebi. I would also like to thank all the representatives of the Member States, who have provided invaluable feedback on our work.

References

Aksoy, U. and Tuncay, S. (2007) Short communication: investigation of intestinal coccidia in patients with diarrhea. *Mikrobiyoloji Bülteni* 41, 127–131.

Andersson, Y. and Ekdahl, K. (2006) Wound infections due to *Vibrio cholerae* in Sweden after swimming in the Baltic Sea, summer 2006. *Eurosurveillance* 11, E060803 2.

Anyamba, A., Chretien, J.P., Small, J., Tucker, C.J., Formenty, P.B., Richardson, J.H., *et al.* (2009) Prediction of a Rift Valley fever outbreak. *Proceedings of the National Academy of Sciences* 106, 955–959.

Bi, P., Cameron, A.S., Zhang, Y. and Parton, K.A. (2008) Weather and notified Campylobacter infections in temperate and sub-tropical regions of Australia: an ecological study. *Journal of Infection* 57, 317–323.

Boxall, A.B., Hardy, A., Beulke, S., Boucard, T., Burgin, L., Falloon, P.D., *et al.* (2009) Impacts of climate change on indirect human exposure to pathogens and chemicals from agriculture. *Environmental Health Perspectives* 117, 508–514.

Brownstein, J.S., Holford, T.R. and Fish, D. (2005) Effect of climate change on Lyme disease risk in North America. *Ecohealth* 2, 38–46.

Christensen, J.H., Hewitson, B., Busuioc, A., Chen, A., Gao, X., Held, I., *et al.* (2007) *Regional Climate Projections*. Cambridge University Press, Cambridge, UK, and New York.

Craig, D.L., Fallowfield, H.J. and Cromar, N.J. (2003) Effectiveness of guideline faecal indicator organism values in estimation of exposure risk at recreational coastal sites. *Water Science and Technology* 47, 191–198.

Ebi, K.L., Lindgren, J.E., Suk, J.E. and Semenza, J.C. (2013) Adaptation to the infectious disease impacts of climate change. *Climatic Change* 118, 355–365.

ECDC (European Centre for Disease Prevention and Control) (2008) Workshop on linking environmental and infectious diseases data (http://www.ecdc.europa.eu/en/publications/Publications/0805_MER_Linking_environmental_and_infectious_diseases_data.pdf, accessed 21 March 2014).

ECDC (European Centre for Disease Prevention and Control) (2010) Handbook for national vulnerability, impact and adaptation assessments. ECDC, Stockholm (http://ecdc.europa.eu/en/publications/Publications/1003_TED_handbook_climatechange.pdf, accessed 21 March 2014).

EEA (European Environment Agency) (2008) Impacts of Europe's changing climate – 2008 indicator-based assessment. Joint EEA-JRC-WHO Report. EEA-JRC-WHO, Copenhagen.

Eisen, R.J., Glass, G.E., Eisen, L., Cheek, J., Enscore, R.E., Ettestad, P., *et al.* (2007) A spatial model of shared risk for plague and hantavirus pulmonary syndrome in the southwestern United States. *American Journal of Tropical Medicine and Hygiene* 77, 999–1004.

Fleury, M., Charron, D.F., Holt, J.D., Allen, O.B. and Maarouf, A.R. (2006) A time series analysis of the relationship of ambient temperature and common bacterial enteric infections in two Canadian provinces. *International Journal of Biometeorology* 50, 385–391.

Giorgi, F., Bi, X.Q. and Pal, J. (2004) Mean, interannual variability and trends in a regional climate change experiment over Europe. II: climate change scenarios (2071–2100). *Climate Dynamic* 23, 839–858.

Glass, G.E., Yates, T.L., Fine, J.B., Shields, T.M., Kendall, J.B., Hope, A.G., *et al.* (2002) Satellite imagery characterizes local animal reservoir populations of Sin Nombre virus in the southwestern United States. *Proceedings of the National Academy of Sciences* 99, 16817–16822.

Haas, C.N., Rose, J.B. and Gerba, C.P. (1999) *Quantitative Microbiological Risk Assessment.* Wiley and Sons, Inc, New York.

Hearnden, M., Skelly, C., Eyles, R. and Weinstein, P. (2003) The regionality of campylobacteriosis seasonality in New Zealand. *International Journal of Environmental Health Research* 13, 337–348.

Hoek, M.R., Oliver, I., Barlow, M., Heard, L., Chalmers, R. and Paynter, S. (2008) Outbreak of *Cryptosporidium parvum* among children after a school excursion to an adventure farm, south west England. *Journal of Water and Health* 6, 333–338.

IPCC (Intergovernmental Panel on Climate Change) (2007) Fourth Assessment Report; Climate Change 2007; Synthesis Report. Cambridge University Press, Cambridge, UK.

Jagai, J.S., Castronovo, D.A., Monchak, J. and Naumova, E.N. (2009) Seasonality of cryptosporidiosis: a meta-analysis approach. *Environmental Research* 109, 465–478.

Kovats, R.S., Edwards, S.J., Hajat, S., Armstrong, B.G., Ebi, K.L. and Menne, B. (2004) The effect of temperature on food poisoning: a time-series analysis of salmonellosis in ten European countries. *Epidemiology and Infection* 132, 443–453.

Kovats, R.S., Edwards, S.J., Charron, D., Cowden, J., D'Souza, R.M., Ebi, K.L., *et al.* (2005) Climate variability and campylobacter infection: an international study. *International Journal of Biometeorology* 49, 207–214.

Lindgren, E., Andersson, Y., Suk, J.E., Sudre, B. and Semenza, J.C. (2012) Monitoring EU emerging infectious disease risk due to climate change. *Science* 336, 418–419.

Martinez-Urtaza, J., Saco, M., de Novoa, J., Perez-Pineiro, P., Peiteado, J., Lozano-Leon, A., *et al.* (2004) Influence of environmental factors and human activity on the presence of Salmonella serovars in a marine environment. *Applied and Environmental Microbiology* 70, 2089–2097.

Mills, J.N., Gage, K.L. and Khan, A.S. (2010) Potential influence of climate change on vector-borne and zoonotic diseases: a review and proposed research plan. *Environmental Health Perspectives* 118, 1507–1514.

Miossec, L., Le Guyader, F., Haugarreau, L. and Pommepuy, M. (2000) Magnitude of rainfall on viral contamination of the marine environment during gastroenteritis epidemics in human coastal population. *Revue d'Epidémiologie et de Santé Publique* 48, 2S62–2S71.

Nakazawa, Y., Williams, R., Peterson, A.T., Mead, P., Staples, E. and Gage, K.L. (2007) Climate change effects on plague and tularemia in the United States. *Vector-Borne and Zoonotic Diseases* 7, 529–540.

Naumova, E.N., Jagai, J.S., Matyas, B., DeMaria, A. Jr, MacNeill, I.B. and Griffiths, J.K. (2007) Seasonality in six enterically transmitted diseases and ambient temperature. *Epidemiology and Infection* 135, 281–292.

Nichols, G., Lane, C., Asgari, N., Verlander, N.Q. and Charlett, A. (2009) Rainfall and outbreaks of drinking water related disease and in England and Wales. *Journal of Water and Health* 7, 1–8.

Palmer, S.R., Gully, P.R., White, J.M., Pearson, A.D., Suckling, W.G., Jones, D.M., *et al.* (1983) Water-borne outbreak of campylobacter gastroenteritis. *The Lancet* 1, 287–290.

Patrick, M.E., Christiansen, L.E., Waino, M., Ethelberg, S., Madsen, H. and Wegener, H.C. (2004) Effects of climate on incidence of *Campylobacter* spp. in humans and prevalence in broiler flocks in Denmark. *Applied and Environmental Microbiology* 70, 7474–7480.

Pebody, R.G., Ryan, M.J. and Wall, P.G. (1997) Outbreaks of campylobacter infection: rare events for a common pathogen. *Communicable Disease Report – CDR Review* 7, R33–37.

Pedersen, K., Dalsgaard, I. and Larsen, J.L. (1997) *Vibrio damsela* associated with diseased fish in Denmark. *Applied and Environmental Microbiology* 63, 3711–3715.

Savill, M.G., Hudson, J.A., Ball, A., Klena, J.D., Scholes, P., Whyte, R.J., *et al.* (2001) Enumeration of Campylobacter in New Zealand recreational and drinking waters. *Journal of Applied Microbiology* 91, 38–46.

Schijven, J., de Roda Husman, A.M., Bouwknegt, M., Sudre, B., Suk, J.E., Rutjes, S., *et al.* (2013) A decision support tool to compare waterborne and foodborne infection and/or illness risks associated with climate change. *Risk Analysis* 33, 2154–2167.

Schmid, D., Lederer, I., Much, P., Pichler, A.M. and Allerberger, F. (2005) Outbreak of norovirus infection associated with contaminated flood water, Salzburg, 2005. *Eurosurveillance* 10, E050616 3.

Semenza, J.C. and Menne, B. (2009) Climate change and infectious diseases in Europe. *Lancet Infectious Diseases* 9, 365–375.

Semenza, J.C. and Nichols, G. (2007) Cryptosporidiosis surveillance and water-borne outbreaks in Europe. *Eurosurveillance* 12, E13–14.

Semenza, J.C., Ploubidis, G.B. and George, L.A. (2011) Climate change and climate variability: personal motivation for adaptation and mitigation. *Environmental Health* 10, 46.

Semenza, J., Herbst, S., Rechenburg, A., Suk, J., Höser, C., Schreiber, C., *et al.* (2012a) Climate change impact assessment of food and waterborne diseases. *Critical Reviews in Environmental Science and Technology* 42, 857–890.

Semenza, J.C., Höser, C., Herbst, S., Rechenburg, A., Suk, J.E., Frechen, T., *et al.* (2012b) Knowledge mapping for climate change and food and waterborne diseases. *Critical Reviews in Environmental Science and Technology* 42, 378–411.

Semenza, J.C., Suk, J.E., Estevez, V., Ebi, K.L. and Lindgren, E. (2012c) Mapping climate change vulnerabilities to infectious diseases in europe. *Environmental Health Perspectives* 120, 385–392.

Semenza, J.C., Sudre, B., Oni, T., Suk, J.E. and Giesecke, J. (2013) Linking environmental drivers to infectious diseases: the European environment and epidemiology network. *PloS Neglected Tropical Diseases* 7, e2323, doi:10.1371/journal.pntd.0002323.

Sudre, B., Rossi, M., van Bortel, W., Danis, K., Baka, A., Vakalis, N., *et al.* (2013) Mapping environmental suitability of malaria transmission in Greece. *Emerging Infectious Diseases* 19, doi:10.3201/eid1905.120811.

Suk, J.E. and Semenza, J.C. (2011) Future infectious disease threats to Europe. *American Journal of Public Health* 101, 2068–2079.

Vose, D. (2000) *Risk Analysis: A Quantitative Guide*. John Wiley and Sons, West Sussex, UK.

World Health Organization (2008) *Protecting Health in Europe from Climate Change*. World Health Organization, Geneva.

Zhou, X.N., Yang, G.J., Yang, K., Wang, X.H., Hong, Q.B., Sun, L.P., *et al.* (2008) Potential impact of climate change on schistosomiasis transmission in China. *American Journal of Tropical Medicine and Hygiene* 78, 188–194.

21 Climate Change and Health in the Arctic

Alan J. Parkinson[1] and Birgitta Evengård[2]
[1]*Arctic Investigations Program, Centers for Disease Control and Prevention, Anchorage, Alaska, USA;* [2]*Arctic Research Center (ARCUM), Division of Infectious Diseases, Umeå University, Sweden*

21.1 Background

The Arctic can be defined generally as the region that extends above 60°N latitude and borders the Arctic Ocean, including all or northern parts of eight nations: the USA, Canada, Greenland, Iceland, Norway, Finland, Sweden and the Russian Federation (RF). The climate in the Arctic varies geographically from severe cold in arid, uninhabited regions, to temperate forests bordering coastal agrarian regions. Some lands are completely located within this region, such as Iceland, Greenland and the Faroe Islands. Russia, Canada, the USA, Norway, Sweden and Finland have small proportions of their overall populations residing within their respective Arctic areas (Einarsson *et al.*, 2004). For practical reasons, it is often more convenient to define the Arctic by jurisdictional or administrative boundaries and the availability of demographic and other data (see Plate 7). Using these boundaries, we can determine that the Arctic is home to 4 million people, of whom almost half reside in the northern part of the RF. People in the Arctic live in social and physical environments that differ from their more southern-dwelling counterparts. Approximately 400,000 (10%) of persons are of indigenous ancestry, half of whom live in the northern part of the RF.

Arctic 'indigenous peoples' occupy the northern reaches of Eurasia, from the Saami in Scandinavia to the Chukchi at the easternmost tip of Russia, and various groups in between. The Inuit today live under four national flags: Chukotka (RF), Alaska (USA), northern Canada and Greenland (Denmark). Some countries define and recognize specific indigenous groups. In Alaska, the collective term 'Alaska Natives' is used, which includes Inuit, Eskimo, Aleut and several Indian groups. Canada identifies First Nations, Inuit and Metis as Aboriginal. In the RF, they use the term 'numerically small peoples of the North, Siberia and Far East', where some 38 groups exist and where 12 different languages are spoken.

The indigenous populations of northern Canada, Alaska, Greenland and the northern RF generally reside in remote, isolated communities consisting of 150 to several thousand inhabitants. In some regions, the only access to communities is by small aircraft or boat in summer and by small aircraft and snow machine in winter. Arctic communities, once isolated, are now very much a part of the global village in which we all live and are as vulnerable to health threats as any other community on the globe. Through their unique relationship with nature, many of these peoples are more vulnerable to health threats generated by climate change.

These communities often have little economic infrastructure and are still largely dependent on subsistence harvesting of wildlife resources from terrestrial, freshwater and marine ecosystems for a significant proportion of their diet. Food security is often dependent on subsistence wild-life migration patterns, predictable weather and some method of food storage. In these remote regions, access to public health and acute-care systems is often marginal and poorly supported.

The health of the indigenous peoples of the circumpolar region has improved over the past 50 years or so. Much of this improvement can be attributed not only to the implementation of prevention and treatment activities that have resulted in reductions in morbidity and mor-tality from infectious diseases, such as tubercu-losis, and the vaccine-preventable diseases of childhood, but also to the provision of safe water supplies and sewage disposal in many communities.

However, life expectancy of the indigenous populations of Alaska, northern Canada, Greenland and the northern RF is lower than that of the respective national populations. Infant mortality remains higher than respective popula-tions of the USA, northern Canada, Greenland and northern RF (Young and Bjerregaard, 2008).

Mortality rates for heart disease and cancer were once lower among the indigenous popula-tions of the USA, Canada and northern European countries, but are now similar to their national rates. According to the US Centers for Disease Control and Prevention (CDC), the cancer rate is 1.5 times higher for Alaska Natives than for the general US population.

However, it is evident that the indigenous populations of Alaska, Canada and Greenland have much higher rates of unintentional injury and suicide. Unintentional injuries have always been a fact of life for those who live close to the land. These have often been related to hunting (animal attacks, shootings and boating acci-dents) and hypothermia. With modernization, motor vehicle accidents and house fires have recently assumed more importance. Many acci-dents are now alcohol related, and misuse of alcohol is a major determinant of ill health and social problems among indigenous communities of the north. Suicides were uncommon before the 1950s, though they did occur among the elderly and infirm. This contrasts sharply with the pattern of adolescent suicide (peak rates 15–24 years), which often occurs in clusters and more often in males. This is an ongoing and difficult behavioural health problem in many communities across the circumpolar north.

Other health concerns of indigenous popu-lations of the north include the high prevalence of certain infectious diseases such as hepatitis B, tuberculosis, *Helicobacter pylori*, respiratory syncytial virus (RSV) in infants, influenza and sexually transmitted infections, as well as the potential health impacts associated with exposure to environmental pollutants through the trad-itional food supply, the health impacts of rapid economic change and modernization and now the health impacts of climate change.

The Arctic has warmed substantially more than the rest of the world over the past century, principally in recent decades. Models indicate that the global temperature will increase between 1.4 and 5.5°C by 2100, compared to the average tem-perature between 1850 and 1900 (IPCC, 2014).

Arctic climate models project continued warming, with a 3–7°C increase by 2100. The greatest warming will occur in the winter months. The mean annual precipitation will also increase, and continued melting of land and sea ice is expected to increase river discharge and con-tribute to a 1 m sea level rise by 2100, which will impact many low-lying coastal communities greatly, not only in the Arctic but also worldwide (IPCC, 2014).

Climate change is already impacting indi-genous communities of the Arctic. Some of the expected health-related impacts of climate change in the Arctic are described next.

21.2 Primary Health Effects

The primary health impacts of climate change are those that result directly from changes in ambient temperature, extreme weather events and ultraviolet B (UV-B) radiation (Berner *et al.*, 2005). Increased precipitation is expected to result in an increase in the magnitude, fre-quency and unpredictability of extreme weather conditions and other natural disasters (i.e. flood-ing), which create risks for injury in communi-ties during evacuation attempts as well as for air, overland, waterway and coastal travel.

Indigenous people have reported that the weather is less predictable than in recent memory (Huntington *et al.*, 2005; Furberg *et al.*, 2011). As ambient temperature increases, the incidence of hypothermia-associated morbidity and mortality may decrease. However, it is also predicted that some areas of the Arctic will experience colder ambient temperatures, and cold exposure has been shown to increase the frequency of certain injuries such as frostbite accidents and respiratory, circulatory and cardiovascular diseases (Hassi *et al.*, 2005; Messner, 2005). Conversely, hyperthermia may increase, particularly among the very young and the elderly (Nayha, 2005). However, because of the low mean temperature in many Arctic regions, the likelihood of such events having a large public health impact is low. Nevertheless, some Arctic residents are reporting respiratory distress associated with extreme warm summer days; an unprecedented experience (Furgal *et al.*, 2002), and it is well known that exposure to extreme high temperature can cause heatstroke, dehydration, heat exhaustion and respiratory problems (Epstein and Ferber, 2011). Increased cardiovascular and non-accidental mortalities can increase during both heatwave and cold spells even in Arctic regions (Revich and Shaposhnikov, 2010). More significantly, unintentional injuries related mostly to travel, substance hunting and fishing activities are already a significant cause of morbidity and mortality among Arctic residents, and these may increase (Berner *et al.*, 2005; Curtis *et al.*, 2005; Ford *et al.*, 2008; Fleischer *et al.*, 2013). The reduction in sea ice thickness, shorter ice season, reduced snow cover and permafrost thawing will make hunting and gathering more difficult, dangerous and less successful, thereby increasing the risk of injury and death by drowning. Increasing exposure to UV-B radiation among Arctic residents has the potential to affect the response of the immune system to disease and to influence the development of skin cancer and non-Hodgkin's lymphoma, as well as the development of cataracts. However, as the incidence rates of these conditions are already low in small Arctic communities, it is difficult to detect or predict any trends in their future incidence (Berner *et al.*, 2005; De Fabo, 2005).

21.3   Secondary Effects

Secondary health effects of climate change arise as a result of environmental or human behavioural change that is affected by climate. These include social and mental stress related to changes in environment or lifestyle brought about by changes in climate. Examples include: village/community relocation; decreased access to subsistence foods and impacts on diet and nutrition; restricted access to adequate and safe water supply; increase in infectious diseases related to failure of water and sanitation infrastructure; changes in human incidence of zoonotic infectious diseases; and increase in respiratory problems related to poor air quality (dust, smoke, pollens).

Impacts on seasons

The meaning of seasons in northern communities is different than in the western context (Brubaker *et al.*, 2010). Although tied to weather and natural landscape, it is also tied to the timing of subsistence activities and cultural events, both of which have a significant impact on community health and well-being (e.g. whaling, seal hunting, waterfowl hunting, fishing, caribou hunting). Small changes in weather can have a large impact on many of these traditional activities, which are essential for a sustainable food supply and good nutrition. Observed changes include: later freeze-up, early thaw and change in the timing of wildlife migrations. Health impacts include dangerous travel conditions and changes in wildlife availability.

Impacts on sea and coastline

There is already a considerable impact on the sea and coastline in many regions of the Arctic. The average extent of sea ice in summer has declined by 15–20% over the past 30 years. The sea ice is getting thinner. There has been a significant loss of multi-year ice over the past decade (http://nsidc.org/arcticseaicenews/). Ice is important for hunting and fishing. It provides easy access to open water and a platform to hunt from; not only for indigenous people but also for the prey they hunt. Thin ice means fewer seals, walrus and whales to hunt, impacting diet, nutrition and cultural well-being (Brubaker *et al.*, 2011). Thin ice also may increase unintentional injury and death by drowning to those travelling the surface to hunting grounds (Fleischer *et al.*, 2013).

Reduction in sea ice will have widespread effects on the marine ecosystem, coastal climate, human settlements and subsistence activities.

The reduction in sea ice has for the first time created ice-free shipping lanes, to the north-west, from Labrador to the Bering Sea, and to the north-east, from the Bering Sea to Norway, representing fuel-saving shortcuts for transportation by sea and access to oil, gas and mineral reserves once inaccessible (Arctic Council, 2009). This traffic is projected to increase rapidly. Benefits to isolated communities will include the construction of military bases and the development of other industrial and commercial ventures such as tourism, which will result in infrastructure support and employment. Public and private services will increase to support emerging economies. However, these ventures will affect population distribution, dynamics, culture and local environments, and will challenge the traditional subsistence way of life for many communities and lead to accelerated and long-term cultural change, which will create additional stress on an already vulnerable population.

Delayed freeze-up, lack of the sea ice barrier and increased storm intensity has accelerated coastal erosion and damage to water and sanitation systems, forcing some communities in north-western Alaska to evacuate during storm events and to consider eventual permanent relocation (Brubaker *et al.*, 2010). Such storm events also place residents at higher risk for unintentional injury and for chronic stress. Fear for safety and security will have long-term effects on mental and behavioural health in these communities. The movement of rural residents to urban centres is occurring in some regions of the circumpolar north (Einarsson *et al.*, 2004; Driscoll *et al.*, 2010). This is currently being driven by economic, educational and access to health-care opportunities. However, this trend may accelerate, due to the impact of climate change.

Impacts on the river and lake systems

For many communities, the river is a byway that connects a community to inland-based subsistence resources such as caribou, salmon trout, waterfowl and wild berries.

With warming has come widespread thawing of shallow permafrost, resulting in the collapse of tundra into the rivers. This increased erosion changes river flow, increases turbidity and restricts access to upriver hunting grounds. Rivers are becoming wider, shallower, warmer and dirtier, affecting navigation, critical fish habitat and water quality and quantity when used as a community water source (Brubaker *et al.*, 2011; Evengård *et al.*, 2011).

Lakes are also changing. Some are expanding with water from thawing permafrost, others are shrinking as an underlying ice lens thaws and drains the lake (Alaskool, 2013). Warmer water allows new vegetation to grow; algae, aquatic plants and mosses flourish, creating problems if the lake is used as a community water source.

Impact on water and sanitation infrastructure

In many communities in the north, the built infrastructure is supported on permafrost. Loss of this support will result in damage to water intake systems and pipes, and may result in the contamination of community water supplies and damage to water and sanitation infrastructures and distribution systems, forcing communities to rely more on untreated (or traditional) water sources (Brubaker *et al.*, 2011; Evengård *et al.*, 2011). This may result in an increase in clinic visits and hospitalizations for various 'water washed' infectious diseases, those commonly prevented by hand washing, such as gastroenteritis, respiratory infections caused by RSV, influenza, skin infections, impetigo, boils caused by methicillin-resistant *Staphylococcus aureus* (MRSA) (Hennessy *et al.*, 2008; Wenger *et al.*, 2010; Brubaker *et al.*, 2011).

Impact on the food supply

Traditional food storage methods often include aboveground drying of fish and meat at ambient temperature, belowground cold storage using ice cellars dug into the permafrost, and fermentation (Brubaker *et al.*, 2011). As temperatures rise, traditional methods of food preparation and storage are less likely to prevent food-borne illness and will force indigenous communities to depend increasingly on non-traditional and often less healthy Western foods. This will result in increased rates of modern diseases associated with processed foods, such as obesity, diabetes and cardiovascular diseases (Young and Bjerregaard, 2008). Climate change is already altering the food web structures in some areas, including influence on the exposure of wildlife and humans to persistent organic pollutants (POPs) (Hansen and Van Oostdam, 2009). A warmer climate may result in increased use of pesticides globally as insect populations increase, and together with the accelerated release of POPs from melting sea ice and permafrost, the result might be an increase in bioaccumulation in plants and animals in the Arctic, posing a significant risk to the health of Arctic residents who depend on those plants and animals for food. The interaction between climate change and contaminants is complex. Increasing temperature could result in the increased toxicity and increased mobility of POPs, but also in increased degradation. Climate change could affect food safety, as increasing temperatures may result in alterations in the food web structures, lipid profiles and carbon flux that in turn affect the concentrations of POP water, soil and biota (Rylander et al., 2010). Potential health threats include nervous system and brain development problems, interference with hormones and sexual function, weakened immune systems, organ damage, cardiovascular diseases and cancer (Arctic Monitoring and Assessment Programme, 2003).

Impacts on air quality

Temperature and humidity are important factors in mould production in homes and on improperly dried fish and meat. Other airborne pollutants expected to increase due to a warming Arctic include dust, smoke from wildfires and allergens such as pollens. With an increase in CO_2 in the atmosphere, some trees will produce much more pollen than they do today. This may lead to more allergies among humans and later chronic pulmonary diseases. The increase in precipitation and flooding events will create favourable conditions for mould growth in homes. This indoor exposure may result in an increase in mould-related respiratory disorders and allergies. Exposure to these airborne pollutants increases the risk of respiratory diseases, incites asthma attacks and bronchitis and compromises people with respiratory disease, the elderly and mothers and newborns (Rylander et al., 2010). Warmer temperatures have also resulted in an increase in the ranges and longevity of insects, including those that sting or bite. The Allergy and Immunology Center of Alaska reports a threefold increase in patients suffering from allergies from stinging insects from 1999–2002 to 2003–2007 (Demain et al., 2008).

Impacts on the land, flora and fauna

As warming temperatures move northwards, associated plants and wildlife will follow. Biologic responses to a warming Arctic are expected to outpace those at lower latitudes. Spring will occur earlier and the growing season will be longer. Recent studies in Greenland, for example, show that flowering in six plant species, median emergence of twelve taxa of arthropods and clutch initiation in three species of birds have advanced in some cases by 30 days during the past decade. Most long-term records of phenological events from northern temperate environments show advancements of 2.7 days/decade for plants and 5.1 days/decade for animals and plants globally (Høye et al., 2007). The treeline is projected to reach the Arctic Ocean in most of Asia and western North American by the end of this century. This is likely to lead to a near loss of tundra vegetation in these areas, with important consequences for many types of wildlife (Weller et al., 2005). The dominant response of current Arctic species to climate change, as in the past, will be relocation rather than adaptation. Terrestrial Arctic animals are likely to be the most vulnerable to: high summer temperatures; changes that affect migration routes, staging areas for long-distance migration; climate events that affect snow cover, freeze–thaw cycles; changes that affect the timing of reproduction and development that are linked to seasonal queues or peaks in food source availability; influxes in new competitors, predators, parasites and diseases. Thus, climate change is likely to have a significant impact on key terrestrial species used as subsistence food by shifting the range and abundance of key species such as caribou, moose, waterfowl and seabirds. The health impacts of a decline in the proportion of traditional food consumed by the indigenous population may be significant. A shift away from a traditional diet to a more Western diet, higher in carbohydrates and sugars, is associated with increased levels of cardiovascular disease, diabetes, vitamin-deficiency disorders, dental cavities, anaemia, obesity and lower resistance to infection.

21.4   Impact on Infectious Diseases

It is well known that climate and weather affect the distribution and risk of many vector-borne diseases such as malaria, Rift Valley fever, plague and dengue fever in tropical regions of the globe. Weather also affects the distribution of food- and waterborne diseases and emerging infectious diseases such as West Nile virus, hantavirus and Ebola haemorrhagic fever (Haines *et al.*, 2006). Less is known about the impact of climate change and the risk and distribution of infectious diseases in Arctic regions. It is known that Arctic populations have a long history of both endemic and epidemic infectious diseases (Parkinson and Butler, 2005; Parkinson, 2008). However, with the introduction of antimicrobial drugs, vaccines and public health systems, morbidity and mortality due to infectious diseases have been reduced greatly. The impact of climate on the incidence of these existing infectious disease challenges is unknown. In many Arctic regions, however, inadequate housing and sanitation are already important determinants of infectious disease transmission. The cold northern climate keeps people indoors, amplifying the effects of household crowding, smoking and inadequate ventilation. Crowded living conditions increase person-to-person spread of infectious diseases and favour the transmission of respiratory and gastrointestinal diseases and skin infections. Many homes in communities across the Arctic lack basic sanitation services (e.g. flush toilet, shower or bath, kitchen sink). Providing these services is difficult in remote villages where small, isolated populations live in a harsh, cold climate. A study in western Alaska demonstrated two to four times higher hospitalization rates among children less than 3 years of age for pneumonia, influenza and childhood RSV infections in villages where the majority of homes had no in-house piped water compared with villages where the majority of homes had in-house piped water service. Likewise, outpatient *S. aureus* infections and hospitalization for skin infections among persons of all ages were higher in villages with no in-house piped water service compared to villages without water service (Hennessy *et al.*, 2008; Wenger *et al.*, 2010). Damage to the sanitation infrastructure by melting permafrost or flooding may therefore result in increased rates of hospitalization among children for respiratory infections, as well as an increased rate of skin infections and diarrhoeal diseases caused by bacterial, viral and parasitic pathogens.

Some infectious diseases are unique to the Arctic and lifestyles of the indigenous populations and may increase in a warming Arctic. For example, many Arctic residents depend on subsistence hunting, fishing and gathering for food, and on a predictable climate for food storage. Food storage methods often include aboveground air-drying of fish and meat at ambient temperature, belowground cold storage on or near the permafrost, and fermentation. Changes in climate may prevent the drying of fish or meat, resulting in spoilage. Similarly, loss of the permafrost may result in spoilage of food stored below the ground. Outbreaks of gastroenteritis caused by *Vibrio parahaemolyticus* have been related to the consumption of raw or inadequately cooked shellfish collected from seawater at temperatures of higher than 15°C. Prior to 2004, the most northerly outbreak occurred in northern British Columbia in 1997. However, in July 2004, an outbreak of gastroenteritis caused by *V. parahaemolyticus* was documented among cruise ship passengers consuming raw oysters while visiting an oyster farm in Prince William Sound, Alaska (McLaughlin *et al.*, 2005). The outbreak investigation documented an increase of 0.21°C/year in the July–August water temperature since 1997, and reported that 2004 was the first year that the oyster farm water temperature exceeded 15°C in July. This event provides direct evidence of an association between rising seawater temperature and the onset of illness.

Warmer temperatures may allow infected host animal species to survive winters in larger numbers, increase in population and expand their range of habitation, thus increasing the opportunity to pass infections to humans. For example, milder weather and less snow cover may have contributed to a large outbreak of Puumala virus infection in northern Sweden in 2007. Puumala virus is endemic in bank voles and causes haemorrhagic fever with renal syndrome in humans (Pettersson et al., 2008). Similar outbreaks have been noted in Finland and in the RF (Revich, 2008; Makary et al., 2013). The climate-related northern expansion of the boreal forest in Alaska and northern Canada has favoured the steady northward advance of the beaver, potentially extending the range of giardia, a parasitic infection of beaver that can infect other mammals, including humans, who consume untreated surface water. Elevated runoff from snowmelt and increased precipitation could exacerbate contamination of water supplies with giardia and cryptosporidium cysts and oocysts (Davidson et al., 2011). The association between infection and increased precipitation is well recognized. In a recent outbreak in two towns in northern Sweden, more than 50,000 residents developed cryptosporidium-related gastroenteritis after drinking contaminated municipal water following heavy rainfall that overwhelmed water purification systems (Evengård et al., 2011).

Climate change could exacerbate the potential for the food- and/or waterborne transmission of toxoplasmosis in the Arctic. Toxoplasmosis is caused by infection with *Toxoplasma gondii*, a widespread protozoan parasite of mammals and birds. Members of the cat family are the only known definitive host for the sexual infectious stages (oocysts) of *T. gondii*; however, the asexual encysted stage is found in the muscle tissues of animals and can serve as the main reservoirs of infection in cat-free areas. Humans become infected by ingesting raw or insufficiently cooked meat or foods that have come into contact with infected meat; or indirectly or directly ingesting cysts from soil, such as on items that have come into contact with cat faeces (unwashed vegetables); or transplacentally in humans from mother to her fetus.

A serosurvey conducted among the Inuit of Nunavik showed a seroprevalence of 60%

(Messnier et al., 2009). Because of the absence of a felid host in Nunavik, it is unclear how Toxoplasma infection would be maintained in this region of the Arctic. It would appear to require a non-felid definitive host, possibly rodents or various migratory species (barren-ground caribou, birds, marine mammals), with terrestrial runoff feeding into a marine cycle. It is also possible that it could be maintained by carnivores and/or by vertical transmission (marine mammals, herbivores). The recent discovery of Toxoplasma in polar bears and Arctic foxes in Svalbard underscores the widespread nature of this infection (Elmore et al., 2012). It has been hypothesized that this infection was introduced to this region by migratory birds. The prevalence in polar bears in Svalbard, the Barents Sea region and eastern Greenland areas has doubled in the past decade (now 46%), and detection in ring seals for the first time highlights the predator–prey cycles in this region and the increasing risk to populations that rely on marine mammals for food. In the Arctic, the consumption of undercooked meat from marine mammals seems a much more important risk factor for human infection than drinking water. A recent serosurvey among Inuit in northern Quebec showed that 80% of Inuit with a dietary preference for dried meat from sea mammals were seropositive compared to 10% among ethnic Cree in the same community, who preferred cooked terrestrial mammals (Messnier et al., 2009).

Another important meat-borne parasite in the Arctic is Trichinella, commonly responsible for outbreaks related to the consumption of undercooked bear or walrus meat (Davidson et al., 2011). The most common species is *Trichinella nativa*, which, unlike other Trichinella species (such as Trichinella T6), survives freezing. The geographical distribution of cold-tolerant versus freeze-tolerant *Trichinella* sp. follows the January isothermal lines ($-5°C$ for *T. nativa*). Thus, shifts in host diversity and environmental temperature could lead to altered distribution.

Similarly, warmer temperatures in the Arctic and sub-Arctic regions could support the expansion of the geographical range and populations of foxes and voles, common carriers of *Echinococcus multilocularis*, the cause of alveolar echinococcus in humans (Jenkins et al., 2011). Alveolar echinococcus was common in two regions of north-western Alaska prior to 1997 (Hueffer *et al.*, 2013). Disease in humans was associated with contact with dogs; however, improvements in housing and dog-lot management have largely eliminated dog-to-human transmission in Alaska. Climate change will influence the transmission of *E. multilocularis* through the effects on the distribution and abundance of rodent intermediate hosts and sylvatic definitive hosts such as the Arctic fox. Increased precipitation might lead to the increased stability and density of rodent populations, facilitating the transmission of *E. multilocularis*. However, the increased frequency of severe weather events may decrease overall transmission. These events are occurring in parallel with other drivers of disease emergence such as landscape change and the translocation of hosts. *E. multilocularis* was recently detected in Svalbard in the Norwegian Arctic following the introduction of a suitable intermediate host, likely from shipping (Henttonen *et al.*, 2001). Such events will increase in frequency with the opening of ice-free shipping lanes across the Arctic, bringing increasing cargo, tourist traffic and other flora and fauna to regions once inaccessible to invasion from the sea. In a warming Arctic, the red fox may become more abundant, outcompeting the Arctic fox for food and den sites. In Sweden, red foxes carrying *E. multilocularis* were detected for the first time in 2011.

The northern strain cystic hydatid disease is caused by *Echinococcus granulosus*, which maintains a cycle that includes an adult cestode stage in the definitive host, such as wolf, coyote, fox or dog, and a larval cestode stage in an intermediate host in cervids such as moose, deer, caribou and reindeer. Humans usually acquire the infection via exposure to eggs that are shed in canid faeces and are an accidental host. Populations of cervid hosts are already being affected by climate change. Caribou populations in the Arctic are declining due to the secondary effects of climate and landscape changes (Kerby and Post, 2013). Increased precipitation and extreme weather events will likely contribute to further declines in population and parasite transmission in Arctic regions. However, several cervid species such as moose and deer, which are also competent hosts for *E. granulosis*, may move further north in response to climate change and thus could fill the current role of caribou in the

life cycle of *E. granulosis* in the Arctic. It was endemic among the Swedish Saami until the 1970s.

In Sweden, the incidence of tick-borne encephalitis (TBE) has increased substantially since the mid-1980s (Lindgren and Gustafson, 2001). This increase corresponds to a trend of milder winters and an earlier onset of spring, resulting in an increase in the tick population (*Ixodes ricinus*) that carries the virus responsible for TBE and other potential pathogens (Skarphédinsson *et al.*, 2005). Similar movement of TBE has been documented in northern north-western Russia, where *Ixodes persculcatus* is the predominant vector. This movement corresponds to the estimated climate-induced changes in the *Ix. persulcatus* habitat (Revich *et al.*, 2012). In north-eastern Canada, climate change, among other factors, is projected to result in a northward expansion in the range of *Ixodes scapularis*, a tick that carries *Borrelia burgdorferi*, the aetiologic agent of Lyme disease. The current northern limit of *Ix. scapularis* is southern Ontario, including the shoreline of Lake Erie, and the southern coast of Nova Scotia. Some temperature-based models show the potential for a northward expansion of *Ix. scapularis* above 60°N latitude and into the Northwest Territories by 2080 (Ogden *et al.*, 2005) However, tick distribution is influenced by additional factors such as habitat suitability and dispersal patterns, which can affect the accuracy of these predictions. The contribution of climate change-induced alterations in vector range to human disease thus depends on many other factors, such as land-use practices, human behaviour (suburban development in wooded areas, outdoor recreational activities, use of insect repellents, etc.) and human population density, as well as the adequacy of the public health infrastructure. Precise attribution of climate change is thus difficult, but is unlikely to be zero.

Past outbreaks of anthrax among cattle and reindeer have resulted in more than 13,000 burial grounds in Russia containing the carcasses of infected animals. More than half of these are located on permafrost in Siberia. There is concern that with a warming of the Arctic, melting permafrost in these regions will expose many of these burial sites together with the Anthrax spores, which will result in an epizootic among grazing animals and increase the risk of infection in humans who come into contact with infected animal products (undercooked meat, hides, bone; Revich *et al.*, 2012).

21.5 Tertiary Health Effects

Climate change will bring fundamental changes in the way land is used, infrastructure is protected and natural resources are accessed (Hess *et al.*, 2008). These changes will in turn impact population distribution, dynamics and culture.

In particular, warming will reduce the extent of ice, fundamental to human settlement and Arctic ecosystems alike. The recent decades of warming have been accompanied by extensive permafrost thawing in many areas, causing riverbank erosion, ground subsidence and damage to community buildings and infrastructure.

Reductions in sea ice pose other challenges. The sea ice retreat will increase marine transport and access to vast oil, gas and mineral reserves. Tourism will increase, as will public sector services to support the expanded oil, gas and mineral economies. While these new ventures will bring employment opportunities, they will also affect population distribution and challenge the traditional subsistence way of life for many communities. The reduction in traditional food may lead indigenous communities to depend increasingly on non-traditional Western foods, resulting in increasing rates of the diseases associated with processed foods, including obesity, diabetes, cardiovascular diseases and food-borne infectious diseases associated with imported fresh and processed foods. As the Arctic experiences increased globalization, changes in demographics will become an important issue. For peoples wishing to sustain their traditional way of life, migration from small communities to urban areas poses a multitude of challenges.

Dislocated families and communities will have to adapt to new ways of living, may face unemployment and will have to integrate and create new social bonds. Relocation may also lead to rapid loss of traditional culture, exacerbating stressors and mental health challenges. The stress of acculturation to new adoptive communities can be lonely and difficult, precipitating depression, anxiety, substance abuse and suicide. Alternatively, relocation may provide opportunities. Greater access to education, employment, health care and less expensive foods, including fresh fruits and vegetables, may contribute to improved overall health for individuals and communities that manage the transition well.

21.6 Adaptation

The potential impact of climate change on human health will differ from place to place depending on regional differences, even local differences, in climate alterations, as well as variations in health status and the adaptive capacity of different populations (Hess *et al.*, 2008). Similarly, the response to these effects will also vary. Actions may include the identification of communities and segments of the population at greatest risk. These should be targeted for assessment of the existing or potential health risks, vulnerabilities and engagement in the design of community-based monitoring and the formulation of intervention and adaptation strategies. Actions could include: (i) enhanced surveillance of health conditions that may be climate related and that are likely to have the most impact on human health; (ii) conducting research into the relationships between weather, climate and human health to help guide early detection and intervention; (iii) developing communication strategies targeting health-care providers and indigenous communities.

A key aspect of the response to climate change in Arctic regions will be the formation of community-based partnerships with indigenous groups to identify potential threats to the community and to develop strategies to address those threats. Communities at greatest risk should be targeted for education, outreach and the assessment of existing or potential health risks and vulnerabilities, and engagement in the design of community-based monitoring and the formulation of intervention strategies. The identification, selection and monitoring of basic indicators for climate change and community health will be important for any response to climate change at the community level (Furgal *et al.*, 2002; Symon *et al.*, 2005; Brubaker *et al.*, 2011). The selection of site- or village-specific indicators should be guided by local concerns and may include activities such as the surveillance of a key wildlife or insect species in a region where climate changes may contribute to the emergence of new zoonotic diseases, or the measurement of weather (i.e. precipitation and temperature), water quality (i.e. turbidity, pathogens) and gastrointestinal illness (i.e. clinic visits) in a community. Linking communities across regions and internationally should facilitate the sharing of standard protocols, data-collection instruments and data for analysis. These linkages will be important for the detection of trends over larger geographic regions, should enhance a community's ability to detect changes that impact health and will allow the development of strategies to minimize the negative health impacts of climate change on Arctic residents in the future.

Note

[1] The findings and conclusions in this chapter are those of the authors and do not necessarily represent the official position of the US Centers for Disease Control and Prevention.

References

Alaskool (2013) Alaska Regional Profiles, Northwest Region (http://www.alaskool.org/resources/regional/nw_reg_pro/permafrost.html, accessed 6 November 2013), pp. 74–83.

Arctic Council (2009) Arctic Marine Shipping Assessment 2009 Report, second printing. Arctic Council (http://www.arctic.noaa.gov/detect/documents/AMSA_2009_Report_2nd_print.pdf, accessed 7 April 2014).

Arctic Monitoring and Assessment Programme (2003) *AMAP Assessment 2002: Human Health in the Arctic*. Arctic Monitoring and Assessment Program, Oslo.

Berner, J., Furgal, C., Bjerregaard, P., Bradley, M., Curtis, T., De Fabo, E., *et al.* (2005) Human health. In: Symon, C., Arris, L. and Heal, B. (eds) *Arctic Climate Impact Assessment*. Cambridge University Press, Cambridge, UK, pp. 863–906.

Brubaker, M., Berner, J., Bell, J. and Warren, J. (2010) Climate Change in Kivalina, Alaska. Strategies for Community Health. ANTHC (http://www.anthc.org/chs/ces/climate/upload/Climate-Change-in-Kivalina-Alaska-Strategies-for-Community-Health-2.pdf, accessed 7 April 2014).

Brubaker, M., Berner, J., Chavan, R. and Warren, J. (2011) Climate change and health effects in Northwestern Alaska. *Global Health Action* 4, 8445, doi:10.3402/gha.v4i0.8445.

Curtis, T., Kvernmo, S. and Bjerregaard, P. (2005) Changing living conditions lifestyle and health. *International Journal of Circumpolar Health* 64, 442–450.

Davidson, R., Simard, M., Kutz, S., Kapel, C., Hamnes, I.S. and Robertson, L.J. (2011) Arctic parasitology; why should we care? *Trends in Parasitology* 27, 239–245.

De Fabo, E.C. (2005) Arctic stratspheric ozone depletion and increased UV-B radiation; potential impacts to human health. *International Journal of Circumpolar Health* 64, 509–522.

Demain, J., Gessner, B., McLaughlin, J., Sikes, D. and Foote, T. (2008) Increasing insect reactions in Alaska: is this related to changing climate? *Allergy and Asthma Proceedings* 30, 238–243.

Driscoll, D., Dotterrer, B., Miller, J. and Voorhess, H. (2010) Assessing the influence of health on rural outmigration in Alaska. *International Journal of Circumpolar Health* 69, 528–544.

Einarsson, N., Larsen, J.N., Nilsson, A. and Young, O.R. (eds) (2004) AHDR (Arctic Human Development Report), Akureyri, Iceland (www.svs.is/AHDR/AHDR%20chapters/English%20version/AHDR_chp%201.pdf, accessed May 2012).

Elmore, S.A., Jenkins, E.J., Huyvaert, K.P., Polley, L., Root, J.J. and Moore, C.G. (2012) *Toxoplasma gondii* in circumpolar people and wildlife. *Vector Borne and Zoonotic Diseases* 12, 1–9.

Epstein, P.R. and Ferber, D. (2011) *Changing Planet, Changing Health: How the Climate Crisis Threatens Our Health and What We Can Do about It*. University of California Press, Oakland, California.

Evengård, B., Berner, J., Brubaker, M., Mulvad, G. and Revich, B. (2011) Climate change and water security with a focus on the Arctic. *Global Health Action* 4, 8449, doi:10.3402/gha.v4i0.8449.

Fleischer, N.L., Melstrom, P., Yard, E., Brubaker, M. and Thomas, T. (2013) The epidemiology of falling-through-the-ice in Alaska, 1990–2010. *Journal of Public Health* 1–8, doi:10.1093/pubmed/fdt081.

Ford, J.D., Pearce, T., Gilligan, J., Smit, B. and Check, O. (2008) Climate change and hazards associated with ice use in Northern Canada. *Arctic, Antarctic and Alpine Research* 40, 647–659.

Furberg, M., Evengård, B. and Nilsson, M. (2011) Facing the limit of resilience: perception of climate change among reindeer herdering Sami in Sweden. *Global Health Action* 4, 8417, doi:10.3402/gha.v4i0.8417.

Furgal, C., Martin, D. and Glosselin, P. (2002) Climate change and health in Nunavik and Labrador: lessons from Inuit knowledge. In: Krupnik, I. and Jolly, D. (eds) *The Earth is Faster Now: Indigenous Observations of Arctic Environmental Change*. Arctic Consortium of the United States, Fairbanks, Alaska, pp. 266–300.

Haines, A., Kovats, R.S., Campbell-Lendrum, D. and Corvalán, C. (2006) Climate change and human health: impacts, vulnerability, and mitigation. *The Lancet* 360, 2101–2109.

Hansen, J.C. and Van Oostdam, J. (2009) *AMAP Assessment 2009: Human Health in the Arctic*. Arctic Monitoring and Assessment Program (AMAP), Oslo.

Hassi, J., Rytkonen, M., Kotaniemi, J. and Rintamaki, H. (2005) Impacts of cold on human heat balance, performace and health in circumpolar areas. *International Journal of Circumpolar Health* 64, 459–467.

Hennessy, T.W., Ritter, T., Holman, R.C., Bruden, D.L., Yorita, K., Bulkow, L.R., *et al.* (2008) The relationship between in-home water service and the risk of infections on the lung, skin and gastrointestinal tract among Alaska Native persons. *American Journal of Public Health* 98, 1–7.

Henttonen, H., Fuglei, E., Gower, C.N., Haukisalmi, V., Ims, R.A., Niemimaa, J., *et al.* (2001) *Echinococcus multilocularis* on Svalbard: introduction of an intermediate host has enabled the local life-cycle. *Parasitology* 123, 547–552.

Hess, J.J., Malilay, J.N. and Parkinson, A.J. (2008) Climate change: the importance of place. *American Journal of Preventive Medicine* 35, 468–478.

Høye, T.T., Post, E., Meltofte, H., Schmidt, N.M. and Forchhammer, M.C. (2007) Rapid advancement of spring in the high Arctic. *Current Biology* 17, R449–R451.

Hueffer, K., Parkinson, A.J., Gerlach, R. and Berner, J. (2013) Zoonotic infections in Alaska: disease prevalence, potential impact of climate change and recommended actions for earlier disease detection, research, prevention and control. *International Journal of Circumpolar Health* 72, doi:10.3402/ijch.v72i0.19562.

Huntington, H., Weller, G., Bush, E., Callaghan, T.V., Kattsov, V.M. and Nuttall, M. (2005) Introduction. In: Symon, C., Arris, L. and Heal, B. (eds) *Arctic Climate Impact Assessment*. Cambridge University Press, Cambridge, UK, pp. 1–20.

IPCC (Intergovernmental Panel on Climate Change) (2014) Summary for Policymakers. In: Stocker, T.F., Qin, D., Plattner, G.-K., Tignor, M.M.B., Allen, S.K., Boschung, J., *et al.* (eds) *Climate Change 2013: The Physical Science Basis: Working Group I Contribution to the Fifth Assessment Report of the Intergovernmental Panel on Climate Change*. Cambridge University Press, Cambridge, UK and New York, pp 3–32.

Jenkins, M.J., Schurer, J.M. and Gesy, K.M. (2011) Old problems on a new playing field: helminthzoonoses transmitted among dogs, wildlife and people in a changing northern climate. *Veterinary Parasitology* 182, 54–69.

Kerby, J.T. and Post, E. (2013) Advancing plant phenology and reduced herbivore production in a terrestrial system associated with sea ice decline. *Nature Communications* 4, doi:10.1038/ncomms3514.

Lindgren, E. and Gustafson, R. (2001) Tick-borne encephalitis in Sweden and climate change. *The Lancet* 358, 16–18.

McLaughlin, J.B., DePaola, A., Bopp, C.A., Martinek, K.A., Napolilli, N.P., Allison, C.G., *et al.* (2005) Outbreak of *Vibrio parahaemolyticus* gastroenteritis associated with Alaskan oysters. *New England Journal of Medicine* 353, 1463–1469.

Makary, P., Kanerva, M., Ollgren, J., Virtanen, M.J., Vapalahti, O. and Lyytikäinen, O. (2013) Disease burden of Puumala virus infection 1995–2008. *Epidemiology and Infection* 138, 1484–1492.

Messner, T. (2005) Environmental variables and the disease risk. *International Journal of Circumpolar Health* 64, 523–533.

Messnier, V., Levesque, B., Proulx, J.F., Rochette, L., Libman, M., Ward, B.J., *et al.* (2009) Seroprevalence of *Toxoplasma gondii* among Nunavik Inuit (Canada). *Zoonoses and Public Health* 56, 188–197.

Nayha, S. (2005) Environmental temperature and mortality. *International Journal of Circumpolar Health* 64, 451–458.

Ogden, N.H., Bigras-Poulin, M., O'Callaghan, C.J., Barker, I.K., Lindsay, L.R., Maarouf, A., *et al.* (2005) A dynamic population model to investigate effects of climate on geographic range and seasonality of the tick *Ixodes scapularis. International Journal for Parasitology* 35, 375–389.

Parkinson, A.J. (2008) The International Polar Year, 2007–2008. An opportunity to focus on infectious diseases in Arctic regions. *Emerging Infectious Diseases* 14, 1–3.

Parkinson, A.J. and Butler, J.C. (2005) Potential impact of climate change on infectious disease emergence in the Arctic. *International Journal of Circumpolar Health* 64, 478–486.

Pettersson, L., Boman, J., Juto, P., Evander, M. and Ahlm, C. (2008) Outbreak of Puumala virus infection, Sweden. *Emerging Infectious Diseases* 14, 808–810.

Revich, B., Tokarevich, N. and Parkinson, A.J. (2012) Climate change and zoonotic infectious diseases in the Russian Arctic. *International Journal of Circumpolar Health* 71, 18792 - http://dx.doi.org/10.3402/ijch.v71i0.18792.

Revich, B.A. (2008) Climate change alters human health in Russia. *Studies on Russian Economic Development* 19, 311–317.

Revich, B.A. and Shaposhnikov, D.A. (2010) Extreme temperature episodes and mortality in Yakutsk, East Siberia. *Rural and Remote Health* 10, 1338, doi:10.3402/ijch.v71i0.18792.

Rylander, C., Odland, J.Ø. and Sandanger, T.M. (2010) Climate change and environmental impacts on maternal and newborn health with a focus on Arctic populations. *Global Health Action* 4, doi:10.3402/gha.v4i0.8452.

Skarphédinsson, S., Jensen, M. and Kristiansen, K. (2005) Survey of tickborne infections in Denmark. *Emerging Infectious Diseases* 11, 1055–1061.

Symon, C., Arris, L. and Heal, B. (eds) (2005) *Arctic Climate Impact Assessment.* Cambridge University Press, Cambridge, UK.

Weller, G., Bush, E., Callaghan, T.V., Corell, R., Fox, S., Furgal, C., *et al.* (2005) Summary and synthesis. In: Symon, C., Arris, L. and Heal, B. (eds) *Arctic Climate Impact Assessment.* Cambridge University Press, Cambridge, UK, pp. 991–1020.

Wenger, J., Zutz, T., Bruden, D., Bruce, M., Bulkow, L., Parks, D., *et al.* (2010) Invasive pneumococcal disease in Alaskan children: impact of the 7-valent pneumococcal vaccine and the role of water supply. *Pediatric Infectious Disease Journal* 29, 251–256.

Young, K.T. and Bjerregaard, P. (eds) (2008) *Health Transitions in Arctic Populations.* University of Toronto Press Inc, Toronto, Canada.

Young, O.R. and Einarsson, N. (2004) Introduction. In: Einarsson, N., Larsen, J.N., Nilsson, A. and Young, O.R. (eds) *AHDR (Arctic Human Development Report).* Stefansson Arctic Institute, Akureyri, Iceland (www.svs.is/AHDR/AHDR%20chapters/English%20version/AHDR_chp%201.pdf, accessed May 2012).

22 Climate Change and Health in Africa

Colin D. Butler,[1,2] Andrew Mathieson,[3] Devin C. Bowles,[2] Godson Ana[4] and Guéladio Cissé[5]

[1]*Faculty of Health, The University of Canberra, Australia, and Benevolent Organisation for Development, Health and Insight (BODHI);* [2]*National Centre for Epidemiology and Population Health, The Australian National University, Canberra, Australia;* [3]*The Australian National University, Canberra, Australia;* [4]*Department of Environmental Health Sciences, University of Ibadan, Nigeria;* [5]*Swiss Tropical and Public Health Institute, University of Basel, Switzerland*

The continent is too large to describe. It is a veritable ocean, a separate planet, a varied, immensely rich cosmos. Only with the greatest simplification, for the sake of convenience, can we say 'Africa'. In reality, except as a geographical appellation, Africa does not exist.

(Kapuściński, 1998)

22.1 The African Continent

Africa, the 'cradle of humanity', is the home of numerous languages, peoples, customs and cultures. Its size is enormous – almost ten times that of India. Like India, it has a deep history of imperialism and colonialism, still visible today (Rodney, 1972). At a time when the people of the continent had already been raided for centuries, by both Europeans and Arabs, often in collusion with and assistance from people born in Africa, the Berlin conference in the mid-1880s formally 'divided' most of the continent among just a few European powers; principally, the UK, France, Germany, Italy and the King of Belgium.

Today, it has a population exceeding 1 billion, in many parts still growing rapidly, with consequences feared by some as of Malthusian, poverty and other forms of entrapment (King, 1990; Butler, 2004). Africa is home to almost 60 separate states, most of them independent, yet still cartographically captured by lines (44% of them straight) drawn by Europeans, with scant regard for local cultures, languages or rivalries.

22.2 Climate Change in Africa

The fraction of the world's total emissions of greenhouse gases (GHGs) that originate from African nations, mainly from power generation and transport, is minor, whether in the past, present or projected into the future (Collier *et al.*, 2008). Deforestation, a major source of GHGs from some developing countries such as Indonesia, is also occurring in Africa, both in the Congo, for valuable timber, and the Sahel, for energy. Its tropical belt harbours one of the largest and most biodiverse rainforests in the world: it is vulnerable to clearing at an increasing rate (Nogherotto *et al.*, 2013).

The methane emissions of Africa's large and expanding herd of cattle (like in the Sudan) probably constitute a significant proportion of the continent's total climate footprint (Thorpe, 2009).

All regions and populations of the world are vulnerable to climate change, but numerous authors, including some based in Africa, agree that many African developing countries are particularly vulnerable. This is due principally to the continent's widespread poverty (Collier *et al.*, 2008; Hope, Sr, 2009; Kula *et al.*, 2013). But this is added to, and in some parts worsened by, its high birth rate and its history of conflict and poor governance.

Some analysts believe that the impacts of climate change on Africa are already apparent, and also that the physical changes due to climate change in Africa will be unusually severe (Collier *et al.*, 2008). Climate change affects different regions in different ways: East Africa may become wetter and South Africa drier. The Intergovernmental Panel on Climate Change (IPCC) reports with 'medium' confidence that, since the 1950s, West Africa has experienced a trend towards more intense and longer droughts. Sea level rise will be problematic in some regions, especially the Nile delta (Bohannon, 2010), Lagos and Banjul, the capital of The Gambia (Douglas *et al.*, 2008). Africa is considered, with small island states, as the region least likely to be able to afford to protect its coasts (Nicholls *et al.*, 2011).

Development in many sectors (particularly agriculture, water and health) will be an important counterweight to climate change. Before describing the various likely health effects, we outline climate change and agriculture in Africa, as the capacity to grow and protect more food is crucial to positive African development scenarios. Without enough food for all, many other problems will be more difficult to solve. If hunger and undernutrition worsen, then it is likely that many other problems will be aggravated, including those that are health related.

22.3   Climate Change and Agriculture in Africa

Home to 15% of the world's population, Africa, especially its sub-Saharan region, houses (with India) one of the largest and greatest concentrations of undernourished children and adults in the world. Few African nations are net food exporters. In fact, Africa is a major grain importer,

responsible for importing about one-third of the world's internationally traded rice and one-quarter of the wheat (Seck *et al.*, 2013). Its main staple is maize, a grain native to Central America. Comparatively little maize is imported, but in much of southern Africa the rain-fed maize crop already experiences frequent drought stress (Collier *et al.*, 2008). This is thought likely to worsen, particularly in parts of Zimbabwe and South Africa. Wheat yields in North Africa are also likely to be threatened, including because of salinity and sea level incursion to the Nile delta, not only from sea level rise but also from subsidization due to loss of silt trapped by the Aswan High Dam (Bohannon, 2010).

A substantial part of Africa's agricultural potential is considered highly vulnerable to climate change (Collier *et al.*, 2008), from East Africa (Grace *et al.*, 2012) to West Africa (Jalloh *et al.*, 2013) and the conflict-prone Horn of Africa (Funk, 2011). Fisheries are also vulnerable, impacted by overfishing (including predation of resources by wealthier nations) and by climate change, as reported of Lake Tanganyika (Verschuren, 2003). In the scenario of an average increase of 4°C of warming globally, then agricultural prospects in Africa may, without exaggeration, be dire (New *et al.*, 2011).

Patterns of livestock use are already changing, partly driven by climate change. Increasingly, the Maasai, traditionally cattle herders, in Kenya and Tanzania are raising chickens. These changes in the patterns of people, animals and land use can, in turn, lead to novel encounters between hosts and pathogens and the creation of new and re-emerging illnesses.

However, it remains to be seen how effective these strategies will be for reducing poverty, especially in drought areas, where water is precious and the effort to acquire it is becoming more and more risky and extremely time-consuming under climate change (Ndenyele and Badurdeen, 2012).

'Land and water grabs'

Adding to these problems is the issue of the acquisition and control of land and its associated water (Rulli *et al.*, 2013), whether by purchase or lease, in developing countries by nations in East Asia, the Persian Gulf and elsewhere.

Often pejoratively called 'land grabs', these lease-holds are claimed to be of mutual benefit. Poor nations gain financial capital and the promise of the transfer of technology and expertise. Leaseholders gain the potential for additional food supplies, which could be very useful if climate or other forms of adverse global environmental change cause agricultural deterioration in their own nations. But, poor nations lose, at least temporarily, use of the natural capital, that is, of the land and its water. It is unknown whether food and fuel produced by such areas will be exported in future, even if there is a major domestic shortage of such commodities.

However, in the 19th century, British India exported food to Europe, even during severe famine. Already, in Sudan, land acquired by foreign interests along the Blue Nile (the best source of water in an otherwise dry region) is used increasingly for the large-scale export of food commodities. Yet Sudanese smallholder access to water and land is diminishing, and an increasing number of Sudanese are becoming dependent on food aid and subsidies (Rulli *et al.*, 2013). Further illustrating such concerns, the author, Robin Palmer, reports how African leaders from Mozambique, Ethiopia or Zambia have claimed that many millions of hectares can be used to grow biofuels 'without threatening food production'. The Ethiopian Minister for Agriculture suggested that pastoralists displaced by land grabbing 'can just go somewhere else' (Palmer, 2011).

Another form of land grab also occurred in Zimbabwe, particularly in the 1990s under President Mugabe, as the 6000 white commercial farmers gradually lost control of their land, culminating in 2000 with the introduction of the 'fast-track' programme, resulting in the coercive acquisition of about 90% of commercial farms still owned by white farmers. Exacerbated by drought, HIV/AIDS and hyperinflation, Zimbabwe agriculture regressed from southern Africa's breadbasket to an impoverished land of hunger and undernutrition (Gwatirisa and Manderson, 2012).

Mycotoxins and pests

Both drought stress and floods increase the risk of mould and fungi in staple crops, including stored grain. Such food can be contaminated by mycotoxins. People ingesting such contaminated food will usually not see or taste anything abnormal, and can develop liver failure and hepatocellular carcinoma. Though a simple fluorescence test can usually determine aflatoxin contamination, it is rarely used in developing countries. Climate change may increase the burden of disease from mycotoxins, including through increased rainfall events during or immediately after harvest (Tefera, 2012).

Many insect pests may be worsened by climate change, including maize, an important African staple (Diffenbaugh *et al.*, 2008). On the positive side, some analysts think that conditions for growing the root crop, cassava, will be improved by climate change (Jarvis *et al.*, 2012).

Possible improvements to African agriculture

Though progress is being made, the Green Revolution has not yet fully reached the continent, due mainly to a lack of human expertise and institutional capacity (Ejeta, 2010). Only 4% of Africa is irrigated, compared to 40% in South Asia (Costello *et al.*, 2011). This at least suggests a possibility of great improvement; irrigation is also thought to be a good buffer for climate change, at least for crops (Hassan, 2010). Another possible improvement is to use more fertilizer (Editorial, 2012), including with phosphorus. African soils are deficient in this vital element, and yet Morocco, including occupied Western Sahara, is the world's major phosphate exporter, hardly any of which is used in sub-Saharan Africa (Cordell *et al.*, 2009).

More optimistically, 'evergreen agriculture', which integrates fertilizer (nitrogen fixing), fruit and fodder tree species with food cropping, is being promoted as a fresh way to improve rain-fed agricultural productivity for smallholders, in regions as diverse as Malawi, Niger and Burkina Faso (Garrity *et al.*, 2010). Advocates of these methods hope that these techniques can be scaled up across the entire continent. Advocates of an African Green Revolution, African aquaculture and genetically modified organisms targeted for the African environment also remain hopeful. For example, Monsanto is trying to develop maize with increased drought tolerance, and has made its work available to WEMA – the Water-Efficient Maize for Africa consortium (Jones, 2011).

22.4 Climate Change and Health in Africa

Consistent with the general framework of this book, this chapter will discuss three main categories of health effect: *primary* (e.g. heatwaves and perhaps reduced coldwaves in highland areas, flooding and disasters); *secondary*, especially vector-borne diseases and changes to other infectious diseases; and *tertiary*, including famine, conflict and large-scale migration. While the latter two phenomena are not often considered *health* problems, they each clearly have important health manifestations and consequences.

22.5 Primary Health Effects of Climate Change

Heatwaves

In June 2010, Climate Signals reported that Chad and Niger set new maximum temperature records (www.climatesignals.org). In Sudan, the mercury peaked at 49.6°C (121.3°F) at Dongola. Heatwaves cause acute and chronic stress, not only for humans but also for animals, including cattle, birds and bats. Impacts on these may have many indirect effects to human well-being. Extreme high air temperatures contribute directly to deaths from cardiovascular and respiratory disease, particularly among elderly people (see Chapter 3, this volume).

The number of African megacities is growing (e.g. Cairo, Lagos, Nairobi and Johannesburg), and these are likely to be vulnerable to heatwaves, including their interaction with air pollution. However, there is as yet little literature on heatwaves and health in Africa (Capuano *et al.*, 2012).

Flooding and disasters

Climate change may be altering the frequency of intense precipitation in Africa (Douglas *et al.*, 2008), as in most of the world. The vulnerability of the poor to flooding is enhanced by urbanization and inadequate regulation. The poor are frequently forced to build low-quality shelters on river flood plains, steep hillsides (at the risk of erosion and landslides) and other locations considered undesirable by people with more means. Both circumstances are compounded by a lack of insurance and limited or no chance of compensation. Due to flooding, the poor experience higher rates of trauma and property loss and some infectious diseases, such as gastroenteritis and leptospirosis. The poor are also likely to experience increased rates of mental stress from their experience of such events in comparison to those who are better off.

In recent years, extremely heavy rainfall events have occurred in several parts of Africa, particularly in East and West Africa (Cissé *et al.*, 2011). In 2000, floods in Mozambique affected up to 1 million people (Douglas *et al.*, 2008). In 2002, floods in East Africa displaced thousands of people from their homes in Kenya, Rwanda and other regions. Due to insufficient data, the IPCC recently concluded with only low confidence that heavy precipitation had increased in this region, though they found, with medium confidence, that it would increase in future, due to climate change (Field *et al.*, 2012, p. 18). If so, this will contribute to many disasters, particularly in lesser cities and areas with limited infrastructure and reduced adaptive capacities for disaster risk reduction (Cissé, 2013).

22.6 Secondary Health Effects of Climate Change

Secondary effects are less direct than primary ones, but more causally obvious than tertiary health effects. An important mechanism for this group is an altered distribution of arthropod vectors, intermediate hosts and pathogens, due to a shift in environmental parameters that make areas more or less welcoming to hosts and pathogens. This in turn will produce changes in the epidemiology of many infectious diseases. Some diseases may expand their range, but others are predicted to decrease.

Irrigation, if used more widely in Africa, is likely to contribute to local environmental change, including the temperature and humidity.

This could favour vector-borne diseases; however, the wealth from irrigation, combined with public health measures, could reduce such diseases. This has been called the 'paddies paradox' (Ijumba and Lindsay, 2001). However, recent studies in India show that the wealth-related reduction of malaria may need decades to take hold (Baeza *et al.*, 2013).

Malaria and other febrile arthropod-transmitted illnesses

Many parts of Africa, especially away from the highlands, have very high rates of infectious diseases. West Africa was long called the 'fever coast' by European invaders, and mysterious febrile illnesses delayed European incursions and conquest. While most of these fevers were caused by malaria, many illnesses mimic malaria. According to the historians, McNeill and Crosby, both malaria and yellow fever were introduced to the New World due to the slave trade. Malaria has shaped the genome of many Africans, with red cell variants including sickle cell disease, thalassaemia and G6PD deficiency (Eridani, 2011).

In general, the incidence of malaria is declining in Africa, due to public health measures, especially the prompt treatment of malarial fevers with the artemisinin combination therapies (ACTs) (that also inhibit transmission) and the use of insecticides (particularly pyrethroids in relation to treated bed nets) (Feachem *et al.*, 2010). Numerous non-climatic factors influence the epidemiology of malaria, though it is implausible that changes in temperature, rainfall and humidity will not also play a role (see Chapter 7, this volume). These factors not only influence the distribution, longevity and activity of arthropod vectors, but in some cases accelerate the life cycle of the parasite within the vector. There is good evidence that climate change is already contributing to the increased survival of *Anopheles* vectors at higher altitudes (Chen *et al.*, 2009) and of malaria transmission (Alonso *et al.*, 2011). In some cases, the changes wrought to malaria by climate change may be beneficial (Meyrowitsch *et al.*, 2011), but on the whole, they are likely to be harmful. It is also plausible that already existing resistance to ACTs and pyrethroids (Trape *et al.*, 2011) will spread

in Africa, again expanding the malarial disease burden (Gosling and Chandramohan, 2008).

Other fevers, such as dengue and chikungunyah,[1] may be on the increase in Africa (Simon *et al.*, 2008), especially as pools and containers of water in urban settlements allow greater breeding opportunities for vectors, and as population density increases.

Meningococcal meningitis

Epidemics of this vaccine-preventable infection occur worldwide, but the highest incidence is observed in the 'meningitis belt' of sub-Saharan Africa, from Senegal on the Atlantic to Ethiopia on the Red Sea (Greenwood, 2006). Populations in northern Nigeria are particularly vulnerable. Other areas in the Rift Valley, the Great Lakes and southern Africa are also affected.

Seasonal, climate-related epidemics occur during the dry season, associated with the dusty Harmattan, which is postulated to impair nasopharyngeal defences (Greenwood, 2006). Changes in the Harmattan have been linked to land-use change and alterations in regional climate (Molesworth *et al.*, 2003).

Other infectious diseases in Africa

HIV/AIDS could worsen in future, in Africa, as a consequence of falling morale and failing public health, for example in a 'four-degree world' with large-scale undernutrition, conflict and declining governance (McMichael *et al.*, 2008). In turn, poorly controlled HIV/AIDs would increase the risk of deepening systemic failure, due to high mortality and consequent lack of cultural transmission and diminished expertise.

The pattern of numerous other infectious diseases is potentially altered in Africa by climate change. These include geohelminths and schistosomiasis (see Chapter 10), African trypanosomiasis, Rift Valley fever virus, cholera and diseases borne by ticks and rodents. Stress from deforestation could weaken immunity in bats, potentially enhancing bat-borne diseases such as Henipavirus and even Ebola. No doubt there are others (we cannot be exhaustive).

Several studies have confirmed and quantified the effects of high temperatures on common forms of food poisoning, such as salmonellosis (Fleury *et al.*, 2006). Contact between food and pest species, especially flies, rodents and cockroaches, is also temperature sensitive. Fly activity is driven largely by temperature rather than by biotic factors (Goulson *et al.*, 2005).

Climate change-related alterations in rainfall, surface water availability and water quality could affect the burden of water-related diseases, including cholera. At least 50 people were killed in a cholera outbreak in Nigeria's northern state of Jigawa bordering Niger and over 100 in Ibadan following different flood episodes in 2011.

22.7 Tertiary Health Effects

The easiest way to think of the tertiary health effects of climate change may be as a 'threat multiplier'. Famine, large-scale migration and conflict can all occur without climate change, but climate change increases the risk of all of these phenomena. While climate change is neither necessary nor sufficient for these effects, it is increasingly likely to be a factor, and in some cases already is. In the long run, the health effects of these tertiary conditions are likely to exceed those of the others, perhaps by two or more orders of magnitude. For this reason, all possible steps to minimize their risk should be taken. A 'four-degree world' should remain theoretical, not something which the next generation has to endure.

Climate change, food security, Africa and the Food and Agricultural Organization (FAO)

Climate change and agriculture in Africa has been discussed above, but food security also depends critically on governance (local, regional and global) and purchasing power (see Chapter 13). Numerous high-level food agencies, especially the FAO, have been slow to comprehend the risk to food security represented by climate change. For example, while the first models which expressed concern about global food insecurity induced by climate change were published in leading journals in the early 1990s, it took more than a decade for a high-level response from the FAO, when it belatedly highlighted climate change at the 29th session of the FAO Committee (Butler, 2009). Some of this lag seems likely to reflect sources of funding and thinking which are biased toward people and institutions with pre-existing power, few of whom have been based in Africa and few of which lobby on behalf of her people. However, this situation has changed markedly (Jalloh *et al.*, 2013); recognition is 'better late than never'.

Climate change, food security and famine

Africa has experienced several severe famines in recent decades, particularly in the Sahel, including Ethiopia, Somalia and Niger. These have arisen through drought, poor governance and conflict. In the recent Somali famine, Islamist militants called Al Shabaab attacked famine relief workers (Moszynski, 2011). On the positive side, global governance measures, especially famine relief, have been able to reduce the most severe effects partially. But, chronic undernutrition is widespread in many parts of Africa, even without overt famine. This reflects and creates vulnerabilities, including the risk of being exploited by those who are better fed and stronger.

Climate change was not a factor in the Rwandan genocides of the early 1960s and again in 1994, but food insecurity and the anticipation that this would worsen due to land scarcity, in the absence of sufficient other sources of income, was (André and Platteau, 1998). Today, the population in Rwanda is higher than in 1994. The birth rate of this substantially Catholic country is declining, but political tensions are returning. However, its human carrying capacity today is also higher, due in part to extensive foreign aid (25% of its GNP in 2006) (Beswick, 2010). But this has been complemented by a large increase in income from ecotourism, particularly associated with the observing of habituated gorillas in the Virunga National Park, a legacy of the fierce and controversial conservationist Dian Fossey.

Climate change and migration

The refugee camp of Dadaab, with almost half a million people, is the world's largest (Polonsky *et al.*, 2013). In large part, the outflow of migrants from Somalia and regional countries that generated this settlement (founded in 1991) has been caused by a combination of ecological and social factors, most recently the interaction of drought and the lack of a functional government in part of Somalia, creating a vacuum which has been filled in part by the militant Islamist group, Al Shabaab.

Already, poverty and insecurity at home and the hope for a better life has motivated hundreds of thousands of Africans to migrate, including to Europe. Climate change is likely to increase the number of attempted migrants and refugees, perhaps by a large factor. In 2012, the African Union Convention on the Protection of and Assistance to Internally Displaced Persons in Africa (the Kampala Convention) came into force. For the first time, this treaty recognizes that displacement can occur because of climate change, as well as several other factors (UNHCR, 2013). However, as with all tertiary manifestations (and primary and secondary ones, too), the scale of this can be modified by other forms of development and by mitigation. Climate change adaptation, including through migration (Black *et al.*, 2011) is also vital, but should not be used as a substitute for mitigation. Adapting to a four-degree world may be impossible.

Climate change, resource scarcity and conflict in Africa

Climate change is also likely to multiply the risk of conflict, especially by diminishing resource availability. This effect could be substantial in Africa over coming decades, as resources such as water, food, productive cropland and animal pasture are already scarce.

Two socio-political characteristics make conflict a particularly likely consequence of resource scarcity in many African countries. One of these is low levels of institutional inclusivity, which impede other routes to redress grievances (Kahl, 2006). The degree to which a society is divided, whether along religious, ethnic, cultural or class lines, also influences the risk that scarcity will lead to violence. In societies with high levels of such cross- and subnational 'groupness', identity and security is associated strongly with membership to a single group. At times of stress, people turn to these groups alone for support, rather than the collective state or a network of groups, in so doing often excluding other groups, who become competitors. Such fault lines are also prone to manipulation by leaders to increase support and distract from more fundamental grievances (Kahl, 2006).

Weak states, of which Africa has a disproportionate number, heighten the potential for violent spillover, including cross-border diffusion of military skills and grievances. War in Darfur (Sudan) spread into Chad and the Central African Republic. Inconsistency between international and cultural borders and the presence of failing states can both create safe havens for insurgents and internationalize civil conflicts.

22.8   Conclusion

Perhaps more than anywhere except small island states, African health is hostage to the increasingly accepted myth that adaptation can be a sufficient solution to climate change. Africa is responsible for only a small part of the annual global emission of heat-trapping gases. Climate change mitigation in Africa may therefore seem of little importance. However, some forms of 'technological leapfrogging' offer hope, both locally and globally. One is the rapid spread of mobile telephony and the Internet, which is particularly strong in Kenya (Aker and Mbiti, 2010). Another may be the widespread replacement of polluting and increasingly expensive kerosene lamps by highly efficient light-emitting diodes.[2]

Photovoltaic technology, the cost of which is declining rapidly, offers hope to provide electricity in remote locations, where it is currently lacking. Allied with access to the Internet, this could accelerate literacy, human development and reduce market asymmetries, in ways which reduce poverty and vulnerability. Infrastructure such as railways and roads, designed and built by Chinese, principally to allow the export of raw materials, may also be used by Africans for trade and travel, further reducing isolation. However, improved governance and slower population growth in many parts of Africa are also essential if the numerous challenges of climate change are to be met. Recently, Goodluck Jonathan, the president of Nigeria, has hinted at an understanding that slowed population growth will lessen poverty (Rosenthal, 2012). To date, his calls have met resistance (Population Media Center, 2012). On the other side of the continent, President Musevini of Uganda continues to be pro-natalist, despite the existence, within his nation, of the Lords Resistance Army, a group whose barbaric practices suggest a continent where a heart of darkness still beats.

Somehow, the population of more than 1 billion people on this enormous continent need to find ways to reduce their poverty and to improve their food security and nutrition, and also to find ways to collaborate effectively with people of goodwill everywhere, whether from China, Europe or the Middle East. Religious and ethnic differences must become subsidiary to the recognition that we are all human. But rich populations, represented by almost all who read this chapter, must reduce their own dependency on coal and other forms of 'Earth poison', to give not only Africans a chance to flourish but also the whole world.

Notes

[1] First isolated in the early 1950s after an epidemic of polyarthritis in Tanganyika (now Tanzania). In Kimakonde, a language spoken in Tanzania and Mozambique, it means 'that which contorts or bends up'. In the Congo, people call it 'buka-buka' or 'broken-broken' (Simon *et al.*, 2008).

[2] The World Bank is financing a project called 'Lighting Africa'.

References

Aker, J.C. and Mbiti, I.M. (2010) Mobile phones and economic development in Africa. *Journal of Economic Perspectives* 24, 207–232.

Alonso, D., Bouma, M.J. and Pascual, M. (2011) Epidemic malaria and warmer temperatures in recent decades in an East African highland. *Proceedings of the Royal Society B* 278, 1661–1669.

André, C. and Platteau, J.-P. (1998) Land relations under unbearable stress: Rwanda caught in the Malthusian trap. *Journal of Economic Behavior and Organization* 34, 1–47.

Baeza, A., Bouma, M.J., Dhiman, R.C., Baskerville, E.B., Ceccato, P., Yadav, R.S., *et al.* (2013) Long-lasting transition toward sustainable elimination of desert malaria under irrigation development. *Proceedings of the National Academy of Sciences* 110, 15157–15162.

Beswick, D. (2010) Managing dissent in a post-genocide environment: the challenge of political space in Rwanda. *Development and Change* 41, 225–251.

Black, R., Bennett, S.R.G., Thomas, S.M. and Beddington, J.R. (2011) Climate change: migration as adaptation. *Nature* 478, 447–449.

Bohannon, J. (2010) The Nile delta's sinking future. *Science* 327, 1444–1447.

Butler, C.D. (2004) Human carrying capacity and human health. *Public Library of Science Medicine* 1, 192–194.

Butler, C.D. (2009) Food security in the Asia-Pacific: Malthus, limits and environmental challenges. *Asia Pacific Journal of Clinical Nutrition* 18, 577–584.

Capuano, P., Sellerino, M., Ruocco, A.D., Kombe, W. and Yeshitela, K. (2012) Climate change induced heat wave hazard in eastern Africa: Dar Es Salaam (Tanzania) and Addis Ababa (Ethiopia) case study. Climate Change and Urban Vulnerability in Africa (CLUVA) (http://www.cluva.eu/index. php?option=com_content&view=article&id=164:climate-change-induced-heat-wave-hazard-in-eastern-africa-dar-es-salaam-tanzania-and-addis-ababa-ethiopia-case-study&catid=60:egu-2013-posters& Itemid=109, accessed 27 March 2014).

Chen, H., Githeko, A.K., Zhou, G., Githure, J.I. and Yan, G. (2009) New records of *Anopheles arabiensis* breeding on the Mount Kenya highlands indicate indigenous malaria transmission. *Malaria Journal* 5, 17, doi:10.1186/1475-2875-5-17.

Cissé, G. (2013) Water-related disaster management and adaptation to climate change: bridges and challenges? *Water International* 38, 11–16.

Cissé, G., Koné, B., Bâ, H., Mbaye, I., Koba, K., Utzinger, J., *et al.* (2011) Ecohealth and climate change: adaptation to flooding events in riverside secondary cities, West Africa. In: Otto-Zimmermann, K. (ed.) *Resilient Cities: Cities and Adaptation to Climate Change: Proceedings of the Global Forum 2010.* Springer, Dordrecht, the Netherlands, pp. 55–67.

Collier, P., Conway, G. and Venables, T. (2008) Climate change and Africa. *Oxford Review of Economic Policy* 24, 337–353.

Cordell, D., Drangert, J.-O. and White, S. (2009) The story of phosphorus: global food security and food for thought. *Global Environmental Change* 19, 292–305.

Costello, A., Maslin, M., Montgomery, H., Johnson, A.M. and Ekins, P. (2011) Global health and climate change: moving from denial and catastrophic fatalism to positive action. *Philosophical Transactions of the Royal Society* 369, 1866–1882.

Diffenbaugh, N.S., Krupke, C.H., White, M.A. and Alexande, C.E. (2008) Global warming presents new challenges for maize pest management. *Environmental Research Letters* 3, 044007, 9, doi:10.1088/1748-9326/3/4/044007.

Douglas, I., Alam, K., Maghenda, M., McDonnell, Y., McLean, L. and Campbell, J. (2008) Unjust waters: climate change, flooding and the urban poor in Africa. *Environment and Urbanization* 20, 187–205.

Editorial (2012) Food for thought. *Nature* 483, 510–510.

Ejeta, G. (2010) African green revolution needn't be a mirage. *Science* 327, 831–832.

Eridani, S. (2011) Sickle cell protection from malaria. *Hematology Reports* 19, e24.

Feachem, R., Phillips, A.A., Hwang, J., Cotter, C., Wielgosz, B., Greenwood, B.M., *et al.* (2010) Shrinking the malaria map: progress and prospects. *The Lancet* 376, 1566–1578.

Field, C.B., Barros, V., Stocker, T.F., Qin, D., Dokken, D., Ebi, K.L., *et al.* (eds) (2012) *Intergovernmental Panel on Climate Change Special Report on Managing the Risks of Extreme Events and Disasters to Advance Climate Change Adaptation.* Cambridge University Press, Cambridge, UK, and New York.

Fleury, M., Charron, D.F., Holt, J.D., Allen, O.B. and Maarouf, A.R. (2006) A time series analysis of the relationship of ambient temperature and common bacterial enteric infections in two Canadian provinces. *International Journal of Biometeorology* 50, 385–391.

Funk, C. (2011) We thought trouble was coming. *Nature* 476, 17.

Garrity, D.P., Akinnifesi, F.K., Ajayi, O.C., Weldesemayat, S.G., Mowo, J.G., Kalinganire, A., *et al.* (2010) Evergreen agriculture: a robust approach to sustainable food security in Africa. *Food Security* 2, 197–214.

Gosling, R. and Chandramohan, D. (2008) Tackling malaria today: beware resurgence of malaria where incidence has fallen. *BMJ* 337, a1592.

Goulson, D., Derwent, L.C., Hanley, M., Dunn, D. and Abolins, S. (2005) Predicting calyptrate fly populations from the weather, and the likely consequences of climate change. *Journal of Applied Ecology* 42, 784–794.

Grace, K., Davenport, F., Funk, C. and Lerner, A.M. (2012) Child malnutrition and climate in sub-Saharan Africa: an analysis of recent trends in Kenya. *Applied Geography* 35, 405–413.

Greenwood, B. (2006) Editorial: 100 years of epidemic meningitis in West Africa – has anything changed? *Tropical Medicine and International Health* 11, 773–780.

Gwatirisa, P. and Manderson, L. (2012) 'Living from day to day': food insecurity, complexity, and coping in Mutare, Zimbabwe. *Ecology of Food and Nutrition* 51, 97–113.

Hassan, R.M. (2010) Implications of climate change for agricultural sector performance in Africa: policy challenges and research agenda. *Journal of African Economies* 19, ii77–ii105.

Hope, K.R. Sr (2009) Climate change and poverty in Africa. *International Journal of Sustainable Development and World Ecology* 6, 451–461.

Ijumba, J.N. and Lindsay, S. (2001) Impact of irrigation on malaria in Africa: paddies paradox. *Medical and Veterinary Entomology* 15, 1–11.

Jalloh, A., Faye, M.D., Roy-Macauley, H., Sérémé, P., Zougmoré, R., Thomas, T.S., *et al.* (2013) Summary and conclusions. In: Institute, I.F.P.R. (ed.) *West African Agriculture and Climate Change: A Comprehensive Analysis.* International Food Policy Research Institute, Washington, DC.

Jarvis, A., Ramirez-Villegas, J., Herrera Campo, B. and Navarro-Racines, C. (2012) Is cassava the answer to African climate change adaptation? *Tropical Plant Biology* 5, 9–29.

Jones, J.D.G. (2011) Why genetically modified crops? *Philosophical Transactions of the Royal Society A* 369, 1807–1816.

Kahl, C. (2006) *States, Scarcity, and Civil Strife in the Developing World.* Princeton University Press, Princeton, New Jersey.

Kapuściński, R. (1998) *The Shadow of the Sun.* Penguin, Melbourne, Australia.

King, M. (1990) Health is a sustainable state. *The Lancet* 336, 664–667.

Kula, N., Haines, A. and Fryatt, R. (2013) Reducing vulnerability to climate change in sub-Saharan Africa: the need for better evidence. *PLoS Medicine* 10, e1001374, doi:10.1371/journal.pmed.1001374.

McMichael, A.J., Butler, C.D. and Weaver, H.J. (2008) Climate Change and AIDS: A Joint Working Paper. United Nations Environment Program, United Nations AIDS, Geneva, Switzerland (http://data. unaids.org/pub/BaseDocument/2008/20081223_unep_unaids_joint_working_paper_on_cca_en.pdf. accessed 27 March 2014).

Meyrowitsch, D.W., Pedersen, E.M., Alifrangis, M., Scheike, T.H., Malecela, M.N., Magesa, S.M., *et al.* (2011) Is the current decline in malaria burden in sub-Saharan Africa due to a decrease in vector population? *Malaria Journal* 10, 188, doi:10.1186/1475-2875-10-188.

Molesworth, A.M., Cuevas, L.E., Connor, S.J., Morse, A.P. and Thomson, M.C. (2003) Environmental risk and meningitis epidemics in Africa. *Emerging Infectious Diseases* 9, 1287–1293.

Moszynski, P. (2011) Militant attacks are jeopardising famine relief in Horn of Africa. *BMJ* 343, d6729.

Ndenyele, W.O. and Badurdeen, F.A. (2012) Looking beyond gender in humanitarian interventions: a study of a drought-stricken region of Kenya. *Gender and Development* 20, 323–336.

New, M., Liverman, D., Schroder, H. and Anderson, K. (2011) Four degrees and beyond: the potential for a global temperature increase of four degrees and its implications. *Philosophical Transactions of the Royal Society A* 369, 6–19.

Nicholls, R.J., Marinova, N., Lowe, J.A., Brown, S., Vellinga, P., de Gusmão, D., *et al.* (2011) Sea-level rise and its possible impacts given a 'beyond 4°C world' in the twenty-first century. *Philosophical Transactions of the Royal Society A* 369, 161–181.

Nogherotto, R., Coppola, E., Giorgi, F. and Mariotti, L. (2013) Impact of Congo Basin deforestation on the African monsoon. *Atmospheric Science Letters* 14, 45–51.

Palmer, R. (2011) Would Cecil Rhodes have signed a Code of Conduct? Reflections on global land grabbing and land rights in Africa, past and present. International Conference on Global Land Grabbing. LDPI, International Conference on Global Land Grabbing, IDS Sussex.

Polonsky, J.A., Ronsse, A., Ciglenecki, I., Rull, M. and Porten, K. (2013) High levels of mortality, malnutrition, and measles, among recently-displaced Somali refugees in Dagahaley camp, Dadaab refugee camp complex, Kenya. *Conflict and Health* 7, doi:10.1186/1752-1505-7-1.

Population Media Center (2012) Nigerian President Hears Cacophony After Ignoring Population Taboo (http://www.populationmedia.org/2012/07/05/nigerian-president-hears-cacophony-after-ignoring-population-taboo/, accessed 27 March 2014).

Rodney, W. (1972) *How Europe Underdeveloped Africa.* Zimbabwe Publishing House, Harare.

Rosenthal, E. (2012) Nigeria tested by rapid rise in population. *New York Times* (http://www.nytimes. com/2012/04/15/world/africa/in-nigeria-a-preview-of-an-overcrowded-planet.html, accessed 7 April 2014).

Rulli, M.C., Saviori, A. and D'Odorico, P. (2013) Global land and water grabbing. *Proceedings of the National Academy of Sciences* 110, 892–897.

Seck, P.A., Diagne, A. and Bamba, I. (2013) Designing innovative agriculture policies in Africa. In: Coudel, E., Devautour, H., Soulard, C.T., Faure, G. and Hubert, B. (eds) *Renewing Innovation Systems in Agriculture and Food. How To Go Towards More Sustainability?* Wageninen Academic Publishers, the Netherlands, pp. 205–220.

Simon, F., Savini, H. and Parola, P. (2008) Chikungunya: a paradigm of emergence and globalization of vector-borne diseases. *Medical Clinics of North America* 92, 1323–1343.

Tefera, T. (2012) Post-harvest losses in African maize in the face of increasing food shortage. *Food Security* 4, 267–277.

Thorpe, A. (2009) Enteric fermentation and ruminant eructation: the role (and control?) of methane in the climate change debate. *Climatic Change* 93, 407–431.

Trape, J.-F., Tall, A., Diagne, N., Ndiath, O., Ly, A.B., Faye, J., *et al.* (2011) Malaria morbidity and pyrethroid resistance after the introduction of insecticide-treated bednets and artemisinin-based combination therapies: a longitudinal study. *The Lancet Infectious Diseases* 11, 925–932.

United Nations High Commissioner for Refugees (UNHCR) (2013) *Displacement. The New 21st Century Challenge. Global Trends 2012.* UNHCR, Geneva, Switzerland.

Verschuren, D. (2003) Global change: the heat on Lake Tanganyika. *Nature* 424, 731–732.

23 Zoonotic Diseases and Their Drivers in Africa

Delia Grace and Bernard Bett

International Livestock Research Institute, Nairobi, Kenya

23.1 Introduction

The reliance of our species on animals, both domestic and wild, affects our lives profoundly, including some of the risks we encounter and our resilience to face risk. Currently, around 24 billion livestock (19 billion of them in poor countries) provide food, income and other products and services for around 7 billion people (5 billion of whom are in poor countries) (Grace *et al.*, 2012a).

While billions of animals are owned and reared for food, hundreds of millions are kept for work, sport, companionship and research. Millions more animals are hunted, trapped or scavenged for food or other use. Livestock and wild animal products and by-products provide clothing, shelter, utensils, tools, fuel and building material; they have medicinal, social and ritual uses. Other important human–animal interactions include killing animals considered 'pests', conserving animals considered to have aesthetic or other values and inadvertently changing animal populations as we alter ecosystems for agriculture, habitation, natural resource exploitation and leisure. Increasing human populations are generally accompanied by decreases in biodiversity and wildlife numbers, but also by increases in farmed, companion, pest and peri-domestic (synanthropic) animals. This relation may not hold true in post-industrial societies, which may be associated with increases in wild animals as agricultural land reverts from farms to forest and reserves.

Animals provide income, nutrition, security and psychological benefits. But food that nourishes can contain pathogens that kill; manure that grows crops can poison water; and livestock that survive better than crops in unpredictable climates produce greenhouse gases (GHGs) that contribute to climate change. The current numbers of both livestock and people are of orders of magnitude greater than ecosystems have ever previously been required to support, and nowhere are numbers changing faster than in Africa, the cradle of humanity.

The relationships between livestock keeping and human well-being are multiple and complex (see Fig. 23.1). This chapter focuses on one of the most important relationships; that is, the link between livestock and zoonoses in the context of sub-Saharan Africa.

23.2 The Unique African Context and Its Implication for Zoonoses

Endowed with rich natural resources, sub-Saharan Africa none the less bears a heavy burden of human poverty. Poverty can be defined as a pronounced deprivation in well-being, and while no single measure captures all its dimensions, internationally comparable metrics, such as the US$1 a day (the 'dollar a day' poverty line,

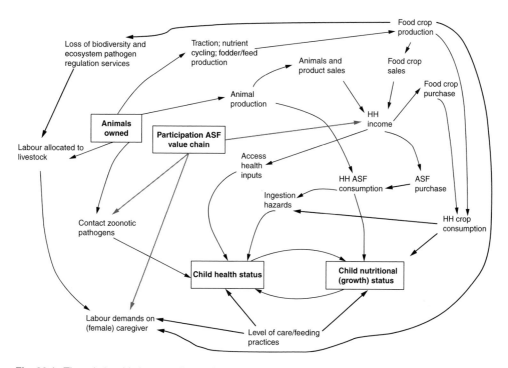

Fig. 23.1. The relationship between livestock and human well-being (modified from Randolph *et al.*, 2007). Hypothesized causal linkages between livestock keeping and human nutrition and health outcomes among the poor (adapted from Nicholson *et al.*, 2003). ASF = animal-source food; HH = household.

established in 1999, was recalibrated to US$1.25 a day in 2008), are useful for spatial and temporal comparisons. These measures show that while, globally, poverty has been declining over recent decades, the gains have been restricted mostly to South, East and South-east Asia and Latin America. In sub-Saharan Africa, the number of poor people is stable or increasing. In 2008, 48% of people in this region were living in poverty (United Nations, 2012), a higher proportion than in any other part of the world.

Around one in three of poor African families keeps livestock. There is debate as to whether the popularity of livestock keeping among the poor is a pathway out of poverty or a mechanism and expression of it (Perry and Grace, 2009), destined to be replaced by highly intensive, and impersonal, livestock systems (Dijkman, 2009), as has happened in other continents. Meanwhile, in contrast to most of the rest of the world, livestock keeping in Africa continues to be dominated by smallholder farming[1] and extensive pastoral systems which, while requiring low external inputs, have lower productivity per livestock unit (Table 23.1).

The persistence of poverty in Africa, the high proportion of poor families who keep livestock and the relative low level of agricultural intensification and input use have implications for the emergence and persistence of zoonoses in this continent. The rest of this chapter focuses on aspects of zoonoses in ecosystems of most importance to Africa. These include the high vulnerability to endemic zoonoses and the relative protection from the outbreak zoonoses associated with livestock intensification. Also distinct to the populations of this enormous and diverse region are the high dependence on wild-caught 'bushmeat' and the exceptional vulnerability to a deadly trifecta of climate change, high rates of demographic increase and uncontrolled disease.

23.3 Living with the Enemy: Proximity to Livestock and Other Animals Fosters Pathogen Access

Poor people in Africa live in close association with the animals they rely on for income, food, work and sociocultural functions. Livestock are valuable

Table 23.1. Number of animals and people (million) in different farming systems in Africa (from Herrero et al., 2013). Around one in three of poor African families keep livestock (Thornton et al., 2003).

Farming system	Agropastoral/ pastoral	Mixed extensive	Mixed intensifying	Others (e.g. urban)
Bovines	62	100	21	10
Goats	72	78	27	10
Sheep	84	79	22	8
Pigs	3	5	4	4
Poor LKs[a]	38	84	36	16

[a]LKs = livestock keepers.

assets, and in sedentary agricultural systems, livestock are often kept beside or inside houses, increasing the risk of the transmission of pathogens between livestock and people. For example, even in urban Nairobi, the capital of Kenya, one in 80 households keeps dairy cattle and one in 8 households keeps pigs (Kang'ethe et al., 2012). In many systems, children have an important role in herding and looking after young animals, making them especially vulnerable to disease transmission. Women are at risk from their role in milking, feeding, taking care of young animals and the preparation of animal-source foods (Grace et al., 2008).

Compounding the high contact rate is the high susceptibility of the human host: poor people are often undernourished and/or have intercurrent infections that make them more susceptible to zoonoses. Particularly dangerous are immune-suppressive conditions such as HIV/AIDS, and possibly aflatoxicosis (Wu et al., 2010). Moreover, poor people have less access to health services for diagnosis and treatment, and the underfunded health services that poor people use have less capacity to differentiate zoonotic diseases from diseases with similar symptoms (Schelling et al., 2007). When your only tool is chloroquine, then everything looks like malaria.

While keeping animals is an important risk factor, killing and consuming animals are two additional pathways for contacting the pathogens that animals harbour. The poor consume less livestock products than do richer people, but their killing and consumption habits are often more risky. The process of slaughter and butchery exposes humans to the bodily tissues and fluids of animals, and is a known risk factor for Rift Valley fever (RVF), brucellosis, anthrax, cysticercosis and other diseases. A study of butchers in Ibadan, Nigeria, reported an exceptionally high incidence

of gastrointestinal disease – one a fortnight. This may be linked to business practices – butchers ate pieces of raw meat in front of customers to demonstrate their 'safeness' and routinely ate the meat which they were unable to sell and which was starting to decay or 'go off' (Grace et al., 2012a). Cultural practices can increase exposure to viable pathogens. Cooking is effective at inactivating many pathogens; however, women who prepare household meals often lack awareness of cross-contamination, and thus place themselves and their families at risk. Eating raw meat is considered a delicacy in Ethiopia; as a consequence, meat-borne disease is prevalent. 'Who told you having a tapeworm is a disease: a man without a tapeworm is not really a man', was the reply of a respondent to a researcher in one of our studies. Moreover, poverty drives the consumption of livestock products that people suspect to be unsafe: a recent study in Tanzania found that nearly two-thirds of people killed and ate poultry which they perceived to be sick (Young et al., 2012).

Bats also interact closely with humans, livestock and wildlife, and are natural reservoirs for high-impact zoonotic viruses, many of which are present in Africa, including rabies, Hendra, Ebola, Marburg, Nipah and SARS-corona viruses (Wood et al., 2012). Bats often live in high-density colonies and have co-evolved over many millions of years with many pathogens; the majority of these are RNA viruses that have a short generation interval and a short genome that can adapt to multiple hosts over a short time.

Various rodent species are also usually found in dump sites and other filthy areas like marketplaces, where they can come into direct or indirect contact with humans. The most well-known zoonotic disease associated with rodents is bubonic plague caused by a bacterium,

Yersinia pestis. There are, however, many other diseases that rodents can transmit. These include leptospirosis, hantavirus diseases, salmonellosis and monkey pox. It is also believed that rodents can act as reservoirs of RVF virus in various parts of Africa (Chevalier *et al.*, 2012). A high mortality rate among rodents (*Arvicanthis abyssinicus nairobae* and *Rattus rattus kijabius*) was observed on farms affected during the 1930 RVF epizootic[2] in Kenya, suggesting that they were involved in the disease transmission. In Egypt, RVF virus was detected by reverse transcriptase polymerase chain reaction (RT-PCR) in the blood of 29 *Rattus rattus* individuals among 300 sampled. Serological evidence of RVF virus infection was also observed in *Mastomys* spp., *Arvicanthis niloticus* and *Aethomys namaquensis* in Senegal, in West Africa. In South Africa, serological and experimental studies suggested that *A. namaquensis*, another rodent, could act as an amplifying host during inter-epizootic periods.

Given greater contact with animal carriers and inadequate health services for diagnosis and treatment, it is not surprising that people in poor countries are much more at risk from zoonoses: a recent study estimated that in least developed countries (mostly African), the relative burden of zoonoses and diseases recently emerged from animals was 500 times greater than the burden of these diseases in rich countries (Grace *et al.*, 2012b).[3]

23.4 Where Poverty Protects

Some aspects of the farming systems common in Africa may reduce the risk of disease emergence and spread. Bird flu put zoonoses on the world's front page and changed the way people thought about the unforeseen and unwanted health consequences (or as economists would say, negative externalities) of a cheap chicken in every pot. Highly pathogenic avian influenza (HPAI; bird flu) is much researched (though much less managed) and can be seen as an exemplar of the type of outbreak disease that is characteristic of rapidly growing, intensified animal agriculture (Lebarbenchon *et al.*, 2010).

When bird flu arrived in Indonesia in 2003, it found the perfect stage for rapid, destructive spread in the country's growing poultry industry. Population growth, wealth, urbanization and changes in diet led to greatly increased demand for animal-source protein. Monogastric animals (pigs and poultry) can be scaled up much more rapidly than digastric (multiple stomachs or ruminant) cattle and sheep, and because of cultural and religious reasons, poultry is strongly preferred in Indonesia. Commercial breeds of low genetic diversity were crammed densely in substandard housing, with poor biosecurity and constant contact with wildlife, including pests. Methods of keeping poultry on a massive scale were novel to the region, having been adopted rapidly from America and Europe. Oversight by government, the private sector and consumers was limited, and high levels of antibiotics were probably used to mask poor husbandry and to enable animals to produce despite stress and immunosuppression. These conditions provided the tinder for the wildfire epidemics of HPAI epidemics that spread throughout Asia and along the Nile delta in Africa but, despite several incursions, did not establish in sub-Saharan Africa (Bett *et al.*, 2012).

We can think of rapidly intensifying and unregulated livestock systems as 'hot spots', where development and rapid change surpass the speed of the Industrial Revolution, threatening animal and perhaps human health in ways never seen before. Prosperity has stimulated a rapid expansion of concentrated animal-feeding operations in Asia and South America and has fostered a 'wild west' intensification mentality, where corners may be cut. Operations are conducted on a massive scale, but with minimal regulation and with limited transparency, even if epidemics occur. But the Western countries that have developed intensive animal agriculture are not fully immune from its drawbacks. In a globalized world, pathogens in animal products, animals, humans or fomites[4] can travel around the world in less time than the incubation period of disease. While previous livestock pandemics were linked with intensive livestock production in Western Europe and the USA, the past decade has seen more disease emergence in South America and South-east Asia (Grace *et al.*, 2012b). The World Bank estimates that pandemic zoonoses between 1997 and 2009 cost the world at least US$80 billion (World Bank, 2012), with most of the losses incurred in rich countries. The chickens are coming home to roost.

In contrast, an entirely different set of health problems threaten 'cold spots' throughout much of Africa, areas locked in a time warp of underdevelopment and crippled by old diseases. Here, livestock keeping is still predominately extensive and low input. As a result, the 'crowd diseases' of livestock intensification may be less likely to establish. For example, recent studies showed that avian flu was not likely to establish in Nigeria, one of the most populous countries in Africa, because the poultry population was too low and poultry carriers too few to allow disease to establish (Bett *et al.*, 2012). This implies that avian influenza probably was not controlled in sub-Saharan Africa, rather the disease burnt out unassisted (Grace and McDermott, 2011). Yet policy and funding is often driven by the concerns of the 'worried well' in rich countries, and even in Africa, diseases of intensification that are unlikely to establish receive more attention and support than the silent epidemic of old-fashioned zoonoses that reliably kill and sicken on a large scale. This inevitably diverts funding and scarce human resources from the diseases of the poor.

23.5 The Risks of Eating Your Relatives

Africa is especially vulnerable to diseases at the livestock/wildlife interface. Bushmeat (by which we include wild animals hunted, trapped or found dead in unfarmed forest and savannah) has traditionally been an important and valued part of the African diet, especially for poor people, for whom it is a 'free' item in their increasingly expensive food budget (poor people spend a much higher percentage of their income on food than rich people). Current use of wild animals for meat is both demonstrably unsustainable and largely illegal, but these characteristics paradoxically drive greater and more irresponsible use in a predictable manifestation of the 'tragedy of the commons', where people engage in a rush to the bottom because they lack confidence that their neighbours will refrain from exploitation even if they do.

Especially risky as a source of new diseases are the animals that are related most closely to humans. An analysis of human disease emergence found that primates, along with bats and rodents,

were over-represented as sources of new diseases (Grace *et al.*, 2011). The consumption of meat from chimpanzees found dead has been associated with Ebola outbreaks in humans; although this may be due to handling the dead bodies of infected chimpanzees rather than via ingestion of contaminated meat. Simian foamy virus has been found infecting people in direct contact with fresh, non-human primate meat. Monkey pox, though rarely fatal, is an emerging disease in central Africa, and a threat to a rapidly growing young population that has not been vaccinated against smallpox (Rimoin *et al.*, 2010). The adaptation of non-human primate lentiviruses to humans was probably the most significant pathogen species jump in human history, affecting more people than any other disease, and it occurred in Africa. The phylogeny of current circulating human immunodeficiency virus (HIV) strains is consistent with at least two such species-jumping events (HIV 1 and HIV 2) over the past from chimpanzee and sooty mangabey. (Other studies suggest a role for gorillas in transmission to people.) This is believed to have occurred through regular infection of humans with simian immunodeficiency virus (SIV) through the consumption of bushmeat and subsequent viral adaptation through mutation or reassortment.

However, it has been argued that while SIV may have spilled over many times to people, these were dead-end events because social, cultural and economic conditions did not allow the establishment of the disease in the human population. Colonialization has been identified as an additional factor, along with the growth of towns with large populations, as well as prostitution, essential for the establishment of HIV/AIDS (Worobey *et al.*, 2008).

23.6 The Deadly Trifecta: Climate Change, Demography and Disease

The Horsemen of the Apocalypse do not ride alone, and historically some of the most deadly plagues have been associated with major societal dislocations. A classic case is the neolithic transition from hunter-gathering to agriculture, which most likely led to both increases in zoonoses and also livestock pathogens evolving into human diseases. These 'lethal gifts of livestock' probably include measles, smallpox, diphtheria and influenza. Tuberculosis appears to have gone the other way: originating in humans, it spilled over to become a deadly disease of cattle (Gagneux, 2012).

Other civilization-altering crises that have been implicated in the epidemic, and pandemic plagues that closely followed them include:

- Mediaeval: bubonic plague associated with massive crop failures resulting from the Little Ice Age;
- 16th century: smallpox, measles and other diseases killed up to 90% of Native Americans following the European 'conquests'.
- 19th century: novel diseases along with the dislocation of colonization wipe out Tasmanians, almost eradicating their culture.
- 19th century: consumption (tuberculosis) epidemics associated with industrialization, overcrowding and poverty.
- 20th century: Spanish influenza associated with the war of 1914–1918.

Africa, home to 15% of the world's population, is a major importer of grains, already responsible for 32% of the world's rice and 25% of the world's wheat imports (Seck et al., 2013). Although it has a comparatively low population density (though a high population growth rate), a substantial part of Africa's agricultural potential is considered highly vulnerable to climate change (Collier et al., 2008), especially if there is an average of four degrees of warming (New et al., 2011). Some coastal cities of Africa are likely to become partially submerged by sea level rise (Nicholls et al., 2011), and extreme weather events are likely to increase, including flooding.

Changes in disease distribution in Africa, caused by climate change, are also likely. Vector-borne diseases are especially climate sensitive. In Africa, arthropod-borne zoonotic diseases such as sleeping sickness (transmitted by tsetse flies), West Nile virus (transmitted by a wide range of mosquito species) and Crimean-Congo haemorrhagic fever (transmitted by ticks) have recurrent seasonal patterns influenced by the population dynamics of their arthropod vectors. Population densities of these vectors shrink during the dry seasons as breeding sites contract and preferred sources of blood meals relocate to areas where water and pasture can be found. This is particularly true in the range or pastoral areas, where

livestock and wildlife move between dry and wet grazing sites as a survival strategy against drought. At the same time, harsh climatic conditions such as high temperatures and low humidity increase the mortality rates of these vectors. These effects have important implications on infectious disease transmission, given that high mortality rates translate to a reduction in the overall lifespan of a vector; yet, to transmit a pathogen successfully, a vector has to live longer than the intrinsic incubation period of that pathogen.

Climate variability also has many other indirect effects on the transmission of infectious diseases. In arid and semi-arid areas, livestock, wildlife and people utilize common sources of water during the dry periods when seasonal rivers dry up. These sources often comprise stagnant water pans constructed to trap surface run-off. However, the microbial quality of the water obtained from such pans is usually low (especially during the dry season), because of frequent contamination with animal faeces, urine and the carcasses of sick or emaciated animals that die while taking water (Shivoga and Coppock, 2003). Some of the zoonotic diseases that can be acquired from drinking such water include cryptosporidiosis, giardiasis, leptospirosis, collibacilosis and haemolytic uraemic syndrome caused by Escherichia coli (E. coli) O157:H7.

Climate change is inducing, and will increasingly induce, land-use changes, leading to disruption of the composition and functioning of ecosystems, and hence stress and probable decline in the range of animal and plant species in such areas. A reduction in biodiversity can lead to an increase in disease transmission, if the animal species that persist or thrive as biodiversity is lost are more efficient amplifying hosts (Keesing et al., 2010).

Despite its many achievements, Africa continues to experience underdevelopment, poor governance and high rates of population growth that render the continent especially vulnerable to the emergence and spread of new plagues. In this context, added stressors from anthropogenic climate change can only make things worse. A warmer, wetter, poorer world is likely to be a sicker world, and Africa, whose livestock produces less than 3% of the total GHGs from ruminants, is especially vulnerable to climate change.

Box 23.1   The Coming Plague? Rift Valley Fever

There is great uncertainty about how climate might change in different regions of Africa and how this will interact with other factors. However, there is already some evidence that the African climate is becoming more variable and that temperature and rainfall are shifting in ways that are harming agriculture.

Rift Valley fever (RVF) is a viral zoonosis with an interesting and imperfectly understood epidemiology, first detected in the Rift Valley of Kenya as a plague among imported sheep; indigenous breeds appear much less susceptible. Like West Nile virus, which is emerging in Europe and the USA, RVF is an arbovirus transmitted mainly by mosquitoes. Before 1977, it was considered mainly an animal health problem, causing abortions in pregnant ewes and death in lambs. Since then, for unclear reasons, humans have been increasingly affected, with epizootics resulting in the death of millions of animals and thousands of people, with major impacts on poverty and well-being.

The epidemiology of RVF depends on land use and climate, and several patterns exist. In forested zones of Africa, the disease is endemic, but in the plateau grasslands of East Africa, epidemics in wildlife occur in 3- to 15-year cycles, closely associated with excess rainfall driven by the El Niño–Southern Oscillation. Currently, both demography and climate change are leading to greatly expanded irrigation in regions of East Africa previously occupied mainly by pastoralists, as well as the introduction of new and more susceptible livestock breeds. Basic biology predicts that these changes in farming system may lead to the increased transmission and possibly pathogenicity of RVF.

Because many vectors can transmit the virus, it appears to have a high potential for emergence from Africa to other regions of the world.

23.7   Conclusion

African agriculture differs quantitatively and qualitatively from that practised in other continents, and this also has implications for the emergence and spread of zoonotic disease. Climate change is just one of the drivers that are changing African farming systems: notable trends of the past decades include increasing irrigation, movement of farmers to new regions, urban agriculture, increased use of exotic livestock breeds and increase in plantation farms funded by foreign investment.

Evidence suggests that emerging diseases are unlikely to have devastating impacts in stable, wealthy societies with good access to health care, but may be much more deadly in impoverished and immunosuppressed societies undergoing rapid growth in degraded and polluted environments. Given continuing population growth in the poorest countries, unplanned urbanization, likely failure to meet most Millennium Development Goals (MDGs) and vulnerability to climate change, there are real risks we are creating in Africa an ideal milieu for the emergence and propagation of new diseases.

On the other hand, Africa has achieved notable health victories in recent years. Infant mortality has plummeted, and more children than ever are completing primary education. The latest MDG report found that the absolute number of people living in extreme poverty had, for the first time since reporting began, declined in Africa. A recent review of poverty and livestock disease concluded that human and economic development was essential for disease control, and increasing wealth could trump increasing warmth (Perry and Grace, 2009). The history of humanity started in Africa, and Africa has a key role, not only as a custodian of irreplaceable resources but also as a tipping point from where the world's future can get dramatically worse or start to improve.

Acknowledgements

This work 'Dynamic Drivers of Disease in Africa: Ecosystems, livestock/wildlife, health and well-being: NE-J001570-1' was partly funded with support from the Ecosystem Services for Poverty Alleviation Programme (ESPA). The ESPA programme is funded by the Department for International Development (DFID), the Economic and Social Research Council (ESRC) and the Natural Environment Research Council (NERC). Other funding was provided by CGIAR Research Program Agriculture for Nutrition and Health.

Notes

[1] Low-intensity agriculture also persists in parts of Asia and Central America, and in the Andes.
[2] Epizootics are disease outbreaks affecting many animals at the same time.
[3] Schistosomiasis is an important parasitic disease in Africa, but is not discussed in this chapter as it is not a zoonosis.
[4] Inert material which can carry infectious material, at least for a limited time.

References

Bett, B., Henning, J., Abdu, P., Okike, I., Poole, J., Young, J., *et al.* (2012) Transmission rate and reproductive number of the H5N1 highly pathogenic avian influenza virus during the December 2005–July 2008 epidemic in Nigeria. *Transboundary and Emerging Diseases* doi:10.1111/tbed.12003.

Chevalier, V., Rakotondrafara, T., Jourdan, M., Heraud, J.M., Andriamanivo, H.R., Durand, B., *et al.* (2012) An unexpected recurrent transmission of Rift Valley Fever virus in cattle in a temperate and mountainous area of Madagascar. *PLoS Neglected Tropical Diseases* 5, e1423.

Collier, P., Conway, G. and Venables, T. (2008) Climate change and Africa. *Oxford Review of Economic Policy* 24, 337–353.

Dijkman, J. (2009) Innovation capacity and the elusive livestock revolution. In: Dijkman, J. (ed.) *LINK News Bulletin*. LINK, UNU-MERIT, Maastricht, the Netherlands.

Gagneux, S. (2012) Host–pathogen coevolution in human tuberculosis. *Philosophical Transactions of the Royal Society B: Biological Sciences* 367, 850–859.

Grace, D. and McDermott, J. (2011) Livestock epidemics and disasters. In: Wisner, B., Gaillard, J.C. and Kelman, I. (eds) *Handbook of Hazards and Disaster Risk Reduction*. Routledge, London, pp. 372–383.

Grace, D., Randolph, T., Olawoye, J., Dipelou, M. and Kang'ethe, E. (2008) Participatory risk assessment: a new approach for safer food in vulnerable African communities. *Development in Practice* 18, 611–618.

Grace, D., Jones, B., McKeever, D., Pfeiffer, D., Mutua, F., Njuki, J., *et al.* (2011) Zoonoses: wildlife/livestock interactions. A report to the Department for International Development, UK. Submitted by The International Livestock Research Institute, Nairobi, and Royal Veterinary College, London.

Grace, D., Dipeolu, M., Olawoye, J., Ojo, E., Odebode, S., Agbaje, M., *et al.* (2012a) Evaluating a group-based intervention to improve the safety of meat in Bodija Market, Ibadan, Nigeria. *Tropical Animal Health and Production* 44, 61–66.

Grace, D., Gilbert, J., Randolph, T. and Kang'ethe, E. (2012b) The multiple burdens of zoonotic disease and an ecohealth approach to their assessment. *Tropical Animal Health and Production* 44, 67–73.

Herrero, M., Grace, D., Njuki, K., Johnson, N., Enahoro, D., Silvestri, S., *et al.* (2013) The roles of livestock in developing countries. *Animal: An International Journal of Animal Bioscience* 7 (Supplement s1), 3–18.

Kang'ethe, E.K., Kimani, V.N., McDermott, B., Grace, D., Lang'at, A.K., Kiragu, M.W., *et al.* (2012) A transdisciplinary study on the health risks of cryptosporidiosis from dairy systems in Dagoretti, Nairobi, Kenya: study background and farming system characteristics. *Tropical Animal Health and Production* 44(S1), 3–10.

Keesing, F., Belden, L.K., Daszak, P., Dobson, A., Harvell, C.D., Holt, R.D., *et al.* (2010) Impacts of biodiversity on the emergence and transmission of infectious diseases. *Nature* 468, 647–652.

Lebarbenchon, C., Feare, C.J., Renaud, F., Thomas, F. and Gauthier-Clerc, M. (2010) Persistence of highly pathogenic avian influenza viruses in natural ecosystems. *Emerging Infectious Diseases* 16, 1057–1062.

New, M., Liverman, D., Schroder, H. and Anderson, K. (2011) Four degrees and beyond: the potential for a global temperature increase of four degrees and its implications. *Philosophical Transactions of the Royal Society A* 369, 6–19.

Nicholls, R.J., Marinova, N., Lowe, J.A., Brown, S., Vellinga, P., Gusmão, D.d., *et al.* (2011) Sea-level rise and its possible impacts given a 'beyond 4°C world' in the twenty-first century. *Philosophical Transactions of the Royal Society A* 369, 161–181.

Nicholson, C.F., Mwangi, L., Staal, S.J. and Thornton, P.K. (2003) Cow ownership and child nutritional status in Kenya. *Research Bulletin 11–2003*. Department of Applied Economics and Management, Cornell University, Ithaca, New York.

Perry, B. and Grace, D. (2009) The impacts of livestock diseases and their control on growth and development processes that are pro-poor. *Philosophical Transactions of the Royal Society B* 364, 2643–2655.

Randolph, T.F., Schelling, E., Grace, D., Nicholson, C.F., Leroy, J.L., Cole, D.C., *et al.* (2007) Invited review:
role of livestock in human nutrition and health for poverty reduction in developing countries. *Journal of
Animal Science* 85, 2788–2800.

Rimoin, A.W., Mulembakani, P.M., Johnston, S.C., Lloyd Smith, J.O., Kisalu, N.K., Kinkela, T.L., *et al.*
(2010) Major increase in human monkeypox incidence 30 years after smallpox vaccination campaigns
cease in the Democratic Republic of Congo. *Proceedings of the National Academy of Sciences* 107,
16262–16267.

Schelling, E., Grace, D., Willingham, A.L. and Randolph, T.F. (2007) Which research approaches for pro-
poor control of zoonoses? *Food and Nutrition Bulletin* (2 Suppl), S345–356.

Seck, P.A., Diagne, A. and Bamba, I. (2013) Designing innovative agriculture policies in Africa. In: Coudel, E.,
Devautour, H., Soulard, C.T., Faure, G. and Hubert, B. (eds) *Renewing Innovation Systems in
Agriculture and Food. How To Go Towards More Sustainability?* Wageninen Academic Publishers, the
Netherlands, pp. 205–220.

Shivoga, W. and Coppock, D.L. (2003) For pastoralists the risk may be in the drinking water: the case of
Kargi, N. Kenya. Research Brief 03-03-PARIMA. Global Livestock Collaborative Research Support
Program. University of California, Davis, California.

Thornton, P.K., Kruska, R.L., Henninger, N., Kristjanson, P.M., Reid, R.S. and Robinson, T.P. (2003)
Locating poor livestock keepers at the global level for research and development targeting. *Land Use
Policy* 20, 311–322.

United Nations (2012) *The Millennium Development Goals Report*. United Nations, New York.

Wood, J.L.N., Leach, M., Waldman, L., MacGregor, H., Fooks, A.R., Kate, E.J., *et al.* (2012) A framework
for the study of zoonotic disease emergence and its drivers: spill-over of bat pathogens as a case
study. *Philosophical Transactions of The Royal Society of Biological Sciences* 367, 2881–2892.

World Bank (2012) *People, Pathogens and our Planet, Volume 2, The Economics of One Health*. World
Bank, Washington, DC.

Worobey, M., Gemmel, M., Teuwen, D.E., Haselkorn, T., Kunstman, K., Bunce, M., *et al.* (2008) Direct evidence of extensive diversity of HIV-1 in Kinshasa by 1960. *Nature* 455, 661–664.

Wu, F., Narrod, C., Tiongco, M. and Liu, Y. (2010) *The Global Burden of Disease Caused by Foodborne
Aflatoxin*. International Food Policy Research Institute, Washington, DC.

Young, J., Grace, D., Young, M., Alders, R.G., Kibaya, A., Msami, H., *et al.* (2012) Newcastle disease
vaccination: from technology to poverty reduction. 13th International Symposium on Veterinary
Epidemiology and Economics, 20–24 August, Maastricht, Netherlands.

24 Climate Change, Food and Energy: Politics and Co-benefits

Ole Faergeman

Department of Cardiology B, Aarhus University Hospital, Aarhus, Denmark

Can we make actual in our minds the sometimes urgent things we say we know?
(Wendell E. Berry, 2012 Jefferson Lecture, National Endowment for the Humanities, 23 April 2012)

24.1 Introduction

The many manifestations of climate change are closely related to how billions of people obtain their food as well as their fuel. One relationship is bidirectional. Obtaining food and fuel contributes to climate change, and climate change and what we do about it affects the production of food and fuel. Another is the causal relationship between climate change and human disease. It has at least two facets. Climate change can be a cause of human diseases such as malnutrition and certain infections, whereas other diseases such as the most common forms of cardiovascular disease share causes with climate change.

If the political will existed, anthropogenic climate disaster might be avoided, and human health improved along the way, by eating more plant foods, planting more trees and by stopping the use of fossil fuels. This agenda is simple in principle, but, at our peril, it is proving difficult to implement in practice.

Agriculture provides almost all of the food required by the huge and growing human population, but farming and the food industry account for about one-quarter, perhaps more, of anthropogenic greenhouse gas (GHG) emissions. A substantial proportion of those emissions is due to raising livestock. Meat and milk are valuable sources of protein, iron, calcium and more, but eating large amounts of some animal products raises the chance of cardiovascular disease and certain cancers. This is a problem especially in affluent countries, whereas farming for feed and fuel rather than for plant food increases food prices and causes undernutrition among the poor in both poor and rich countries.

We farm feed to grow animals, but we also farm crops for fuel, allegedly because this will reduce GHG emissions. Burning the products of recent photosynthesis (biomass) rather than ancient photosynthesis (fossil fuels) is of questionable climatic benefit, however. This chapter is a brief attempt to describe how some of these intersections and networks also affect the nutritional background for human health and disease.

Three questions are considered. First, how much livestock do we need to raise and eat to promote human health and yet reduce the risk of climate disaster? Second, what are the nutritional consequences of trying to substitute agrofuels for fossil fuels? Third, why is it so difficult for us to avert preventable disaster?

24.2   How Much Livestock Do We Need to Raise and Eat?

Intake[1] of meat and dairy as a proportion of total diet varies widely within and between countries. In general, the more affluent, the greater the intake (Steinfeld et al., 2006). For humans, there are no absolute physiological requirements for animal products, however. How much animal food to eat to be healthy depends on the availability of other forms of food, the presence or absence of gastrointestinal parasites, such as hookworm, and a range of determinants including food prices, cultural norms and political and economic forces. The production of livestock is associated with at least two sets of problems, nutritional and ecological/climatological.

Livestock and nutrition

The nutritional question can be considered in terms of foods, nutrients, absorption and metabolic demand.

In developed countries, populated mostly by people of European descent who are able to digest lactose easily, early research taught us to entertain a love–hate relationship with dairy products. Some of us, especially women, drank a lot of cow's milk in the hope of preventing osteoporosis and bone fractures. Others have avoided cow's milk in the hope of avoiding heart attacks. Reviews of the more recent literature indicate that the evidence for both perceptions is weak (Alvarez-Leon et al., 2006; Elwood et al., 2010; Soedamah-Muthu et al., 2011). How drinking and eating dairy products affect the risk of cancer is also not straightforward. While drinking milk probably protects against colorectal cancer, it may increase the risk of prostate cancer. To complicate matters further, milk is associated with a lower risk of bladder cancer, and cheese, but not milk, is associated with increased, not decreased, risk of colorectal cancer (World Cancer Research Fund/American Institute for Cancer Research, 2007). Thus, in thinking about the risk of the major diseases of affluent societies, dairy products come out neither clearly good nor bad for you.

A different picture has emerged when it comes to eating processed meat and 'red meat'.

The latter is an epidemiological term to designate meat from pigs, cows, sheep and goats.[2] Eating processed meat and red meat increases the risk of colorectal cancer (World Cancer Research Fund/American Institute for Cancer Research, 2007; Egeberg et al., 2013). It also seems to increase the risk of cardiovascular disease and death (Sinha et al., 2009; Bernstein et al., 2010). That evidence is considered convincing, although it comes from association studies, particularly cohort studies, rather than from randomized clinical trials. The relative risks associated with eating a lot rather than a little red meat and processed meat are approximately 40% (risk ratio 1.4). That is moderate from the individual perspective, but, given that most people eat meat, it is important in terms of public health. A reasonable comparison concerns environmental tobacco smoke ('passive smoking'), which has a similar or even lower risk (He et al., 1999). A sensible recommendation is to eat at most 500 g red meat/week and to limit the average intake of red meat to less than 300 g/week (World Cancer Research Fund/American Institute for Cancer Research, 2007).

Do we need to eat meat and drink milk at all? Of the four extant hominid genera, both chimpanzees and humans are carnivorous, and meat has been part of the human diet during evolution (Smil, 2002). Northern Europeans and several other ethnic groups, moreover, became able to drink cow's milk after the age of weaning due to a mutation in the gene for intestinal lactase (Bersaglieri et al., 2004).

Populations in developed countries do not in general need to be concerned about inadequate protein intake due to low meat intake (Millward and Garnett, 2010). The daily requirement for dietary protein is conventionally estimated to be 0.66 g/kg body weight. That estimate is based on obligatory daily nitrogen losses equivalent to approximately 0.3 g protein/kg body weight and assumes, conservatively, that the utilization of even good-quality protein is less than 50%. The true daily requirement for protein is therefore likely to be less than 0.66 g/kg. Since old people eat less than the young, protein must account for a higher proportion of energy intake for the elderly. Vegetable protein is less digestible than animal-source protein, but the biological value (a measure of how well the amino acid composition of food matches dietary requirements) is likely to be low only in developing countries with diets based on cereals or starchy roots. Moving from a diet with mainly animal protein to one with all vegetable protein could reduce the overall quality of protein (digestibility × biological value) from 0.95 to 0.80, but this is not a critical reduction, and it is even less important if some animal protein remains in the diet.

The absorption of iron from the intestine is a closely regulated function of body iron stores, of which serum ferritin is a marker. Those stores can fall considerably before there is detectable anaemia. Although haem iron from meat is absorbed more easily than non-haem iron from plants, its absorption improves when the phytate content of some plant food is lowered by cooking or when it contains ascorbate, citrate or certain spices. Thus, meat in the diet may be less important for iron homeostasis than commonly supposed.

Meat and meat products contribute about one-third (and dairy products about 17%) of dietary zinc in developed countries. Legumes, whole grains, seeds and nuts are also sources of zinc. Although they also contain phytic acid, an inhibitor of the absorption of zinc as well as iron, plant-based zinc sources could probably compensate for lower intakes of zinc by adults on a no-meat diet. Whether this conclusion pertains to children is an open and important question, because zinc deficiency is one of several causes of conditions such as stunting, hypogonadism and impaired cognitive development.

In developed countries, we can calculate the requirements for dietary calcium on the basis of calcium balance studies indicating peak bone retention rates of calcium (e.g. 250–300 mg/day) and percentage absorption from the intestine (e.g. 40%). Such requirements (e.g. 625–750 mg/day) are then adjusted upwards to give recommended intakes, for example 1000 mg/day, which are met fairly easily when diets contain large amounts of dairy products. The latter can provide almost half of dietary calcium, yet intake of milk is not related to rates of incident hip fractures in women (Bischoff-Ferrari et al., 2011), and rates of bone fractures are not higher in developing countries with low intakes of calcium (Millward and Garnett, 2010; Kanis et al., 2012). In fact, fracture rates are lower, suggesting that requirements for dietary calcium adapt to some extent to the amount available.

Meat, dairy products, eggs and fish provide most vitamin B12, and milk and other dairy foods provide most riboflavin. In both cases, it would be prudent to fortify foods or provide supplements when people decide or are forced to eat substantially less food from animals. The question of fortification is not straightforward, however, and there continues to be scientific uncertainty about many of these nutritional issues. Furthermore, the reductions in meat and dairy foods necessary to limit environmental damage really do pose nutritional challenges for some key nutrients. They can be met, however, by improved public health advice on alternative dietary sources and by increasing food fortification (Millward and Garnett, 2010). They can be overcome.

Livestock, ecology and climate

There are many humans, and they have many animals. The global wet biomass of 7 billion humans with an average weight of 60 kg is 420 million tonnes (Mt), and the biomass of our domesticated animals has increased from about 180 Mt in 1900 to about 620 Mt in 2000. In comparison, the biomass of all wild terrestrial mammals is probably below 40 Mt (Smil, 2002).

Anthropogenic emissions of GHGs total approximately 49 gigatonnes annually, calculated as carbon dioxide equivalents (CO_2e). The Food and Agriculture Organization (FAO) has estimated that 14.5% of those emissions stem from raising domesticated animals (Gerber et al., 2013). That estimate is considered authoritative, but others have argued that raising livestock contributes as much as 50% of GHG emissions (Goodland and Anhang, 2009).[3]

Central to the arguments in support of the higher estimate is that the FAO, in conforming to methods employed by the Intergovernmental Panel on Climate Change (IPCC), has ignored livestock respiration. Herrero et al., on the other hand, defend the FAO's view that CO_2 expired by livestock is not a net source of CO_2, since livestock eat plant material created by recent photosynthesis, which, as we know, removes CO_2 from Earth's atmosphere (Herrero et al., 2011).

World meat production in 2010 was 293 Mt, and annual per capita meat supply continues to rise in both developed (currently c.90 kg/capita) and developing countries (currently about 32 kg/capita) (Prakash and Stigler, 2012). These figures are in terms of carcass weight, slightly more than half of which is actually eaten (Smil, 2002). An average 42 kg carcass weight/capita (293 billion kg divided by 7 billion people) implies that we eat on average about 23 kg meat/year, or about 440 g/week, substantially more than the 300 g limit recommended by the World Cancer Research Fund (World Cancer Research Fund/American Institute for Cancer Research, 2007). Even the average consumption in developing countries now exceeds that limit.

So, how much livestock should we produce? Despite disingenuous claims by agribusiness to increase production to 'feed the world', we already produce more than we need. If this could be distributed fairly, 161 Mt of meat (carcass weight)/year would be enough to provide 300 g/week of meat on the plate for all 7 billion people. Yet, there is no simple formula or answer. Livestock serve useful purposes apart from becoming food (traction, landscaping, recycling of nutrients, social functions and more), and they pose problems apart from GHGs (pollution, eating food that we could eat, water depletion, deforestation and more) (Janzen, 2011). Pigs and poultry emit less GHGs than cattle, but they eat what we could eat, while cattle and other ruminants eat cellulosic plants we cannot eat, converting them, though at low efficiency, to meat and milk. A safe bet, however, is that less than half of the livestock we produce today is more than enough for 7 billion, and even 9 billion, people.

Industrial agriculture has certain advantages over small-scale farming. Handling of manure can be more efficient, for example, and animals in confinement in industrial agriculture spend less energy on movement than animals on extensive, small-scale farms (Steinfeld et al., 2006). Nevertheless, when factors such as source of feed for pigs are also considered, pigs raised on very small-scale farms ('backyards') emit slightly less CO_2e/kg carcass weight than pigs in industrial systems (Gerber et al., 2013). Moreover, neither harvest yields nor farmer income are necessarily lower on organic or semi-organic farms employing 3- or 4-year crop rotations (Davis et al., 2012; Seufert et al., 2012) than on farms using large amounts of chemical fertilizers and pesticides. Many of the drawbacks of livestock production (e.g. disruption of nutrient cycles, pollution, inhumane treatment) and growing crops (e.g. loss of biodiversity, eutrophication, climate change) are greatest, moreover, with intensive, industrial production of food.

Thus, we would probably be wise to decrease, not increase, the production of livestock and to allow most livestock to be produced by smallholders.

24.3   Agrofuels

According to the FAO and the World Food Programme, almost 1 billion people are undernourished (calorie intake in macronutrients is less than needed for light physical activity and maintenance of body weight). There is undernutrition in Europe and the USA, but the problem is greatest in developing countries in protracted crisis, for complex reasons including war, natural disasters and weak government, particularly in sub-Saharan Africa. Inability to buy food became increasingly important as a cause of undernutrition as food prices rose from 2006 to 2008, and again in 2011. The poor spend as much as 60–70% of their income on food, and when prices rise, poor households try to maintain energy intake at the expense of dietary diversity, protein and micronutrients; for example, by buying cheap rice rather than more expensive eggs and green leafy vegetables. Stunting due to undernutrition afflicts at least 170 million children, causing 7.6 million deaths before 5 years of age (World Cancer Research Fund/American Institute for Cancer Research, 2007).

The FAO, as a branch of the United Nations (UN) dependent on funding from national governments, is cautious in identifying the causes of the latest food price rises (FAO, 2011). In contrast, a study modelling food price data, performed by a non-governmental organization, suggests that the rising price of food in 2007/08 can be ascribed to speculation and to agrofuels. Whereas price volatility is due to speculation in agricultural commodities, the underlying steady rise in food prices results from the increasing production of agrofuels (Lagi et al., 2012). Regulation of speculation and subsidization of the agrofuel ('food fuel') industry are, of course, the prerogative of politicians.

Of the maize harvest in the USA, 42% is now used to produce bioethanol, and more than two-thirds of rapeseed oil produced in Europe is used to make biodiesel (Fischer *et al.*, 2009). Burning agrofuels allegedly releases only carbon to the atmosphere that recently has been taken from the atmosphere for plant growth, and it is thus 'carbon neutral'. This idea is simplistic and increasingly challenged by life-cycle analysis, including the GHG emissions that could otherwise be sequestered on land used to grow agrofuels (EEA, 2011). More accurate calculations from the European Commission indicate that burning several of the most important first-generation agrofuels produces more CO_2 equivalents per energy unit than conventional petroleum. They include rapeseed oil and soybean oil. Palm oil emits almost as much CO_2e/megajoule as does oil obtained from tar sands. Although both first-generation and especially second-generation bioethanol, as well as second-generation biodiesel, do better (Neslen, 2012): current economic and political investments in agrofuels have been, at best, premature.

Furthermore, the 'net energy' from agrofuels, including second-generation fuels, is scarcely worth the trouble (Patzek, 2007). In 2006, Pimentel and Patzek calculated that the 4.5 billion gallons of ethanol which were produced from 18% of the US maize crop, could replace only 1% of US petroleum consumption (Pimentel and Patzek, 2006). In the European Union (EU), consumption of agrofuel oil equivalents is expected to increase from its current 10 Mt to approximately 20 Mt in 2020 (Laborde, 2011). This amounts to only 3% of the 681 Mt of oil (almost 14 billion barrels/day) consumed in 2009 in the EU (NationMaster, 2009).

One gets a sense of the proportions involved by considering that the food available to Americans each year (half of which is wasted) contains approximately 2 exajoules (2×10^{18} joules) of energy, whereas the whole American food system consumes 20 exajoules, and the total energy used by Americans is approximately 105 exajoules (Patzek, 2007). Yet, agribusiness and the fossil fuel industry share an interest in promoting the idea that agrofuels somehow solve the problems caused by burning fossil fuels.

In summary, three major problems attend attempts to mitigate climate change by growing fuel crops. First, growing fuel instead of food exacerbates hunger and undernutrition. Second, making and burning agrofuels can emit as much as or more GHGs than fossil fuels per unit of energy released. Third, the net energy obtainable from agrofuel production is insignificant. Sadly, much of this was clear to critics when the agrofuel era began in earnest in 2006.

24.4 Why Is It So Difficult To Deal With Preventable Disaster?

One answer is biological and psychological. Our brains have not evolved to understand changes to the world that are so big and distant as climate change. We deal more easily with short-term and local problems than with long-term and global problems, and adapting to climate change is typically a local issue. Building a dyke incurs local costs, but it also confers local benefits and hence is easier to sell politically. Reducing emissions of GHGs or conserving forests, on the other hand, have global benefits but incur local costs. They are harder to sell.

Another answer is sociological. The corporate state cannot deal rationally with climate change, because business interests and the state are one body, a corpus, and corporations are responsible at most to shareholders, not to the public and the commons (Bakan, 2005). The Polaris Institute (based in Canada) has documented how corporations, including agribusinesses like Cargill and Monsanto, have become parties to the international UN climate negotiations (Fernandes and Girard, 2011). Oreskes and Conway have shown, moreover, how scientists with corporate ties misinform on a range of ecological and climatic issues (Oreskes and Conway, 2010). Indeed, we have allowed our universities to become closely tied to business interests, not least in the agricultural sector (Food and Water Watch, 2012).

Knowing what to do about climate change is fundamentally not difficult, however. At the risk of oversimplification, politicians should tax, and they should use tax money wisely. They must tax carbon, preferably at source, i.e. they must tax the company extracting domestic fossil fuels or the importer of fossil fuels extracted abroad (severance taxes at 'the mine, the wellhead, or port of entry') (Hansen, 2009). The effect of the tax will then percolate through the system, finally increasing the price of fuel to the consumer and lowering consumption. The tax can then be increased gradually so that prices to the consumer approach the true costs to society of a fossil fuel economy. An alternative is 'cap and trade'. Government limits ('caps') the amount of CO_2 that may be emitted and sells permits for specified emissions to companies. Such permits can then be traded. Schemes for trading emissions, originated in the USA, vary across the planet. On the whole, they are complex and bureaucratic, and so far they have not reduced GHG emissions.

Carbon taxes must be applied by international agreement, and tax revenue must be used wisely. Since a carbon tax is regressive, that is, it favours the rich, who have less trouble paying higher fuel prices, tax money must be redistributed to people with low incomes. Although corporations have generally supported schemes for trading permits to emit CO_2, one of the world's largest energy corporations (Exxon) has, like James Hansen, come out in favour of a carbon tax (Herald, 2009). Tax revenue must also be used wisely to favour industries that do least damage to nature and climate. Subsidization policies therefore need fundamental revision based on an understanding of relevant science: climatology, ecology, nutrition and thermodynamics.

Wise taxation, subsidization and regulation are also necessary for food. A tax on meat and dairy that reflects the true social costs is probably the only effective way to reduce consumption. Tax redistribution schemes must then, as with fuel, be used to promote forms of agriculture that respect ecosystems, and to ensure that low-income groups can buy enough nutritionally adequate foods, including meat and dairy. As important, governments must again regulate trade in agricultural commodities to avoid the extreme fluctuations in the price of food that harm the poor the most.

All of this requires good and strong government. Although there is little evidence that confidence in government has eroded in the general public of democratic states, many intellectuals are pessimistic about the ability of government to deal with the destruction of ecosystems, climate change and the political pressures exerted by corporations and financial institutions. They are also deeply concerned that so many governments are uninterested in the problem of constant proportional economic growth divorced from physical resources. None of these concerns are new. We became aware of ecosystem destruction in the mid-20th century, and the limits to economic growth were well understood in the early 1970s. Yet, general recognition of the gravity of the climate problem happened as recently as 2006, and, having forgotten 1929, most of us understood the dangers of unregulated financial markets only after 2008.

In the rich European and American tradition of criticism of power and government, modern intellectuals have argued that we live best in small states (Kohr, 2001), that the liberal class has largely failed to understand the moral corruption and malfeasance of the corporate state (Hedges, 2012), and that we must return to an understanding of economics that puts nature before consumerism. Constanza *et al.* have reiterated the case for ecological economics in a recent UN document (Constanza *et al.*, 2012). Ecological economics differs fundamentally from the economic thinking of the past half-century, but it is consistent with an earlier understanding of economics from Aristotle to John Stuart Mill. Since exponential economic growth, pursued by government to avoid unemployment, is incompatible with finite biophysical resources, ecological economics rejects gross domestic product as a measure of social progress.

The issue of food and climate thus branches into energy policy, economics, studies of corporatism and political science, and our discussions of how to avoid disaster must be well and broadly informed.

A word about balance, however. The view of informed people, including many in the general public, is that we are causing Earth's ecosystems and climate to change in ways incompatible with a future for human civilization and many of our companion species. It makes no sense to pretend to balance that view, as journalists often do, against those of corporate interests, including agribusiness and the food industry.

24.5 Conclusions

Food production affects both climate and human health. Dietary recommendations should therefore be based on ecological, including climatological, insights, as well as on nutrition science. Just as the taxation of carbon at source is the simplest way to reduce the burning of fossil fuels, so is taxation of foods to reflect the true costs of production the most effective way to affect consumer choices, and thereby food production systems. Government must also be willing to regulate corporations, including agribusiness, more effectively and to entertain an understanding of economics based on biophysical reality.

Notes

[1] The literature on terms such as 'consumption' is confusing. Sometimes called 'economic disappearance', animal product consumption includes bones and offal, the ingestion of which varies with culture. However, even in rich countries, some animal products (especially animal fat) are ingested in disguised forms, such as in biscuits. Not all edible animal parts are eaten, especially in high-income countries; some is converted to fertilizer, and some may also be used for animal feed; this was a major contributor to the epidemic of bovine spongiform encephalopathy.

[2] Game meat, such as venison or kangaroo, is sometimes also classified as red meat.

[3] The contribution by Working Group III to the Fifth Assessment Report of the Intergovernmental Panel on Climate Change (IPCC, 2014) was published during preparation of this book. It stresses a point already made in this chapter, namely that considerable uncertainty exists about the contribution of agriculture and forestry to global greenhouse gas emissions.

References

Alvarez-Leon, E.E., Roman-Vinas, B. and Serra-Majem, L. (2006) Dairy products and health: a review of the epidemiological evidence. *British Journal of Nutrition* 96, S94–S99.

Bakan, J. (2005) *The Corporation*. Constable, London.

Bernstein, A.M., Sun, Q., Hu, F.B., Stampfer, M.J., Manson, J.E. and Willett, W.C. (2010) Major dietary protein sources and risk of coronary heart disease in women. *Circulation* 122, 876–883.

Bersaglieri, T., Sabeti, P.C., Patterson, N., Vanderploeg, T., Schaffner, S.F., Drake, J.A., *et al.* (2004) Genetic signatures of strong recent positive selection at the lactase gene. *The American Journal of Human Genetics* 74, 1111–1120.

Bischoff-Ferrari, H.A., Dawson-Hughes, B., Baron, J.A., Kanis, J.A., Orav, E.J., Staehelin, H.B., *et al.* (2011) Milk intake and risk of hip fracture in men and women: a meta-analysis of prospective cohort studies. *Journal of Bone and Mineral Research* 26, 833–839.

Constanza, R., Alperovitz, G., Daly, H.E., Farley, J., Franco, C., Jackson, T., *et al.* (2012) *Building a Sustainable and Desirable Economy-in-Society-in-Nature*. United Nations Division for Sustainable Development, New York.

Davis, A.S., Hill, J.D., Chase, C.A., Johanns, A.M. and Liebman, M. (2012) Increasing cropping system diversity balances productivity, profitability and environmental health. *PLoS One* 7, e47149.

EEA (European Environment Agency) (2011) *Opinion of the EEA Scientific Committee on Greenhouse Gas Accounting in Relation to Bioenergy*. Scientific Committee, European Environment Agency, European Commisssion, Brussels.

Egeberg, R., Olsen, A., Christensen, J., Halkjær, J., Jakobsen, M.U., Overvad, K., *et al.* (2013) Associations between red meat and risks for colon and rectal cancer depend on the type of red meat consumed. *The Journal of Nutrition* 143, 464–472.

Elwood, P.C., Pickering, J.E., Givens, D.I. and Gallacher, J.E. (2010) The consumption of milk and dairy foods and the incidence of vascular disease and diabetes: an overview of the evidence. *Lipids* 45, 925–939.

FAO (Food and Agriculture Organization of the United Nations) (2011) *The State of Food Insecurity in the World*. FAO, Rome.

Fernandes, S. and Girard, R. (2011) *Corporations, Climate and the United Nations*. Polaris Institute, Ottawa.

Fischer, G., Hizsnyik, E., Prieler, S., Shah, M. and Velizen, H.v. (2009) *Biofuels and Food Security*. International Institute for Applied Systems Analysis, Laxenberg, Austria.

Food and Water Watch (2012) *Public Research, Private Gain. Corporate Influence Over University Agricultural Research*. Food and Water Watch, Washington, DC.

Gerber, P.J., Steinfeld, H., Henderson, B., Mottet, A., Opio, C., Dijkman, J., *et al.* (2013) *Tackling Climate Change Through Livestock – A Global Assessment of Emissions and Mitigation Opportunities.* Food and Agriculture Organization of the United Nations, Rome.

Goodland, R. and Anhang, J. (2009) Livestock and Climate Change. *World Watch Magazine,* 22, 10–19.

Hansen, J. (2009) *Storms of my Grandchildren.* Bloomsbury, London.

He, J., Vupputuri, S., Allen, K., Prerost, M.R., Hughes, J. and Whelton, P.K. (1999) Passive smoking and the risk of coronary heart disease – a meta-analysis of epidemiologic studies. *New England Journal of Medicine* 340, 920–926.

Hedges, C. (2012) The implosion of capitalism. *Truthout* (http://truth-out.org/news/item/8808-when-civilizations-die, accessed 1 May 2012).

Herald, C. (2009) Exxon supports carbon tax. *Calgary Herald,* 9 January 2009 (http://www.canada.com/calgaryherald/news/calgarybusiness/story.html?id=e8aecbbb-16c6-412d-8054-7e64e2b176ef, accessed 7 April 2014).

Herrero, M., Gerber, P., Vellinga, T., Garnett, T., Leip, A., Opio, C., *et al.* (2011) Livestock and greenhouse gas emissions: the importance of getting the numbers right. *Animal Feed Science and Technology* 166–167, 779–782.

IPCC (Intergovernmental Panel on Climate Change) (2014) Summary for Policymakers. In: Edenhofer, O., Pichs-Madruga, R., Sokona, E., Farahani, S., Kadner, K., Seyboth, A. *et al.* (eds) *Climate Change 2014, Mitigation of Climate Change. Contribution of Working Group III to the Fifth Assessment Report of the Intergovernmental Panel on Climate Change.* Cambridge University Press, Cambridge, UK and New York

Janzen, H.H. (2011) What place for livestock on a re-greening earth? *Animal Feed Science and Technology* 166–167, 783–796.

Kanis, J.A., Oden, A., McCloskey, E.V., Johansson, H., Wahl, D.A. and Cooper, C. (2012) A systematic review of hip fracture incidence and probability of fracture worldwide. *Osteoporosis International* 23, 2239–2256.

Kohr, L. (2001) *The Breakdown of Nations.* Green Books in association with New European Publications, London.

Laborde, D. (2011) *Assessing the Land Use Change Consequences of European Biofuels Policies.* International Food Policy Research Institute (http://trade.ec.europa.eu/doclib/html/148289.htm, accessed 2 April 2014).

Lagi, M., Bar-Yam, Y. and Bar-Yam, Y. (2012) *Update July 2012 – The Food Crises: The US Drought.* New England Complex Systems Institute, Cambridge, Massachusetts.

Millward, D.J. and Garnett, T. (2010) Plenary Lecture 3: Food and the planet: nutritional dilemmas of green-house gas emission reductions through reduced intakes of meat and dairy foods. *Proceedings of the Nutritional Society* 69, 103–118.

NationMaster (2009) Energy > Oil > Consumption: Countries Compared (http://www.nationmaster.com/country-info/stats/Energy/Oil/Consumption, accessed 2 April 2014).

Neslen, A. (2012) Biofuels pollute more than oil, leaked data show (http://www.sott.net/article/240690-Biofuels-pollute-more-than-oil-leaked-data-show, accessed 7 April 2014).

Oreskes, N. and Conway, E.M. (2010) *Merchants of Doubt: How a Handful of Scientists Obscured the Truth on Issues from Tobacco Smoke to Global Warming.* Bloomsbury Press, New York.

Patzek, T.W. (2007) How can we outlive our way of life? (http://www.oecd.org/dataoecd/2/61/40225820.pdf, accessed 17 May 2012).

Pimentel, D. and Patzek, T. (2006) Green plants, fossil fuels, and now biofuels. *BioScience* 56, 875.

Prakash, A. and Stigler, M. (2012) *FAO Statistical Yearbook 2012.* Food and Agriculture Organization of the United Nations, Rome.

Seufert, V., Ramankutty, N. and Foley, J.A. (2012) Comparing the yields of organic and conventional agriculture. *Nature* 485, 229–232.

Sinha, R., Cross, A.J., Graubard, B.I., Leitzmann, M.F. and Schatzkin, A. (2009) Meat intake and mortality: a prospective study of over half a million people. *Archives of Internal Medicine* 169, 562–571.

Smil, V. (2002) Eating meat: evolution, patterns, and consequences. *Population and Development Review* 28, 599–639.

Soedamah-Muthu, S.S., Ding, E.L., Al-Delaimy, W.K., Hu, F.B., Engberink, M.F., Willett, W.C., *et al.* (2011) Milk and dairy consumption and incidence of cardiovascular diseases and all-cause mortality: dose–response meta-analysis of prospective cohort studies. *American Journal of Clinical Nutrition* 93, 158–171.

Steinfeld, H., Gerrber, P., Wassenaar, T., Castel, V., Rosales, M. and de Haan, C. (2006) *Livestock's Long Shadow. Environmental Issues and Options.* Food and Agriculture Organization of the United Nations, Rome.

World Cancer Research Fund/American Institute for Cancer Research (2007) *Food, Nutrition, Physical Actitvity, and the Prevention of Cancer: A Global Perspective.* AICR, Washington, DC.

25 Death of a *Mwana*: Biomass Fuels, Poverty, Gender and Climate Change

Bianca Brijnath

*Department of General Practice, School of Primary Health Care,
Monash University, Notting Hill, Victoria, Australia*

We arrived at the Kakula Health Post to find the ambulance waiting. A child had been badly burned. His mother had been feeling cold that morning and had built a fire in their hut. She had stepped away, gone somewhere else – just for a moment – and when she returned, she found her son had fallen into the fire.

The *mwana* (child) was young, perhaps not more than five or six years old. We were told that he had been burned all over his body; his back, his torso, his legs, his arms and his face. I could not see the extent of all his burns when they carried him, wrapped in a blanket, to the ambulance. I did see that his legs were burnt almost to the bone, that his face had a chunk missing from it, and that his hands also had pieces of flesh missing from them. I did see that his eyes were large and glazed, the whites prominent. I noticed that he did not cry or shout but emitted a low moan – an almost inhuman noise that I had never heard before and have not heard since.

The child was loaded into the ambulance and some members of his community, mostly women, clambered in behind him. His mother, weeping noisily, did not accompany him. She had other children to look after and could not leave them. The doors of the ambulance decisively shut and the vehicle jolted away to the district hospital 80 km away.

Later, we met Thomas, the ambulance driver. We asked him about the status of the child. He said, 'The hospital staff asked me, "Why did you bring him here? He is almost dead. Only another 30 minutes or so and he will be dead".'
(Field notes from rural Zambia, 6 February 2011)

25.1 The Fuel That Warms You Thrice

Kirk R. Smith (2008), playing on a New England adage, wrote, 'Wood is the fuel that warms you thrice'. By this he was referring to the capacity of wood to warm a person when she was chopping it, burning it and inhaling the smoke from it. In this chapter, I will apply the three ways wood can warm a person – collection, burning and woodsmoke inhalation – to explicate the relationship between fuelwood, poverty, gender and health. Then I will discuss how climate change will amplify these relationships, leaving the poor, especially women and children, more vulnerable than other segments of the population. I argue that the challenge that confronts us is one of ecology and equity and that the social sciences can help determine how to navigate this relationship. I would like to emphasize that by my choices, I neither deny the importance of other determinants of climate change nor suggest that fuelwood and deforestation is the only nexus at which gender, socio-economic and environmental inequalities meet. Rather, my choice is pragmatic and helps me make a contained and cathartic argument. This is both an attempt to discuss the links between fuelwood and gender, poverty and the environment and to come to terms with what was and still is a very distressing event for me, i.e. witnessing the death of a child.

25.2   Poverty and Biomass Fuels

According to the United Nations Development Programmes' (UNDP) Multidimensional Poverty Index,[1] lack of access to modern cooking fuel, a significant cause of deforestation and overuse of forest wood, is currently the greatest environmental deprivation experienced by the poor (UNDP, 2011). More than 90% of the world's poor (~2.7 billion) presently lack access to modern cooking fuels (UNDP, 2011), and more than 2.4 billion people rely directly on biomass fuels such as wood, crop residues, dung and plant materials for their heating and cooking (OECD and International Energy Agency, 2004). The vast majority of this is burned in traditional hearths without improvements such as a chimney or cooking stove (see Plate 8). In Asia, nearly 70% of wood removed from forests is used as fuel for domestic cooking; in Africa the rate is 90% (FAO, 2011). In contrast, almost 80% of wood consumption in developed countries is for industrial use (FAO, 2002).

Unsurprisingly, women and children, as groups generally with the least power and status, suffer the worst social and health outcomes from the unimproved burning of biomass fuels. Such outcomes manifest in different and particular ways during the three phases of warming – the collection, burning and smoke inhalation of biomass fuels (Smith et al., 2004). Below, I describe how this occurs.

Fuelwood collection

Chopping and collecting fuelwood is time- and energy-consuming work.[2] Across the developing world, women can spend up to half of their day engaged in this activity. In India, for example, women spend as much as 4 hours and travel an average of 4.5 km/day to collect wood (Agarwal, 2010). This represents a high opportunity cost,

lowering time for more productive activities including education, paid employment and health care (Blackden and Wodon, 2006). This cost also extends to children; studies from Malawi and Kenya show that time collecting fuelwood is at the expense of schooling (Nankhuni and Findeis, 2004; Ndiritu and Nyangena, 2010) and that girls are especially vulnerable, helping with gathering fuel, collecting water, cooking, child-rearing and other related tasks. Of the 60.7 million children not currently in school, more than half are girls (The World Bank, 2012).

Collecting and hauling fuel (and water) over great distances can harm health, including by spinal damage and arthritis in later life (Kramarae and Spender, 2000). Collecting fuelwood (and water) also increases the opportunity for injury and sexual violence. This is especially the case for women who have no choice but to live in conflict zones (Hampton, 2005; Marsh et al., 2006).

Paediatric burns

Burning biomass for cooking, lighting and heating in open fires remains a major underlying factor for the high incidence of burns across the developing world. In houses where ventilation is inadequate and a single space may be used for multiple purposes (e.g. sleeping, playing and cooking), the use of unstable stoves, cooking pots and open fires amplifies the risk of severe burns, especially in children (Onuba and Udoidiok, 1987; Ahuja and Bhattacharya, 2004).

The most common type of paediatric burns are flame burns, typically caused by falling into the fire (Sowemimo, 1993). In Abidjan, Cote d'Ivoire, 75% of all paediatric burn patients were younger than 5 years of age (Vilasco and Bondurand, 1995), in Nigeria it was 80% (Albertyn et al., 2006) and in South Africa, burns were the third most common external cause of death in children younger than 18 years (Van Niekerk et al., 2004).

Apart from high mortality as a result of burns, burn survivors experience long-term physical, emotional and psychological illnesses, scarring and cosmetic disfigurement, repeat operations and limited opportunities in nearly all spheres of life (e.g. economic, education and social) (Dissanaike and Rahimi, 2009). Globally, disabilities from burns result in an US$80 billion annual loss in work productivity. This figure excludes medical expenses for treatment (ReSurge International, 2012).

Woodsmoke and indoor air pollution

The third way that Smith describes how biomass can 'warm' a person is by smoke inhalation. Biomass fuels have low combustion efficiency and release significant amounts of carbon as respirable particles, volatile organic chemicals and carbon monoxide (Smith, 2008). In poor household settings, where there is limited space and bad aeration, women and children are exposed to high doses of particulate matter (PM) from this noxious smoke. PM_{10} (inhalable material <10 μm in aerodynamic diameter) and $PM_{2.5}$ often exceed guideline levels, especially during cooking times. Biomass burning is *the* main contributor to indoor air pollution (Po *et al.*, 2011) (emphasis added). It is equivalent to smoking several cigarettes a day, and in households where people are exposed to biomass burning and exhaled cigarette smoke, the health effects are all the more injurious (Bruce *et al.*, 2000).

According to the World Health Organization (WHO, 2012), indoor air pollution contributes 2.7% of the global burden of disease (about 2 million deaths per annum, including a disproportionate number of young people), ranking it as the second environmental cause of ill health after unsafe water and sanitation. In low-income countries, the burden of indoor air pollution accounts for about 4% of the overall disease burden, an effect much greater than that of outdoor air pollution (WHO, 2012). Indoor air pollution contributes to around 900,000 deaths due to pneumonia in children under 5 years of age (WHO, 2012), and a proportion of pneumonia, chronic obstructive pulmonary disease (COPD) and lung cancer mortality among adults (WHO, 2012). Tuberculosis, eye diseases, adverse pregnancy outcomes and ischaemic heart disease cases can all be predisposed to or worsened by biomass smoke exposure (Smith, 2008).

25.3 Thrice-Warming Fuels and the Death of a Child

The thrice-warming capacities of biomass fuels – collection, burning and inhalation of woodsmoke – speak to the broader determinants of poverty, gender inequalities and environmental degradation. Set within this context, the story I narrated at the opening of this chapter, although very distressing, is all too common:

1. Our *mwana* lives in Zambia, which is ranked 164th from 169 on the Human Development Index. He is a poor child, in a poor country, in the poorest region in the world, Africa.
2. He lives in a rural area, in a badly ventilated mud hut, with his siblings and at least one parent, his mother.
3. He lives without electricity, clean water, adequate sanitation and modern cooking fuel.
4. Therefore, his mother will walk long distances each day to gather fuelwood, so she can cook *nshima* (maize porridge), boil water, light the hut and warm her children.
5. In all likelihood this child's mother will be illiterate.
6. When she dies, at approximately 50 years of age, most probably she will have a respiratory illness.
7. Meanwhile, because of the time and energy she expends every day just to complete the basic tasks of daily living, she has had few employment opportunities.
8. She probably also feels physically and mentally tired from this work and the effort to make ends meet.
9. One day, just like any other day, there is dirty fuel and an open fire in her hut. She steps away *just* for a moment.
10. Her son falls into the fire and we see him as he is dying at the Kakula Health Post.

25.4 Climate Change and Biomass Fuels

By 2030, 2.7 billion people are anticipated to rely on biomass fuels, up from today's 2.4 billion (IEA, 2011). Two reasons explain the increase. First, even though modelling predicts a declining proportion of the population using biomass fuels, actual numbers will still rise.

Second, rising prices of fossil fuels (e.g. oil, kerosene, liquefied petroleum gas and coal-fired electricity) might force some people back down this 'energy ladder' (Smith *et al.*, 2004; Smith, 2008).

Reliance on biomass also interacts with environmental quality. Collection of fuelwood is one of the leading causes of forest degradation in many parts of Africa. Overuse of forest resources[3] contributes to deforestation and desertification, and disrupts ecosystem services such as carbon sequestration, nutrient cycling and flood regulation (Millennium Ecosystem Assessment, 2005). Earlier chapters in this volume have discussed aspects concerned with deforestation for vector biology, local ecologies, food scarcity, migration and urbanization. It is worth stressing that deforestation can alter the risk of some infectious diseases (Butler, 2008b) and diminishes Earth's ecological library of plants and herbs, many of which could have potential medicinal benefits (Chivian and Bernstein, 2008).

The direct relationship between biomass burning and climate change is complex and has not yet been fully understood. On the one hand, incomplete burning of carbonaceous material (in conjunction with deforestation) increases warming by its production of black carbon, including from forest and cooking fires. On the other hand, many other air pollutants (e.g. sulfate aerosols) cool the atmosphere, and it possible that air pollution control initiatives, such as reducing greenhouse gas emissions, may actually lead to further warming (Arneth *et al.*, 2009).

While questions remain on the efficacy of air pollution control and reducing biomass, the evidence has shown conclusively that biomass combustion has serious implications for population health and environmental quality in the short, medium and long term. This is an issue of global social justice (Butler, 2008a). Accordingly, programmes to mitigate the negative effects of air pollution and climate change more broadly need to be multifaceted and to include a strong dimension of poverty and gender inequality reduction. Greater gender equity and female education will also have economic and environmental benefits, including by slowing fertility and giving women greater say over their health outcomes (Bryant *et al.*, 2009).

25.5 Ecology, Equity and the Social Sciences

Social science has an important role to play, not just in explicating the localized effects of climate change and poverty on communities and individual households but also in helping policy makers frame sustainable strategies using a social justice filter. Such frameworks must acknowledge the different effects climate change have on men, women and children, rich and poor, in low-, middle- and high-income countries. A social science analysis must apply from the grass roots to high policy, acknowledging the wider sociopolitical and economic forces which govern people's lives.

This is occurring already. At the community level, social science concepts such as participatory decision making and empowerment have been applied with relative success to populations such as poor women and indigenous peoples and in areas including forest management, rural development, agriculture, land and water use (see, for example, Agarwal, 2001, 2009; Tole, 2010). At the policy level, the first recommendation of the WHO Commission on the Social Determinants of Health (2008) is to improve daily living conditions. The Commission's second recommendation is the need to tackle the inequitable distribution of power, money and resources. Similarly, the Millennium Development Goals (MDGs) enshrine the aims of promoting gender equality and empowering women (MDG3), as well as ensuring environmental sustainability (MDG7). Although such aspirations can sound utopian, progress has been made, and in any case, aspirations are necessary if any change is to occur.

To return to biomass fuels, arguably one of the most important tasks facing social scientists is continued agitation for the mass distribution of improved cooking stoves and to hasten the transition to 'cleaner' fuels, especially ones based on renewable energy. The social and economic consequences stemming from such an initiative have the potential to improve the lives of more than 1 billion people. While some worry that such gains may have a minimal environmental benefit (e.g. by increasing the demand for coal-fired electricity, or by increased demand and consumption) (Sorrell, 2009), others disagree.

In working with the realities of finite resources and the ambition of achieving equity, we must recognize that the right to resources and energy, to electricity, gas and clean water, come with responsibilities – to ensure that current and future generations lead lives of quality and dignity with access to these basic resources (Cafaro, 2012). This is the challenge with which we are faced today, and ecological and equitable reasons compel us to act. No more dead *mwanas*, please.

Notes

[1] This index integrates measures of health, education and environmental deprivation. It can be analysed by component or as a whole across regions, ethnic groups, gender and other socio-economic variables.
[2] As is collecting water (see Dar and Khan, 2011).
[3] Globally, the main cause of deforestation is agriculture. However, in Africa, fuelwood collection is the leading cause.

References

Agarwal, B. (2001) Participatory exclusions, community forestry, and gender: an analysis for South Asia and a conceptual framework. *World Development* 29, 1623–1648.

Agarwal, B. (2009) Gender and forest conservation: the impact of women's participation in community forest governance. *Ecological Economics* 68, 2785–2799.

Agarwal, B. (2010) *Gender and Green Governance : The Political Economy of Women's Presence Within and Beyond Community Forest.* Oxford University Press, Oxford, UK, and New York.

Ahuja, R.B. and Bhattacharya, S. (2004) Burns in the developing world and burn disasters. *BMJ* 329, 447–449.

Albertyn, R., Bickler, S.W. and Rode, H. (2006) Paediatric burn injuries in Sub Saharan Africa – an overview. *Burns* 32, 605–612.

Arneth, A., Unger, N., Kulmala, M. and Andreae, M.O. (2009) Clean the air, heat the planet? *Science* 326, 672–673.

Blackden, C.M. and Wodon, Q. (2006) Gender, time use and poverty: an introduction. In: Blackden, C.M. and Wodon, Q. (eds) *Gender, Time Use and Poverty in sub-Saharan Africa.* World Bank, Washington, DC, pp. 1–10.

Bruce, N., Perez-Padilla, R. and Albalak, R. (2000) Indoor air pollution in developing countries: a major environmental and public health challenge. *Bulletin of the World Health Organization* 78, 1078–1092.

Bryant, L., Carver, L., Butler, C.D. and Anage, A. (2009) Climate change and family planning: least developed countries define the agenda. *Bulletin of the World Health Organization* 87, 852–857.

Butler, C.D. (2008a) Environmental change, injustice and sustainability. *Journal of Bioethical Inquiry* 5, 11–19.

Butler, C.D. (2008b) Human health and forests: an overview. In: Colfer, C.J.P. (ed.) *Human Health and Forests: A Global Overview of Issues, Practice and Policy.* Earthscan, London, pp. 13–33.

Cafaro, P. (2012) Climate ethics and population policy. *Wiley Interdisciplinary Reviews: Climate Change* 3, 45–61.

Chivian, E. and Bernstein, A. (eds) (2008) *Sustaining Life. How Human Health Depends on Biodiversity.* Oxford University Press, Oxford, UK.

Commission on the Social Determinants of Health (2008) Closing the gap in a generation: health equity through action on the social determinants of health. Final Report of the Commission on Social Determinants of Health. World Health Organization, Geneva, Switzerland.

Dar, O.A. and Khan, M.S. (2011) Millennium development goals and the water target: details, definitions and debate. *Tropical Medicine and International Health* 16, 540–544.

Dissanaike, S. and Rahimi, M. (2009) Epidemiology of burn injuries: highlighting cultural and socio-demographic aspects. *International Review of Psychiatry* 21, 505–511.

FAO (Food and Agriculture Organization of the United Nations) (2002) World Agriculture: Towards 2015/2030. Summary Report. FAO, Rome.

FAO (2011) *The State of the World's Forests.* FAO, Rome (http://www.fao.org/docrep/013/i2000e/i2000e00.htm, accessed 1 March 2012).

Hampton, T. (2005) Agencies speak out on rape in Darfur. *The Journal of the American Medical Association* 294, 542–544.

IEA (International Energy Agency) (2011) Energy for all: financing access for the poor. *World Energy Outlook 2011*. OECD, IEA, Paris.

Kramarae, C. and Spender, D. (2000) *Routledge International Encyclopedia of Women : Global Women's Issues and Knowledge*. Routledge, New York.

Marsh, M., Purdin, S. and Navani, S. (2006) Addressing sexual violence in humanitarian emergencies. *Global Public Health* 1, 133–146.

Millennium Ecosystem Assessment (2005) *Living Beyond Our Means. Natural Assets and Human Well-being*. Island Press, Washington, DC.

Nankhuni, F.J. and Findeis, J.L. (2004) Natural resource-collection work and children's schooling in Malawi. *Agricultural Economics* 31, 123–134.

Ndiritu, S.W. and Nyangena, W. (2010) *Environmental Goods Collection and Children's Schooling: Evidence from Kenya*. Environment for Development (http://www.rff.org/RFF/documents/EfD-DP-10-18.pdf, accessed 1 March 2012).

OECD and International Energy Agency (2004) *World Energy Outlook*. Organization for Economic Co-operation and Development, International Energy Agency, Paris.

Onuba, O. and Udoidiok, E. (1987) The problems and prevention of burns in developing countries. *Burns* 13, 382–385.

Po, J.Y.T., FitzGerald, J.M. and Carlsten, C. (2011) Respiratory disease associated with solid biomass fuel exposure in rural women and children: systematic review and meta-analysis. *Thorax* 66, 232–239.

ReSurge International (2012) Burns: The Neglected but Solvable Health Crisis. ReSurge International, Mountain View, California (http://www.resurge.org/transforming_lives/story_burns.cfm, accessed 6 March 2012).

Smith, K.R. (2008) Wood: the fuel that warms you thrice. In: Colfer, C.J.P. (ed.) *Human Health and Forests: A Global Overview of Issues, Practice, and Policy*. Earthscan, London and Sterling, Virginia, pp. 97–111.

Smith, K.R., Mehta, S. and Maeusezahl-Feuz, M. (2004) Indoor air pollution from household use of solid fuels. In: Ezzati, M., Rodgers, A.D., Lopez, A.D. and Murray, C.J.L. (eds) *Comparative Quantification of Health Risks: Global and Regional Burden of Disease Attribution to Selected Major Risk Factors*. World Health Organization, Geneva, Switzerland, pp. 1435–1493.

Sorrell, S. (2009) Jevons' paradox revisited: the evidence for backfire from improved energy efficiency. *Energy Policy* 37, 1456–1469.

Sowemimo, G.O.A. (1993) Burn care in Africa: reducing the misery index: the 1993 Everett Idris Evans Memorial Lecture. *Journal of Burn Care and Rehabilitation* 14, 589–594.

Tole, L. (2010) Reforms from the ground up: a review of community-based forest management in tropical developing countries. *Environmental Management* 45, 1312–1331.

UNDP (United Nations Development Programme) (2011) *Human Development Report 2011 – Sustainability and Equity: A Better Future for All*. UNDP, New York.

Van Niekerk, A., Rode, H. and Laflamme, L. (2004) Incidence and patterns of childhood burn injuries in the Western Cape, South Africa. *Burns* 30, 341–347.

Vilasco, B. and Bondurand, A. (1995) Burns in Abidjan, Cote D'Ivoire. *Burns* 21, 291–296.

WHO (World Health Organization) (2012) *Household Air Pollution and Health*. World Health Organization, Geneva, Switzerland (http://www.who.int/mediacentre/factsheets/fs292/en/, accessed 2 March 2012).

World Bank, The (2012) *Girls' Education*. The World Bank Washington, DC (http://web.worldbank.org/WBSITE/EXTERNAL/TOPICS/EXTEDUCATION/0,,contentMDK:20298916~menuPK:617572~pagePK:148956~piPK:216618~theSitePK:282386,00.html, accessed 2 March 2012).

26 Mental Health, Cognition and the Challenge of Climate Change

Colin D. Butler,[1,2] Devin C. Bowles,[2] Lachlan McIver[2] and Lisa Page[3]

[1]*Faculty of Health, The University of Canberra, Australia, and Benevolent Organisation for Development, Health & Insight (BODHI);* [2]*National Centre for Epidemiology and Population Health, The Australian National University, Canberra, Australia;* [3]*Sussex Partnership NHS Foundation Trust, and Brighton and Sussex Medical School, and Mental Health Liaison Team, Royal Sussex County Hospital, Brighton, East Sussex, UK*

26.1 Introduction

In recent decades, there has been increasing recognition of the importance of mental health, though in most low-income countries much work remains to be done to legitimize and deepen this field. Major recognized forms of mental, neurological and substance use disorders include anxiety, dementia, depression, epilepsy, post-traumatic stress disorder (PTSD), psychosis (principally schizophrenia) and substance abuse. The collective burden of these diseases (BOD)[1] exceeds that of cancer or cardiovascular disease, and has been estimated to be about 13% of the global total (Collins *et al.*, 2011). This high level is, in part, because most mental health conditions, other than dementia, arise in early or mid-life; many sufferers thus endure poor health for decades.

Mental and neurological health is more than the absence of these conditions, some of which have clear physical causes. For example, infectious diseases and head injuries are major causes of epilepsy in low-income countries (Singhi, 2011). Some forms of dementia are related to cerebrovascular disease, including atherosclerosis and repeated cerebral infarctions. For others, such as schizophrenia, the major causes

remain unknown. But, mental health also relates to less well-defined internal states, both positive, such as peace of mind, happiness, relaxation and the absence of stress, and negative, such as anxiety, fear and worry. The epidemiology of a surprising number of these cognitive states is and will be altered by climate change, including through the harm to the material, bodily and financial determinants of mental health which will probably result.

In addition to these issues, there are a range of emotions and cognitive states which will also be affected. Without intending to pathologize normal emotions, including negative ones, cognitive states associated with biophilia and solastalgia will also be influenced by climate change. Biophilia, meaning 'love of the bio' (life) (Wilson, 1984) may also be related to 'nature deficit disorder', the theory (backed by accumulating evidence) that some behavioural disorders arise from, or are aggravated by, too little contact with rich nature, especially when young.

Solastalgia, a more recently defined term, refers to the distress that some people experience in response to unwanted local change in a home environment to which they are attached (Albrecht *et al.*, 2007). Some documented causes of solastalgia are major contributors to climate change, such as large-scale open-cut coal mining and the removal of mountain tops in order to extract the underlying coal (Connor *et al.*, 2004; Cordial *et al.*, 2012). There is as yet little research about solastalgia in developing countries, but the burden of psychic pain is likely to be immense in some vulnerable populations, such as tribal people displaced by coal mining in some areas of India (Kujur, 2006). Solastalgia has also been used to describe the loss experienced by older women in some islands of the Torres Strait, whose homes have been placed at risk by sea level rise associated with climate change (McNamara and Westoby, 2011). Risk is increased in cultures with greater attachment to place, such as many indigenous cultures in which spirituality is localized, as climate change puts at risk many of these localities.

Anxiety and regret in response to genuine threats are normal, evolutionarily driven states of mind, with clear survival benefits. Normal levels of anxiety and regret are not usually considered mental diseases, but they are both unpleasant. For many people, climate change is an amorphous, insoluble issue. It may generate anxiety in some (as the fear of nuclear annihilation did in the Cold War), but for others it may generate denial or nihilism. All of these responses represent ripples in the global pool of mental health, even though sufficient anxiety or anger, if channelled effectively, may eventually contribute to needed reforms.

Adverse cognitive effects such as anxiety, fear and loss of serenity are likely to arise through the anticipation of events related to climate change, with the actual experience of those events giving rise to states such as trauma, loss and distress and after the event, including through harm to the physical and financial determinants of mental health such as depression and PTSD. Climate change may also worsen other mental ill-health conditions, such as dementia and depression, by means of heatwaves, which exacerbate chronic diseases.

Climate change will also affect the pathways by which poor nutrition leads to chronic health problems, including depression and other conditions which can themselves cause or exacerbate depression. In the other direction, high-consumption lifestyles which worsen climate change (e.g. high red meat ingestion) are associated with chronic health problems such as

circulatory disease, which can contribute to mental health problems such as depression and fear.

This chapter builds on selected literature that explores the links between mental health and climate change (Fritze *et al.*, 2008; Berry *et al.*, 2010; Page and Howard, 2010; Doherty and Clayton, 2011). It sketches ways in which mental health interacts with the primary, secondary and tertiary manifestations of climate change. There is as yet little empirical work on this topic, but there is a large foundational literature; there can be no doubt that the BOD of mental ill health and neurological disorders attributable to climate change will prove to be significant. This burden may be further added to by increased substance use abuse, consequent to the breakdown in civil society that could occur if current trajectories deepen, leading to enhanced conflict, hunger and migration, with the potential for increased xenophobia. This chapter will focus on the three areas of mental ill health that the authors consider are, or will be, most sensitive to climate change: anxiety, depression/suicide and PTSD. It then explores briefly some of the major cognitive obstacles which have to date contributed to preventing an adequate collective social response to climate change. Barriers which have been suggested include evolutionary novelty (Ornstein and Ehrlich, 1989), cognitive dissonance, denial (Hamilton, 2010; Gifford, 2011) and the growing mistrust of science in some developed countries (Oxley *et al.*, 2008; Otto, 2012). Forces which profit from climate change, especially fossil fuel interests, have worked with and harnessed sophisticated techniques of persuasion, often with powerful public relations companies and advertisers, in ways to promote doubt and to reinforce conservative social norms, especially through right-wing media and 'think tanks' (Oreskes and Conway, 2010). The chapter concludes by suggesting some approaches that might help protect individual and collective mental health, including collective activism, resistance and cooperation.

26.2   Ecosystem Services and Mental Health

The idea that contact with biodiverse green spaces and other pleasing scenery (e.g. a river or the sea) can benefit mental health is intuitively obvious to many, at least for people with biophilia. Myths of fertile Edens or desired oases are common in many cultures and faiths. The intuitive evidence that exposure to nature benefits health (Van Den Berg *et al.*, 2007) is now being strengthened by empirical evidence, controlled for social class (Francis *et al.*, 2012).

Other forms of evidence include the existence of sacred groves (Ramakrishnan *et al.*, 1998) and the preference of most wealthy people to live with some contact with nature, even if just visual or auditory. Groves are considered sacred, not only to protect the biodiversity they harbour but also because they provide a space for contemplation and serenity. The Japanese have a custom of 'forest bathing' (Noss *et al.*, 2012). The Millennium Ecosystem Assessment conceptualized these benefits as 'culturally enriching' ecosystem services (Corvalán *et al.*, 2005).

In the 1950s, the planetary ecologist, Rene Dubos, warned that ecological crises threatened to harm the quality of life, rather than destroy humanity: 'people adapt so unconsciously to their surroundings', he worried, 'that they would no longer mind the stench of automobile exhausts, ugly urban sprawl, starless skies, treeless avenues, shapeless buildings, tasteless bread, joyless celebrations'. He also predicted that loss of sensual perceptions would be compensated for by stimulations from loud noises, bright lights and drugs. ... 'We do not live *on* the planet Earth but *with* the life it harbours and *within* the environment that life creates' (Moberg and Cohn, 1991).

Today, more than half of humanity lives in cities. Increasingly, the connections between people and nature are virtual, including via television and the Internet. This disconnection may be contributing not only to subtle forms of mental disease but also to the ongoing destruction of distant species and ecosystems, whose existence is not widely perceived. In fact, this disconnect may also hinder the as yet too feeble collective response to the planetary crisis that climate change represents. On the other hand, virtual exposure to a much wider range of nature than was once possible may sensitize new audiences to the importance of conservation (Noss *et al.*, 2012), including of the climate.

26.3 Climate Change and Mental Health

The mental health effects of the primary, secondary and tertiary manifestations of climate change are likely to share similarities, but also to be subtly different (see Fig. 26.1). Increased anxiety is likely to be associated with all three categories. Many people, especially those who are frail, vulnerable or otherwise sensitized by adversity, will be apprehensive of summer, likely to be marked in most temperate and tropical regions by an increased rate of primary climate change effects, including heatwaves, fires and floods. Some people will be fearful of secondary effects, such as the encroachment of malaria in

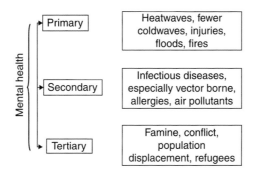

Fig. 26.1. All of the three forms of health effects discussed in this book are likely to have adverse mental health effects, whether as anticipation (anxiety), experience (mental trauma) or sequelae (e.g. depression, anxiety or PTSD). Adverse neurological events are also possible, such as epilepsy following head injuries arising from storm – related trauma.

regions where it was previously rare, as is possible in the future in some East African highland cities.

Beyond anxiety, other likely mental health effects include depression, arising particularly from primary effects (e.g. trauma from floods, fire and extreme weather events) (see Box 26.1). These are discussed in more detail below.

26.4 Anxiety

In early January 2013, during Australia's 'angry summer', an author of this chapter awoke well before light with pain in his left cheek; a nuisance symptom of sinus sensitivity, but which normally never interferes with sleep (see Box 26.1). There is increasing recognition of the relationship between anxiety and climate change (Swim *et al.*, 2011).

26.5 Depression

The causes of depression and suicide are multifactorial and vary with culture. A perhaps surprising number of people kill themselves with no previous diagnosis of clinical depression, though this may also reflect a lack of diagnostic clinical services and, in some places, cultural factors which inhibit open discussion of painful mental states. India and China have high rates of suicide, but very limited psychiatric services, especially for impoverished people and in rural areas. There is considerable anecdotal evidence of high rates of rural suicide in India, associated with agricultural hardship and indebtedness. Many projections for Indian agriculture are bleak due to climate change; the rural suicide rate in that country is therefore likely to rise even higher. Climate change is also likely to intensify regional and ethnic conflict in many developing countries, with numerous, albeit poorly documented, flow-on effects to mental health, including depression and suicide (Doherty and Clayton, 2011). Past examples include Darfur and possibly Syria. Bangladesh, Myanmar and north-east India are possible future examples.

In the Australian state of New South Wales (NSW), where excellent meteorological and health data exist over an extended period, a careful analysis has shown a modest increase in rural suicide associated with drought for the period between 1970 and 2007 (Hanigan *et al.*, 2012). While the authors do not link this with climate change, the findings suggest that any increase in drought associated with climate change is likely to increase the rate of suicide (due, for example, to loss of livelihood and observing the suffering of livestock), unless sufficient countermeasures, such as drought relief, counselling services and antidepressant medication can be

---

Box 26.1

Australia had just experienced 2 consecutive days with an average maximum temperature across the whole continent of over 40°C, the hottest ever recorded, a clear indication of climate change. The regional city of Hobart, Tasmania, had also endured its hottest day, during which fires had destroyed the local primary school and many homes in the nearby hamlet of Dunalley, about 200 miles from where I was staying. Fire is a regular feature of Australian forest ecology, but bushfires appear to be becoming more intense, destructive and fast-spreading, due in part to climate change (Bowman *et al.*, 2013). In 2009, the Black Saturday bushfires killed 173 people and destroyed over 2000 houses in the state of Victoria.

As fires spread through the island state, the taxpayer-subsidized radio provided constant coverage; for example, repeatedly warning people to activate their fire plan or to leave home while there was time to flee. For many people, this was an anxious time, fearing loss of life and property, including of stock. The families of volunteer firefighters (another public good, like the radio) also fretted; almost every year, firefighters are killed in Australia, and 2013 proved no exception. Those with high biophilia also worried about the pain and loss of creatures and perhaps trees and plants in the bush. Wildlife carers prepared for an influx of injured wallabies, possums and other marsupials, for which the state is famous.

I went outside. It seemed unusually hazy – was that smoke, or just my imagination? I went to the Internet, which, like the radio, had become essential for fire information. The forecast for the nearest town was 'smoky'. That was reassuring. Nevertheless, I could not go back to sleep.

In 2013, that part of Tasmania avoided a bushfire. But for how much longer?

Box 26.2   Mental Health and Climate Change in Pacific Atoll Countries

Atolls are coral islands, typically lying at very low elevations above sea level and with very scarce fresh water and arable land. A small handful of countries are comprised exclusively of atolls; most of these are in the South Pacific.

Tokelau, a tiny Pacific island nation of 1500 people spread over three small atolls, is one of the most remote and isolated parts of the planet, accessible only by a 2-day boat journey, which departs every 2 weeks from Samoa. The highest point in Tokelau is 5 m above sea level; perhaps nowhere in the world are the effects of climate change more visible and tangible.

At the Rio+20 summit in Brazil in late 2012, the *ulu* (political leader) of Tokelau, Mr Foua Kerisiano Toloa, made the point – obvious to his compatriots and residents of atolls everywhere but apparently unimportant to the rest of the world – 'We stand to lose the most of any country in the world due to climate change and rising sea levels... we will be among the first to go under water.'

In this same address, the *ulu* chastised the developed world for its apathy and inactivity in relation to climate change mitigation. Leading by example, he also committed his country to a bold but feasible conversion to 100% renewable energy by the end of 2013. Indicative of the ambivalence felt by many island communities with respect to the role of God and religion in relation to climate change and the prospect of migration induced by climate change, he stated: 'We have no intention of leaving. This is a God-given land; we have a culture, a language, an identity and a heritage. We want to preserve Tokelau for future generations.'

People in atoll countries would thus appear to be at high risk from the mental health disorders attributable to climate change discussed elsewhere in this chapter, such as anxiety, depression, excess worry, pathological stress (including PTSD) and solastalgia. Yet, little is known about the mental health of people living on atoll islands. There have been no systematic studies, and baseline data are either scarce or non-existent. For all the attention given to atoll communities, little is known about their psychosocial vulnerability to climate change, nor measures by which these threats could be alleviated. It is essential that this research gap be addressed to further the development of evidence-based policies related to climate change adaptation and strengthening health systems, in order to protect the mental health of atoll, island and coastal communities in the Pacific and around the world.

---

made available. Though such measures have been applied in NSW, resulting in some level of improved outcomes, these countermeasures are less likely to be available in low-income countries, especially if resource availability continues to tighten as limits to growth approach, including the availability of cheap and abundant energy. Approaching limits to growth lift living costs (especially of food and energy) and reduce purchasing power and economic activity, and are thus likely to be associated with reduced 'human service', such as those that might cushion future drought (e.g. financial assistance, disaster relief) and the pain of watching crops or stock perish (e.g. specialized counselling services).

natural disasters (Norris *et al.*, 2002). The prevalence of PTSD following a natural disaster seems highly variable, but is associated with higher exposure to trauma (Galea *et al.*, 2005). The tertiary effects of climate change, arising from conflict and enforced migration (see Box 26.2), are expected to lead to increased incidence of PTSD, particularly if those migrating populations are exposed to torture or other especially traumatic events (Steel *et al.*, 2009). Refugees have high rates of PTSD (Fazel *et al.*, 2005), and prolonged incarceration in camps is likely to further risk of PTSD (Robjant *et al.*, 2009). The health of some receiving populations may decline, due to resentment, resource competition and xenophobia.

26.6   Post-traumatic Stress Disorder

Adverse impacts on mental health, including PTSD, are well documented in the aftermath of

26.7   Children and Intergenerational Mental Health

Children are at risk of many of the same mental health problems from climate change as adults, such as anxiety, depression and PTSD, through similar causal pathways. Their mental health is particularly vulnerable. Childhood and adolescence are characterized by rapid and important psychological development, such that even relatively brief or seemingly small disturbances can cause lifelong psychological problems. Trauma during childhood development can play an important role in the formation of personality disorders (Reich *et al.*, 1997), which can resist treatment and may be lifelong (Bateman and Fonagy, 2008). Natural disasters, food insecurity, social tension and discrimination and conflict all interfere with healthy psychological development.

Children's vulnerability is further heightened by their reliance on adults for emotional and material support. They face double jeopardy from the effects of climate change, which can impact them directly, through their parents' experience, or both. As climate change worsens, many of its effects are likely to test parenting capacity and the ability to insulate children's development from challenges, including those unrelated to climate change. Limits to parenting capacity include enhanced morbidity, mortality and absence of parents, siblings and other relatives and adults, due a range of mechanisms set out in this book, including conflict and migration (see Part IV Tertiary Effects, this volume).

The effects of climate change on mental health will be exacerbated by intergenerational issues, because mental ill health may be self-reinforcing through generations. Not only will climate change make life progressively harder without mitigation but also each generation of adults is likely to carry some of their psychological scars from childhood, sustaining intergenerational stress, through behavioural and perhaps epigenetic mechanisms (Franklin *et al.*, 2010).

As climate change increases its toll on the natural environment, opportunities for, and benefits from, psychological connection with nature will be diminished for many. Already, many urban poor live in total isolation from undisturbed ecosystems. Challenges to personal and cultural identity are increased as ancestral lands are made uninhabitable by climate change. People from low-lying atolls and delta regions are at particular risk (see Box 26.2). Should climate change lead to diasporas, the maintenance of culture may be particularly difficult and may

exacerbate familial and social tension in migrant and, to a lesser extent, in host populations. The mental health impacts of climate change are likely to be cumulative and self-reinforcing.

26.8 Cognitive Obstacles to Dealing with Climate Change

Climate change challenges global cognitive capacity in many ways (Swim *et al.*, 2011). The idea that collective human actions are equivalent to a geological force (reflected in the term the 'Anthropocene'),[2] capable of altering the climate and thus a key determinant of civilization, may rival heliocentrism and natural selection as the third great cognitive revolution of the last thousand years. Furthermore, the most significant health and social burdens from climate change lie well into the future (see Fig. 26.2). At the moment, it is cognitively tempting to hope that recent signs (e.g. Hurricane Sandy) are flukes, rather than precursors of the near-term future. It is also tempting to deny the magnitude of the risks that the tertiary effects pose in the near future.

Only a century ago, most people in the world were illiterate. The theory of evolution so challenged religious doctrine and cultural sensitivities that Charles Darwin, one of its two main proponents, delayed its publication for decades. In comparison, acceptance of the validity of climate change (a major part of the Anthropocene) has been rapid and driven by much higher levels of scientific literacy, by both government and the public. However, even in wealthy countries, strong cultural barriers to the acceptance of science exist (Bloom and Weisberg, 2007).

Furthermore, the degree of acceptance and understanding of climate change is manifestly inadequate compared to the urgency and danger that it represents (Gifford, 2011). The theories of evolution and heliocentricism threaten powerful interests, principally Christianity. Climate change also threatens powerful interests, extending beyond the fossil fuel industry to many people who regard themselves as dependent on them. These forces are so powerful that they have harnessed a vast empire of denial and deception, which collectively has tried to undermine its reality and threat. Denialism is well-known in public health circles; examples other than climate change include the concerted opposition

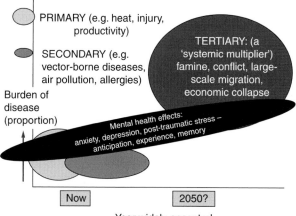

PRIMARY (e.g. heat, injury, productivity)

SECONDARY (e.g. vector-borne diseases, air pollution, allergies)

TERTIARY: (a 'systemic multiplier') famine, conflict, large-scale migration, economic collapse

Burden of disease (proportion)

Mental health effects: anxiety, depression, post-traumatic stress – anticipation, experience, memory

Now

2050?

Year widely accepted

Fig. 26.2. This conceptual diagram compares the likely burden of disease from the primary, secondary and tertiary effects of climate change with the time at which the effects are likely to be widely accepted as causally related by the general and even the scientific community. Two major European heatwaves since 2000 (France and Russia) killed over 100,000 people. Both extreme events are likely to have been contributed to by climate change. Secondary effects, such as changes to vector-borne diseases, probably have a lower burden of disease. There has been greater scientific resistance to their reality, but this is fading. Tertiary effects such as the contribution of anthropogenic climate change to the conflicts in Sudan and Syria are still regarded as speculative by most people, including many scientists. These events have the potential to cause a burden of disease at least of an order of magnitude higher than the others. Waiting for complete consensus is to wait too long.

recognizing the hazards of tobacco, air pollution, asbestos and untreated sewerage (McKee and Stuckler, 2010). There is limited evidence that people resistant to egalitarian beliefs are physiologically different and perhaps more aggressive than those who are more liberal (Oxley *et al.*, 2008). Anecdotal support for this can perhaps be found in the numerous death threats and hate mail received by many leading climate scientists.

26.9 Conclusion: Cognitive Solutions for Addressing Climate Change

For many people, the challenge of climate change is simply too overwhelming, a 'wicked' problem whose scale invites despair. If so, an antidote may be to break the problem into smaller, more manageable fragments (Weick, 1984). Too much bad news about the prospect of climate change is counterproductive (Feinberg and Willer, 2011), but so too is the false reassurance that no

behavioural change is needed. The difficulty is to find a sufficient 'dose' of disturbing information, without destroying all hope. Another useful strategy may be to unite with others who are trying to foster social change, such as with transition town[3] activists (Richardson *et al.*, 2012). Active transport, which involves more exercise, will help mitigate climate change, and is also associated with improved mental health (Richardson *et al.*, 2012). Collaborating with environmental and health reformers and activists may also be beneficial (Stokols *et al.*, 2009).

High social capital is a well-recognized form of insurance against poor mental health (Berry *et al.*, 2010). The connectivity of social media may generate constructive activism, including among the young and the elderly. The effectiveness of this may be complemented by regular contact with nature, and with face-to-face contact with like-minded people. Appropriate leadership by individuals, institutions and governments is important to provide and nurture hope. For example, Los Angeles has announced plans to be a coal-free city by 2025. But broken promises by leaders may harm collective hope.

High social capital should not be accepted uncritically. Antisocial motorcycle and other gangs also have strong group cohesion. A risk exists of increased fragmentation, of the emergence of a fortress or 'enclave' world in which competing groups cooperate closely but which create a future riven by heightened conflict and fear (Swart, 1996).

To minimize this possibility, it is also important that investment is made in technologies which can lower fossil fuel dependency. If the corporations many regard as pariahs could make handsome profits from providing energy efficiency solutions and climate-friendly energy, then we would be well served, though other global problems would nevertheless remain, such as poverty, illiteracy and habitat loss. More suggestions are made in the final chapter of this volume.

Acknowledgements

We thank Dr Susie Burke and Dr Mark Braidwood for their comments.

Notes

[1] BOD accounts not just for years of life lost but also years lived with a disability (disability-adjusted life years).
[2] The modern era, in which collective human actions are considered as equivalent to a geological force.
[3] Transition town initiatives seek to build cohesive sustainable communities to prepare for a future with limited oil and a changing climate.

References

Albrecht, G., Sartore, G.-M., Connor, L., Higginbotham, N., Freeman, S., Kelly, B., *et al.* (2007) Solastalgia: the distress caused by environmental change. *Australasian Psychiatry* 15, S95–S98.

Bateman, A. and Fonagy, P. (2008) 8-year follow-up of patients treated for borderline personality disorder: mentalization-based treatment versus treatment as usual. *American Journal of Psychiatry* 165, 631–638.

Berry, H.L., Bowen, K. and Kjellstrom, T. (2010) Climate change and mental health: a causal pathways framework. *International Journal of Public Health* 55, 123–132.

Bloom, P. and Weisberg, D.S. (2007) Childhood origins of adult resistance to science. *Science* 316, 985–986.

Bowman, D.M.J.S., Murphy, B.P., Boer, M.M., Bradstock, R.A., Cary, G.J., Cochrane, M.A., *et al.* (2013) Forest fire management, climate change, and the risk of catastrophic carbon losses. *Frontiers in Ecology and the Environment* 11, 66–67.

Collins, P.Y., Patel, V., Joestl, S.S., March, D., Insel, T.R., Daar, A.S., *et al.* (2011) Grand challenges in global mental health. *Nature* 475, 27–30.

Connor, L., Albrecht, G., Higginbotham, N., Freeman, S. and Smith, W. (2004) Environmental change and human health in Upper Hunter communities of New South Wales, Australia. *EcoHealth* 1, 47–58.

Cordial, P., Riding-Malon, R. and Lips, H. (2012) The effects of mountaintop removal coal mining on mental health, well-being, and community health in Central Appalachia. *Ecopsychology* 4, 201–208.

Corvalán, C., Hales, S., McMichael, A.J., Butler C.D., Campbell-Lendrum D., Confalonieri U., *et al.* (2005) *Ecosystems and Human Well-Being. Health Synthesis.* World Health Organization, Geneva, Switzerland.

Doherty, T.J. and Clayton, S. (2011) The psychological impacts of global climate change. *American Psychologist* 66, 265–276.

Fazel, M., Wheeler, J. and Danesh, J. (2005) Prevalence of serious mental disorder in 7000 refugees resettled in western countries: a systematic review. *The Lancet* 365, 1309–1314.

Feinberg, M. and Willer, R. (2011) Apocalypse soon?: dire messages reduce belief in global warming by contradicting just-world beliefs. *Psychological Science* 22, 34–38.

Francis, J., Wood, L.J., Knuiman, M. and Giles-Corti, B. (2012) Quality or quantity? Exploring the relationship between Public Open Space attributes and mental health in Perth, Western Australia. *Social Science and Medicine* 74, 1570–1577.

Franklin, T.B., Russig, H., Weiss, I.C., Gräff, J., Linder, N., Michalon, A., *et al.* (2010) Epigenetic transmission of the impact of early stress across generations. *Biological Psychiatry* 68, 408–415.

Fritze, J.G., Blashki, G.A., Burke, S. and Wiseman, J. (2008) Hope, despair and transformation: climate change and the promotion of mental health and wellbeing. *International Journal of Mental Health Systems* 2, 13, doi:10.1186/1752-4458-2-13.

Galea, S., Nandi, A. and Vlahov, D. (2005) The epidemiology of post-traumatic stress disorder after disasters. *Epidemiologic Reviews* 27, 78–91.

Gifford, R. (2011) The dragons of inaction. Psychological barriers that limit climate change mitigation and adaptation. *American Psychologist* 66, 290–302.

Hamilton, C. (2010) *Requiem for a Species. Why We Resist the Truth About Climate Change.* Allen and Unwin, Sydney, Australia.

Hanigan, I.C., Butler, C.D., Kokic, P.N. and Hutchinson, M.F. (2012) Suicide and drought in New South Wales, Australia, 1970-2007. *Proceedings of the National Academy of Sciences* 109, 13950–13955.

Kujur, J.M. (2006) Resettlement debate: continuing the 'historical injustice'. *Social Change* 36, 175–190.

McKee, M. and Stuckler, D. (2010) How cognitive biases affect our interpretation of political messages. *BMJ* 340, 936–937.

McNamara, K. and Westoby, R. (2011) Solastalgia and the gendered nature of climate change. *EcoHealth* 8, 233–236.

Moberg, C.L. and Cohn, Z.A. (1991) René Jules Dubos. *Scientific American* 264, 32–38.

Norris, F., Friedman, M., Watson, P., Byrne, C., Diaz, E. and Kaniasty, K. (2002) 60,000 disaster victims speak: Part I. An empirical review of the empirical literature, 1981–2001. *Psychiatry* 65, 207–239.

Noss, R.F., Dobson, A.P., Baldwin, R., Beier, P., Davis, C.R., Dellasala, D.A., *et al.* (2012) Bolder thinking for conservation. *Conservation Biology* 26, 1–4.

Oreskes, N. and Conway, E.M. (2010) *Merchants of Doubt: How a Handful of Scientists Obscured the Truth on Issues from Tobacco Smoke to Global Warming.* Bloomsbury Press, New York.

Ornstein, R. and Ehrlich, P.R. (1989) *New World, New Mind.* Methuen, London.

Otto, S.L. (2012) Antiscience beliefs jeopardize U.S. democracy. *Scientific American* November (http://www.scientificamerican.com/article.cfm?id=antiscience-beliefs-jeopardize-us-democracy, accessed 1 April 2014).

Oxley, D.R., Smith, K.B., Alford, J.R., Hibbing, M.V., Miller, J.L., Scalora, M., *et al.* (2008) Political attitudes vary with physiological traits. *Science* 321, 1667–1670.

Page, L. and Howard, L.M. (2010) The impact of climate change on mental health (but will mental health be discussed at Copenhagen?). *Psychological Medicine* 20, 177–180.

Ramakrishnan, P.S., Saxena, K.G. and Chandrashekara, U.M. (1998) *Conserving the Sacred: For Biodiversity Management.* UNESCO, Oxford and IBH, New Delhi.

Reich, R.B., Vera, S.C., Marino, M.F., Levin, A., Yong, L. and Frankenburg, F.R. (1997) Reported pathological childhood experiences associated with the development of borderline personality disorder. *American Journal of Psychiatry* 154, 1101–1106.

Richardson, J., Nichols, A. and Henry, T. (2012) Do transition towns have the potential to promote health and well-being? A health impact assessment of a transition town initiative. *Public Health* 126, 982–989.

Robjant, K., Hassan, R. and Katona, C. (2009) Mental health implications of detaining asylum seekers: systematic review. *The British Journal of Psychiatry* 194, 306–312.

Singhi, P. (2011) Infectious causes of seizures and epilepsy in the developing world. *Developmental Medicine and Child Neurology* 53, 600–609.

Steel, Z., Chey, T., Silove, D., Marnane, C., Bryant, R. and van Ommeren, M. (2009) Association of torture and other potentially traumatic events with mental health outcomes among populations exposed to mass conflict and displacement. A systematic review and meta-analysis. *Journal of the American Medical Association* 302, 537–549.

Stokols, D., Misra, S., Runnerstrom, M.G. and Hipp, J.A. (2009) Psychology in an age of ecological crisis: from personal angst to collective action. *American Psychologist* 64, 181–193.

Swart, R. (1996) Security risks of global environmental changes. *Global Environmental Change* 6, 187–192.

Swim, J., Clayton, S., Doherty, T., Gifford, R., Howard, G., Reser, J., *et al.* (2011) Psychology and global climate change: addressing a multi-faceted phenomenon and set of challenges. A report by the American Psychological Association's task force on the interface between psychology and global climate change. *American Psychologist* 66, 265–276.

Van Den Berg, A.E., Hartig, T. and Staats, H. (2007) Preference for nature in urbanized societies: stress, restoration, and the pursuit of sustainability. *Journal of Social Issues* 63, 79–96.

Weick, K.E. (1984) Small wins. Redefining the scale of social problems. *American Psychologist* 9, 40–49.

Wilson, E.O. (1984) *Biophilia.* Harvard University Press, Cambridge, Massachusetts.

27 Climate Change, Housing and Public Health

Jouni J.K. Jaakkola and Adeladza K. Amegah

Center for Environmental and Respiratory Health Research,
University of Oulu, Oulu, Finland

27.1 Importance of Housing in the Context of Climate Change

In the evolution of humanity, housing and shelter has remained a basic need, offering protection against extreme weather events, natural calamities and pest and insect invasion, as well as threats from other species; initially from wild animals, now mostly from other humans. Housing has enabled human survival in a wide range of climatic and environmental conditions for centuries and unsurprisingly represents a significant contributor to climate change through the sector's high energy consumption and associated greenhouse gas (GHG) emissions. Actions in the housing sector thus play an important role, both to mitigate climate change and to adapt to the adverse effects of changing climate. However, some of these strategies are expected to have adverse impact on human health and well-being and yet offer co-benefits for public health.

The carbon footprint from housing is great, both in cold climates, where energy is consumed in heating, and in warm climates, where indoor temperatures are cooled mechanically, assuming the energy is released from the combustion of fossil fuels. In the USA, the average household carbon footprint is approximately 48 t CO_2 equivalents (CO_2e)/year, of which housing is responsible for approximately 13.2 t CO_2e (28%) (Jones and Kammen, 2011). In Finland, buildings consume approximately 40% of the total energy consumption; their contribution to GHGs is approximately 30%. Heating and ventilation require about half of the total energy consumption, with the remainder divided between household electricity and warm water.

As mentioned, housing offers substantial possibilities for climate change mitigation through reduced energy consumption, and thus lower GHG emissions. In cold climates, energy consumption can be reduced by exterior wall and ceiling insulation, the use of low-energy devices, adjustment and timing of heating and minimizing waste. In the future, the carbon footprint of housing can be further lowered by passive solar construction and other designs which optimize energy consumption for heating and ventilation, together with decarbonized energy sources, such as those based on geothermal heat, solar or wind power.

27.2 The Health Effects of Climate Change

The potential adverse effects of climate change on global health have been reviewed extensively (Haines and Patz, 2004; Patz et al., 2005; Patz and Olson, 2006; Haines et al., 2006; McMichael et al., 2006). In his seminal article published in *The Lancet*, McMichael and colleagues presented a framework describing the direct and indirect health impacts of climate change (McMichael *et al.*, 2006). Climate change, influenced by both natural and human phenomena (especially the production of excessive GHGs), manifests in environmental effects such as extreme weather events, expected to have direct effects on mortality and morbidity. Effects on ecosystems, sea level rise and environmental degradation manifest in complex causal webs, which also have various potential long-term health effects. Frumkin and co-workers summarized the principal public health concerns of climate change as including injuries and fatalities related to severe weather events and heatwaves; infectious diseases related to changes in vector biology, water and food contamination; allergic symptoms related to increased allergen production; respiratory and cardiovascular disease related to worsening air pollution; and nutritional shortages related to changes in food production (Frumkin *et al.*, 2008). According to these authors, indirect concerns, for which data to support projections are less available and uncertainties are greater, include mental health consequences, population dislocation and civil conflict. In addition, changes in the patterns of pests, parasites and pathogens affecting wildlife, livestock, agriculture, forests and coastal

marine organisms can alter ecosystem composition and functions, while life-support system erosion also has implications for human health (Epstein, 2005).

27.3 The Housing Life Cycle and GHGs

The contribution of housing to GHG production and the opportunities it offers for mitigating and adapting to climate change are presented in all phases of its life cycle (Fig. 27.1) – the production and haulage of raw materials, construction, occupation, maintenance and renovation, deconstruction and demolition and transportation of waste to a recycling facility or waste disposal site. The mining and industrial processing of building materials are energy-intensive and the largest contributor to CO_2 emissions during the housing life cycle. Heavy-duty vehicles required for the transportation of building materials to construction sites have low fuel efficiency and are an important source of CO_2 emissions. The construction phase tends to have a relatively low impact on GHG emission levels, but the type

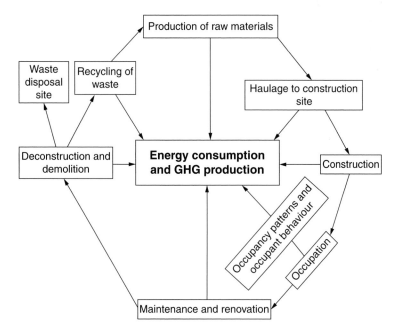

Fig. 27.1. Carbon footprint associated with the life cycle of housing.

of housing and the techniques applied can incur high energy consumption and resultant high CO_2 emission. The occupation of buildings has a higher impact on CO_2 emission due to the large amount of energy required for lighting, cooking, heating and cooling, as mentioned above. Building maintenance also requires energy and GHG emissions, but these tend to be less compared to the construction phase. Renovation and remodelling of buildings can also consume considerable energy, as can deconstruction and demolition, although this is minor compared to the total energy consumption through the housing life cycle. Waste recycling (including of building materials) offers great potential for reducing GHG emissions.

Contribution of the occupation phase of the housing life cycle to climate change

Mitigating and adapting household occupancy to climate change is driven by two main factors – occupancy patterns and behaviour. Occupant behaviour includes the range of attitudes, actions and practices associated with the building's characteristics and with habits used to maximize comfort. Such behaviour influences household energy consumption and associated GHG emission significantly, and includes the following: (i) window and interior door usage and associated heating and cooling methods adopted; (ii) window curtain usage, drawing patterns and associated lighting techniques and choices; (iii) clothing and bedding adjustment and room temperature control; (iv) cooking methods, patterns and fuel choices; (v) household appliance preferences and usage; and (vi) approaches to housekeeping (manual versus use of electrical gadgets).

Occupancy patterns relate to the amount of time spent at home. This is related to the working and social lifestyle of the occupants and is a major determinant of energy consumption. Most people typically spend about 60–70% of their time at home (Klepeis *et al.*, 2001; Schweizer *et al.*, 2007). Ott, for instance, reported that in the USA, people in employment spend about 60% of their time at home indoors, 30% at work indoors and 5% in transit (Ott, 1985). Household energy consumption is linked directly to the amount of occupant time spent indoors: once there, activities and operations are undertaken to put the house in order and to ensure the thermal and overall comfort of the occupant.

27.4 Impact of Housing-related Climate Change Mitigation and Adaptation Strategies on Public Health

Housing-related climate change mitigation and adaptation strategies include changes in building design and characteristics, building operation and occupant behaviour in response to changing climate and its adverse effects. Mitigation strategies are aimed at curtailing anthropogenic influences that produce GHG emissions, whereas adaptation strategies try to assuage climate change effects. Both are intended to reduce health risk.

A major attempt to mitigate climate change in many developed countries has seen buildings being constructed or renovated to become more airtight and insulated. However, compliance with such techniques has compromised ventilation, with consequences for indoor air quality. The effects of poor indoor air quality on respiratory and cardiovascular health are well documented in the epidemiologic literature. Bone and colleagues have highlighted the possible impacts of increased air tightness in the absence of adequate ventilation of buildings on occupant health (Bone *et al.*, 2010). Issues identified by these authors include:

1. Insufficient ventilation increases levels of indoor pollutants such as carbon monoxide (CO), nitrogen dioxide (NO_2), formaldehyde, volatile organic compounds (VOCs), environmental tobacco smoke and radon.
2. Raising indoor temperatures by better insulation in winter reduces relative humidity and condensation, but in the absence of adequate ventilation relative humidity is expected to increase, thereby promoting growth of mould and house dust mite proliferation.
3. Poor ventilation is also likely to increase the risk of CO poisoning, even though reduced reliance on fossil fuel burning in the home lowers this risk.

4. Increased airtightness of homes and associated poor ventilation increases indoor radon levels. In UK homes, double glazing of all windows and full draught-proofing, for instance, have been estimated to increase indoor radon levels by 55% and 9%, respectively (Gunby *et al.*, 1993). This problem is regional and depends on the type of soil and other geological characteristics of the ground underneath the building.

5. Insulation keeps heat out of the home and may lower health risk from intense heatwaves worsened by climate change, but where heat enters (e.g. through unshaded double-glazed windows), insulation may reduce its escape, raising heat-related morbidity and mortality risk.

6. Warmer summers may be associated with higher outdoor levels of ozone, some of which could penetrate the indoor environment, reacting with material surfaces and other air components to produce secondary indoor air pollutants. There is increasing evidence that such products may harm human health (Weschler *et al.*, 2006; Carslaw *et al.*, 2009). Ventilation is necessary to reduce concentrations of pollutants generated indoors, as well as to lessen the time available for chemical reactions between them (Mitchell *et al.*, 2007).

Crump and co-workers have also identified the possible consequences of highly energy efficient homes for the indoor environment and their potentially beneficial or adverse effects on occupant health and well-being (Crump *et al.*, 2009). Building characteristics considered were increased airtightness, higher winter internal temperature, summer internal temperatures, mechanical ventilation, heat recovery aspect, construction materials and ground contaminants. They found that insufficient air infiltration could harm air quality, due to the build-up of stale air and its lack of sufficient refreshment. In addition to increase in air pollution, insufficient air change will increase humidity and condensation, and increase the risks of mould growth, damage to structures and proliferation of house dust mites.

Poor ventilation is also a recognized risk factor for airborne disease transmission, and is also associated with dampness problems. According to the Institute of Medicine (IOM), some modern construction techniques and materials can also increase the risk of dampness (IOM, 2011). Jaakkola and colleagues previously suggested that the major reasons for the high frequency of indoor dampness and mould problems in cold climates might be insufficient building maintenance and excessively airtight buildings to conserve energy, with inadequate ventilation (Jaakkola *et al.*, 2002).

Air conditioning

An adaptation measure in response to elevated indoor temperatures caused by intense heatwaves due to climate change is the increased use of air conditioning (AC). This can trigger or aggravate respiratory illnesses. Occupants in office buildings with AC systems, for instance, have consistently reported increased symptoms of mucous membrane irritation, breathing difficulties, irritated skin and constitutional/neurological symptoms such as headache and fatigue compared to inhabitants of naturally ventilated buildings (Jaakkola and Miettinen, 1995; Mendell, 2004). Mendell identified two possible causes of these non-specific symptoms associated with AC exposure (Mendell, 2004): (i) contaminants disseminated into the indoor air by AC and humidification systems; and (ii) moisture in AC and humidification systems is likely to result in microbiological exposures that cause health effects through mechanisms that are irritant, toxic or allergic. Mendell further stated that surfaces in AC systems that remain constantly moist, such as the cooling coils and drip pans, are supportive environments for the growth of undesirable microorganisms, are directly in the path of all air supplied to the occupants to inhale and are often poorly maintained (Mendell, 2004). These issues can all cause health effects on the occupants. A well-documented public health danger of AC systems is legionnaires' disease, a potentially fatal form of pneumonia caused by the bacteria, *Legionellosis pneumophila*. The presence of AC has also been identified to increase the risk of dampness problem (IOM, 2011). Living in air-conditioned environments also reduces heat acclimatization, thus increasing health risks during power cuts; common occurrences in developing economies (Kjellstrom and McMichael, 2013). AC systems, especially if older and poorly maintained, also generate significant noise, which can disturb the sleep of the occupants and their neighbours. AC systems also contribute to heat island effects, with adverse consequences, especially for poor and vulnerable people.

Fans and fires

Another alternative to maintaining cooling in sweltering conditions is the use of electric fans, but according to a Cochrane review, when temperatures are above 35°C, electric fans can actually contribute to heat gain by blowing hot air over the body, thus leading to heatstroke (Gupta *et al.*, 2012).

An adaptation strategy to extremely low indoor temperatures during winter in Arctic and temperate regions, and during wet seasons in the tropics, is the combustion of solid fuels in fireplaces. But these result in increased levels of combustion-related air pollutants, including carbon monoxide, particulate matter, polycyclic aromatic hydrocarbons, sulfur dioxide and nitrogen oxides, with severe health consequences for the occupants. The health effects of exposure to combustion products on children and adults include acute lower respiratory tract infection (ALRI), nutritional deficiency, chronic obstructive lung disease, lung cancer, tuberculosis, cardiovascular disease, cataracts and adverse pregnancy outcomes (Fullerton *et al.*, 2008).

27.5   Climate Change, Housing Dampness and Mould Problems and Respiratory Health

The projected effects of climate change include an increase of precipitation, flooding and severe hurricanes, which is likely to worsen dampness and mould problems in homes and other buildings, thus exacerbating related health problems. Water damage, leakages and dampness and related mould problems are already common worldwide, constituting one of the most important public health housing problems. In cold climates, the prevalence of water damage and dampness problems in homes has been estimated to range from 5% to 30%, while in moderate and warm climates it has been estimated at 10–60% (Jaakkola and Jaakkola, 2004; World Health Organization (WHO), 2009; Jaakkola *et al.*, 2011). The prevalence of indoor mould

exposure has been estimated as 5–10% in cold climates and 10–30% in moderate and warm climates (Jaakkola *et al.*, 2011). A systematic review and meta-analyses of 16 longitudinal studies (Quansah *et al.*, 2012) provided evidence on the effects of dampness and mould problems on the onset of asthma. The summary effect estimates for the relation between any exposure and onset of asthma was 1.50 (95% confidence interval (CI) 1.25–1.80). The summary effect estimates were elevated significantly for dampness, visible mould and mould odour. A recent systematic review and meta-analysis (Jaakkola *et al.*, 2013) also provides convincing evidence that dampness and mould problems increase the risk of allergic and non-allergic rhinitis. The meta-analysis showed that the risk of rhinitis (Rh), allergic rhinitis (AR) and rhinoconjunctivitis (RC) were increased significantly in relation to home dampness and moulds. The largest risk was observed in relation to mould odour (odds ratio (OR) 2.18, 95% CI 1.56–2.12 for Rh, and 1.87, 0.95–3.68 for AR), and the risk related to visible mould was also consistently increased (OR 1.82, 95% CI 1.56–2.12 for Rh, OR 1.51, 95% CI 1.39–1.64 for AR and OR 1.66, 95% CI 1.27–2.18 for RC).

27.6   Co-benefits for Public Health from Climate Change Mitigation Strategies in the Housing Sector

Several climate change mitigation actions have important health co-benefits, meaning that reduced GHG emissions are achieved with actions that promote and facilitate improved population health (Kjellstrom and McMichael, 2013). Haines and colleagues also previously stated that one favourable aspect of the efforts to mitigate climate change was that local health gains would accrue quickly to populations that undertook such efforts (Haines *et al.*, 2006). We present three case scenarios (see Fig. 27.2):

1. Tightening the building envelope during construction or renovation to improve energy efficiency and reduce GHG emissions also: (i) improves temperature control; (ii) reduces noise exposure; (iii) minimizes the penetration of ambient air pollutants, aeroallergens and dust; (iv) checks or addresses dampness and

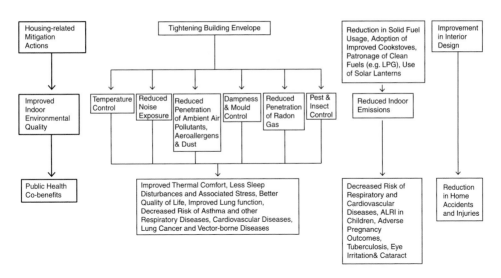

Fig. 27.2. Pathways through which housing-related climate change mitigation leads to public health co-benefits.

mould problems; (v) prevents the penetration of radon gas; and (vi) improves pest and insect control, resulting in improved indoor environmental quality (IEQ). The public health gains accruing from the enhancement of IEQ are improved thermal comfort, less sleep disturbances and associated stress, better quality of life, improved lung function and decreased risk of asthma and other respiratory illnesses, cardiovascular diseases, lung cancer and vector-borne diseases.

2. Reduction in solid fuel use, adoption of improved cooking stoves and use of cleaner fuels such as liquid petroleum gas, especially in developing countries, leads to reduced indoor emissions with a decreased risk of respiratory and cardiovascular diseases, ALRI in children, cataracts and blindness, and adverse pregnancy outcomes. Exchanging kerosene for solar lighting where electricity is lacking will improve indoor air quality and protect against tuberculosis, eye irritation and other health risks.

3. Additionally, some aspects of improved interior design during construction or renovation processes, such as the repairing of uneven and loose floors, provision of slip-resistant flooring, fixing of smoke alarms and fire sprinklers and installation of grab bars in showers and bath tubs, will lower the risk of home accidents and injuries. Avoiding lead-based paints (which causes severe developmental delays in children) and VOC-containing products (known to cause

problems including eye and throat irritations) are further public health gains.

27.7 Controversies in Climate Change Mitigation and Conclusions

In some situations, housing-related climate change mitigation action may generate important public health controversies. Some actions to mitigate climate change and related health impacts in the long term may increase health risk in the short term. We present three such examples.

First, by reducing air change in buildings, energy consumption can be lowered, mitigating climate change and reducing long-term health impact. However, if ventilation is reduced without commensurate lowering of indoor source emissions, indoor air pollution may rise, with short-term ill-health effects. A win–win approach which lessens energy consumption is to reduce indoor emission sources, for example by using low-emitting surface materials or restricting indoor activities that produce pollution in selected areas. In this way, we can lower air change without increasing indoor air pollution and the related adverse health effects.

The second example is related to adaptation to heatwaves. In climate change scenarios, the frequency and intensity of heatwaves will increase. By studying empirically the health effects of heatwaves, we can predict the long-term consequences of climate change. The effects of heatwaves on mortality have been quantified extensively. Bouchama and colleagues have summarized the factors that make some individuals more susceptible to the effects of heatwaves (Bouchama *et al.*, 2007). Clearly, patients with chronic cardiovascular and respiratory diseases face a higher risk. However, the weight of evidence shows that use of AC for home cooling lessens the adverse effects. Thus, reduced AC use will lower energy consumption and mitigate climate change, but may harm heatwave adaptation, thus increasing the health effects of extreme temperature.

The third controversial climate change mitigation action is related to reduced indoor temperatures in the cold season. Reduction of indoor temperatures saves energy and mitigates climate change. However, reducing temperatures below 16–18°C may increase health risks, especially in susceptible populations. Too low an indoor temperature is likely to contribute to more respiratory infections, especially among younger children and the elderly. Low temperatures are also associated with hypertension (Collins, 1986; Brook *et al.*, 2011), which increases the risk of cardiovascular diseases.

In summary, climate change mitigation actions targeted at housing may result in adverse health effects. These risks could be lowered by an increased use of energy produced using renewable means, such as solar and wind. Multidisciplinary collaboration is needed.

Acknowledgements

Supported by the Academy of Finland, Grant No 129419 (SALVE Research Program) and Grant No 266314 (Research Council for Health).

References

Bone, A., Murray, V., Myers, I., Dengel, A. and Crump, D. (2010) Will drivers for home energy efficiency harm occupant health? *Perspectives in Public Health* 130, 233–238.

Bouchama, A., Dehbi, M., Mohamed, G., Matthies, F., Shoukri, M. and Menne, B. (2007) Prognostic factors in heat wave related deaths: a meta-analysis. *Archives of Internal Medicine* 167, 2170–2176.

Brook, R.D., Weder, A.B. and Rajagopalan, S. (2011) 'Environmental hypertensionology' the effects of environmental factors on blood pressure in clinical practice and research. *The Journal of Clinical Hypertension* 13, 836–842.

Carslaw, N., Langer, S. and Wolkoff, P. (2009) New directions: where is the link between reactive indoor air chemistry and health effects? *Atmospheric Environment* 43, 3808–3809.

Collins, K.J. (1986) Low indoor temperatures and morbidity in the elderly. *Age and Ageing* 15, 212–220.

Crump, D., Dengel, A. and Swainson, M. (2009) Indoor air quality in highly energy efficient homes – a review. *NHBC Foundation Report NF18.* National House Building Council, Milton Keynes, UK.

Epstein, P.R. (2005) Climate change and human health. *New England Journal of Medicine* 353, 1433–1436.

Frumkin, H., Hess, J., Luber, G., Malilay, J. and McGeehin, M. (2008) Climate change: the public health response. *American Journal of Public Health* 98, 435–445.

Fullerton, D.G., Bruce, N. and Gordon, S.B. (2008) Indoor air pollution from biomass fuel smoke is a major health concern in the developing world. *Transactions of the Royal Society of Tropical Medicine and Hygiene* 102, 843–851.

Gunby, J.A., Darby, S.C., Miles, J.C.H., Green, B.M.R. and Cox, D.R. (1993) Factors affecting indoor radon concentrations in the United Kingdom. *Health Physics* 64, 2–12.

Gupta, S., Carmichael, C., Simpson, C., Clarke, M.J., Allen, C., Gao, Y., *et al.* (2012) Electric fans for reducing adverse health impacts in heatwaves. *Cochrane Database of Systematic Reviews* 11, CD009888.

Haines, A. and Patz, J.A. (2004) Health effects of climate change. *Journal of the American Medical Association* 291, 99–103.

Haines, A., Kovats, R.S., Campbell-Lendrum, D. and Corvalán, C. (2006) Climate change and human health: impacts, vulnerability and public health. *Public Health* 120, 585–596.

IOM (Institute of Medicine) (2011) *Climate Change, The Indoor Environment, and Health.* The National Academies Press, Washington, DC.

Jaakkola, M.S. and Jaakkola, J.J.K. (2004) Indoor molds and asthma in adults. *Advances in Applied Microbiology* 55, 309–338.

Jaakkola, J.J.K. and Miettinen, P. (1995) Type of ventilation system in office buildings and sick building syndrome. *American Journal of Epidemiology* 141, 755–765.

Jaakkola, M.S., Nordman, H., Piipari, R., Uitti, J., Laitinen, J., Karjalainen, A., *et al.* (2002) Indoor dampness and molds and development of adult-onset asthma: a population-based incident case-control study. *Environmental Health Perspectives* 110, 543–547.

Jaakkola, M.S., Haverinen-Shaugnessy, U., Doewes, J. and Nevalainen, A. (2011) Indoor dampness and mold problems in homes and asthma onset in children. In: Braubach, M., Jacobs, D. and Ormandy, D. (eds) *Environmental Burden of Disease Associated with Inadequate Housing – A Method Guide to the Quantification of Health Effects of Selected Housing Risks in the WHO European Region.* World Health Organization, Geneva, Switzerland, pp. 5–31.

Jaakkola, M.S., Quansah, R., Hugg, T.T., Heikkinen, S.A.M. and Jaakkola, J.J.K. (2013). Association of indoor dampness and molds with the risk of rhinitis: a systematic review and meta-analysis. *Journal of Allergy and Clinical Immunology* 132, 1099–1110.e18.

Jones, C.M. and Kammen, D.M. (2011) Quantifying carbon footprint reduction opportunities for US households and communities. *Environmental Science and Technology* 45, 4088–4095.

Kjellstrom, T. and McMichael, A.J. (2013) Climate change threats to population health and well-being: the imperative of protective solutions that will last. *Global Health Action* 6, doi: 10.3402/gha.v6i0.20816.

Klepeis, N.E., Nelson, W.C., Ott, W.R., Robinson, J.P., Tsang, A.M., Switzer, P., *et al.* (2001) The National Human Activity Pattern Survey (NHAPS): a resource for assessing exposure to environmental pollutants. *Journal of Exposure Analysis and Environmental Epidemiology* 11, 251–252.

McMichael, A.J., Woodruff, R.E. and Hales, S. (2006) Climate change and human health: present and future risks. *The Lancet* 367, 859–869.

Mendell, M. (2004) Commentary: air conditioning as a risk for increased use of health services. *International Journal of Epidemiology* 33, 1123–1126.

Mitchell, C.S., Zhang, J.J., Sigsgaard, T., Jantunen, M., Lioy, P.J., Samson, R., *et al.* (2007) Current state of the science: health effects and indoor environmental quality. *Environmental Health Perspectives* 115, 958–964.

Ott, W. (1985) Total human exposure: an emerging science focuses on humans as receptors of environmental pollution. *Environmental Science and Technology* 19, 880–886.

Patz, J.A. and Olson, S.H. (2006) Climate change and health: global to local influences on disease risk. *Annals of Tropical Medicine and Parasitology* 100, 535–549.

Patz, J.A., Campbell-Lendrum, D., Holloway, T. and Foley, J.A. (2005) Impact of regional climate change on human health. *Nature* 438, 310–317.

Quansah, R., Jaakkola, M.S., Hugg, T.T., Heikkinen, S.A.M. and Jaakkola, J.J.K. (2012) Residential dampness and molds and the risk of developing asthma: a systematic review and meta-analysis. *PLoS One* 7, e47526.

Schweizer, C., Edwards, R.D., Bayer-Oglesby, L., Gauderman, W.J., Ilacqua, V., Jantunen, M.J., *et al.* (2007) Indoor time–microenvironment–activity patterns in seven regions of Europe. *Journal of Exposure Science and Environmental Epidemiology* 17, 170–181.

Weschler, C.J., Wells, J.R., Poppendick, D., Hubbard, H. and Pearce, T.A. (2006) Workgroup report: indoor chemistry and health. *Environmental Health Perspectives* 114, 442–446.

WHO (World Health Organization) (2009) *WHO Guidelines for Indoor Air Quality: Dampness and Mould.* World Health Organization, Copenhagen (http://www.euro.who.int/__data/assets/pdf_file/0017/43325/E92645.pdf, accessed 7 April 2014).

28 Health in New Socio-economic Pathways for Climate Change Research

Kristie L. Ebi

ClimAdapt, LLC, Seattle, Washington, USA

28.1 Background

The magnitude and pattern of the possible future impacts of climate change will depend on:

- how the Earth System will respond to changes in atmospheric composition from emissions of greenhouse gases (GHGs);
- how ecosystems change and human societies evolve and develop, independent of and in interaction with climate change, including demographics, socio-economic conditions, technology, and policies;
- the effectiveness of mitigation and adaptation strategies, policies and measures to avoid, prepare for, cope with and recover from impacts.

The future is, of course, uncertain. Projecting how individual factors could change over the coming decades, such as demographics or global mean surface temperature, provides limited insights into how the wide range of factors that determine human and ecosystem resilience will interact, including with the Earth System, across spatial and temporal scales, and how adaptation and mitigation strategies will interact with environmental and human systems to alter atmospheric conditions and their impacts. Understanding the possible interactions and their associated range of uncertainties will facilitate informed and effective decision and policy making to reduce the human health and well-being risks of climate change.

28.2 Use of Scenarios

Scenarios are a standard approach to explore and evaluate the extensive uncertainties associated with future climate change and development. A *scenario* is a comprehensive description of the future of the human–climate system, including quantitative and qualitative information. Scenarios differ from *pathways* that describe scenario components such as atmospheric concentration or development indicators.

For more than the past decade, projections of the possible impacts of climate change have been based primarily on scenarios developed for the Intergovernmental Panel on Climate Change (IPCC) in the Special Report on Emission Scenarios (SRES) (Nakicenovic *et al.*, 2000). These scenarios were developed to represent a range of driving forces and emissions, reflecting understanding of underlying uncertainties.

Because these scenarios are now dated and greater scientific understanding has been gained, the climate modelling, integrated assessment modelling (IAM), and vulnerability, impacts and adaptation (VIA) communities have developed a process and framework for the creation and use of new scenarios (Moss *et al.*, 2008, 2010). This framework is intended to improve the interdisciplinary analysis and assessment of climate change, the associated risks, and response options, thus informing research and policy. This process includes a set of forcing pathways (the representative concentration pathways, or RCPs) that can be combined with socio-economic development pathways (shared socio-economic pathways, or SSPs) to create a range of scenarios. This chapter reviews the process and development of the RCPs and SSPs and suggests how health pathways could evolve within the SSPs (Ebi, 2013).

28.3   Process

The SRES were developed by first creating internally consistent storylines of possible future worlds (Nakicenovic et al., 2000). Four main storylines described the relationships between the driving forces of GHG and sulfur emissions and their evolution. Underlying each was a 'future history' describing demographic, economic, technological and environmental development over the 21st century. Four families of scenarios were developed to examine the range of possible outcomes associated with similar assumptions about driving forces. The SRES assumed no specifically targeted climate mitigation and adaptation policies and measures. Quantification of the storylines resulted in estimated atmospheric concentrations of GHG and sulfur that were input into climate models to project future temperature and precipitation patterns. These projections have been used extensively to estimate the possible impacts associated with each scenario.

The new process follows a different approach. The goal is to develop scenarios that include GHG emissions (RCPs) over the 21st century and quantitative and qualitative descriptions of socio-economic development pathways (Ebi et al., 2014; Kriegler et al., 2014; O'Neill et al., 2014; van Vuuren et al., 2014). The RCPs are being used to project changing weather patterns over the 21st century, with extensions of projections of global mean surface temperature to 2300. The scenarios developed using the RCPs and

SSPs can be used to project risks and to assess mitigation and adaptation policies.

The emission pathways underlying the RCPs are documented in a special issue of *Climatic Change* (van Vuuren et al., 2011). Climate model simulations based on them are available (Taylor et al., 2011). The IAM and VIA communities have finalized the narratives for the SSPs, including developing qualitative descriptions and quantitative projections. Work is just beginning on developing sectoral and regional extensions of the SSPs, and on assessing policy options.

28.4   Representative Concentration Pathways

To begin the process, four IAM teams (IIASA, JGCRI-PNNL, PBL and NIES) (see Acronyms, page xiii) each calculated the emissions for one RCP; these emissions were used in the Coupled Model Intercomparison Project 5 (CMIP5). The RCPs were selected to span a wider range of emissions than were used in the SRES, taking into consideration the ability of Earth System models to differentiate between different radiative forcing levels. With increased scientific understanding and improved modelling capabilities, emission scenarios now include all GHGs (not just carbon dioxide concentrations). This enables RCPs to be defined in terms of radiative forcing in 2100 and the rate of change (Table 28.1) (van Vuuren et al., 2011).

The RCPs include emissions of GHGs and short-lived species at ½ degree by ½ degree grid scale, and land use and land cover. Only limited assumptions about demographic and socio-economic change were made. The RCPs were developed using idealized assumptions about policy instruments (e.g. no specific policies were assumed) and the timing of participation by the international community. Therefore, the RCPs have limited usefulness to researchers working on

Table 28.1. Representative concentration pathways in 2100. (Source: van Vuuren et al., 2011.)

RCP	Radiative forcing	CO_2 equivalent concentration	Rate of change in radiative forcing
RCP 8.5	8.5 W/m²	1350 ppm	Rising
RCP 6.0	6.0 W/m²	850 ppm	Stabilizing
RCP 4.5	4.5 W/m²	650 ppm	Stabilizing
RCP 2.6	2.6 W/m²	450 ppm	Declining

the possible risks of climate change and those investigating the effectiveness of mitigation and adaptation policies to alter emissions and eco-system and societal resilience to climate change.

28.5 Shared Socio-economic Pathways

Developing new scenarios for use in impacts, adaptation and mitigation research requires more than the emissions of GHGs and the resulting extent of climate change. Scenarios also require a narrative that describes assumptions about socio-economic development patterns, including qualitative and quantitative elements. An insight gained from the SRES (Nakicenovic et al., 2000) and confirmed by van Vuuren et al. (2012) is that the magnitude of GHG emissions is relatively independent of demographic and socio-economic development: multiple demographic and socio-economic development pathways can lead to any particular emission scenario. A relatively wealthy world with high population density could have low GHG emissions because of policies that encourage energy efficiency and sufficient low emission technology. The opposite also is possible. These multiple pathways mean that demographic and socio-economic development pathways can be described separately from the RCPs and then combined, using a matrix architecture, into a broader range of scenarios than was possible with the SRES (see the next section).

Shared socio-economic pathways define the state of human and natural societies at a macroscale. They include a narrative and a set of quantified elements that define the state of so-ciety as it evolves over the 21st century under the assumption of no significant climate feed-back (O'Neill et al., 2014; van Vuuren et al., 2014). This assumption allows the SSPs to be formulated without considering climate change; the interaction between climate change and development will be taken into consideration when scenarios are developed. The possibility that the SSPs will be altered by climate change is considered when an SSP is combined with cli-mate change projections, to generate scenarios.

The key aims of the new scenario process are to facilitate research and assessment that can char-acterize the range of uncertainty in mitigation efforts required to achieve particular radiative forcing pathways, in adaptation efforts that could be undertaken to prepare for and respond to the climate change associated with those pathways, and in residual impacts (Ebi et al., 2014; van Vuuren et al., 2014). Five SSPs are being devel-oped along two axes describing worlds with increasing socio-economic challenges to mitiga-tion (y-axis) and adaptation (x-axis).

Increasing challenges to mitigation include factors that: (i) generate high reference emis-sions; and (ii) reduce social capacity to mitigate emissions. These include factors such as insuffi-cient viable technologies, inadequate national and international policy-making institutions and lack of financial and other resources to sup-port adequate mitigation, such as political will and human and social capital (Klein et al., 2007; Winkler et al., 2007). High reference emissions could result from various combinations of high population growth rates, rapid (conventionally defined) economic growth, energy-intensive eco-nomic systems, carbon-intensive energy supplies and the like. Not all factors need operate in the same direction to result in high (or low) emissions.

Increasing socio-economic challenges to adaptation include factors that reduce social capacity to manage the risks associated with any given level of climate change. These include fac-tors such as increasing inequity, low human and social capital and reduced institutional effective-ness. Adaptation can focus on managing the risks associated with climate hazards, who or what is exposed to those hazards, or on the asso-ciated vulnerability, whether geographic, socio-economic, cultural or other forms (IPCC, 2012).

The five SSPs are described based on O'Neill et al. (O'Neill et al., 2014), along with possible pathways for population health to evolve over the coming decades (Ebi et al., 2014). These pathways may be further modified as the SSPs are used by the scientific community.

Low challenges to adaptation and mitigation. Sustainability; taking the green road

This SSP describes a world making concerted and prolonged efforts toward sustainability, emphasizing more inclusive development within environmental boundaries. Increasing evidence of and accounting for the social, cultural and economic costs of environmental degradation and inequality drive a gradual shift to a more sustainable path. Concerns with economic growth shift consumption toward low material growth, lower resource and energy intensity and environmentally friendly development, even at the expense of somewhat slower growth. Increasingly effective and persistent cooperation and collaboration of local, national and international organizations and institutions, the private sector and civil society facilitate improved management of the global commons. Investments in education and health accelerate the demographic transition, leading to a relatively low global population. At the same time, economic growth in low-income countries reduces the number of people below the poverty line. Improved resource efficiency, driven by investment in environmental technology, changes in tax structures and changing perceptions, reduce overall energy and resource use over the longer term and increase the use of renewables. The improvements in human well-being, along with strong and flexible global, regional and national institutions, imply low challenges to adaptation.

Overall health improves significantly in this world, with reductions in the burden of climate-sensitive health outcomes and increased capacity to adapt to the health risks of climate change as efforts to achieve sustainable development goals (SDGs) increase access to safe water, improved sanitation, medical care and other factors that reduce vulnerability to climate and other global changes. Life expectancies increase in low-income countries with decreasing burdens of the key causes of childhood mortality (undernutrition, diarrhoeal diseases and malaria). However, burdens of non-communicable diseases (NCDs) increase, with more children surviving to adulthood, although changes in dietary patterns and reductions in air pollution from the burning of fossil fuels would lower the burden of some chronic diseases. Funding would increase for public health and health-care organizations and institutions to improve: (i) the monitoring and surveillance of climate-sensitive health outcomes and of adaptation policies and measures; (ii) research on and modelling of the health risks of

climate change; and (iii) the use of iterative management approaches to increase the effectiveness of adaptation policies and measures. Enhanced cooperation across sectors minimizes co-harms from adaptation options implemented in other sectors. Large co-benefits could be achieved with mitigation policies (although specific policies are not included in the SSPs).

Low challenges to adaptation and high challenges to mitigation. Rapid growth; taking the highway

This SSP differs from the one above in the extent of the challenges to mitigation. This is a world characterized by rapid economic growth in industrialized and emerging economies, a strong reliance on fossil fuels and limited concern about global environmental change. The focus is on the integration of global competitive markets, emphasizing maintaining competition and removing institutional barriers for the participation of disadvantaged population groups. Rapid technological progress and the development of human capital support these trends. This world also is characterized by strong investments in health, education and institutions to enhance human and social capital (and thus, low challenges to adaptation). However, economic and social development is coupled with the exploitation of fossil fuel resources and the adoption of resource- and energy-intensive lifestyles around the world. Faith that social and ecological systems can be managed to minimize environmental problems through technological solutions results in relatively little specific proactive efforts to avoid potential global environmental impacts.

Health also improves significantly in this world, but not as much as in a world with low challenges to both adaptation and mitigation, because of the impacts of animal product-rich diets on chronic diseases and the health impacts of exposure to the burning of fossil fuels (e.g. due to exposure to tropospheric ozone and to particulate matter from fossil fuels). Highly engineered urbanization would have positive health effects by increasing access to safe water and improved sanitation, and thus reducing the burden of water- and food-borne diseases.

Medium challenges to adaptation and mitigation. Middle of the road

This is a world where global and national institutions make slow and uneven progress toward the SDGs, with considerable heterogeneity within and across countries. Population growth and persistent income inequality (globally and within economies), societal stratification and limited social cohesion continue challenges to reducing vulnerability to societal and environmental changes and constrain rapid advances in sustainable development. Some environmental systems experience degradation. Although income growth proceeds unevenly, most economies are politically stable. Overall, living conditions and access to education, safe water, improved sanitation and health care improve. However, educational investments are not high enough to accelerate the transition to low fertility rates in low-income countries and rapidly slow population growth. Investments in technological developments result in few fundamental breakthroughs. The intensity of resource and energy use declines, with slowly decreasing fossil fuel dependency.

Health in this world is in many ways a continuation of current trends, with slow gains in reducing the burden of climate-sensitive health outcomes in low-income countries. Climate change, population growth, income inequality and other trends make achievement of the sustainable development and other aspirational health goals more difficult, thus continuing the disproportional impacts of climate change on children in low-income countries. Funding for public health infrastructure and health care continues to fall below requirements, with inadequate resources and international commitment. Adverse health outcomes associated with the burning of fossil fuels increase, particularly in rapidly industrializing economies.

High challenges to adaptation and mitigation. Regional rivalry; a rocky road

This is a world characterized by strong regional identities concerned with competitiveness and security, with poor progress toward sustainability. There is a domestic or, at most, regional focus on strategies and policies to achieve energy and food security, at the expense of broader-based development. There is high resource intensity and fossil fuel dependency. Pockets of extreme poverty exist alongside pockets of moderate wealth, with high levels of inequality, particularly in low-income countries. Impeded development results in many vulnerable groups being unable to adapt. A low priority for environmental stewardship results in environmental degradation in some regions. The few global institutions are comparatively weak, unable to promote coordination and cooperation to address environmental issues. Population growth is low in industrialized countries and high in low-income countries. Many countries struggle to maintain living standards and provide access to safe water, improved sanitation and health care for disadvantaged populations. Vulnerability is exacerbated by the absence of functioning international mechanisms to provide assistance.

This is a world of increasing mortality from climate-sensitive health outcomes, and possibly falling life expectancy in low-income countries because of increased childhood mortality. Developed and developing countries experience a double burden of infectious and chronic climate-sensitive health outcomes, with limited effectiveness of adaptation and mitigation. Limited cooperation and relatively low investments in research and in public health and health-care infrastructure reduce the effectiveness of international, national and local organizations working to improve human well-being. Urbanization mostly fails to improve access to safe water and improved sanitation, thus increasing the burden of infectious diseases in unplanned settlements, and thereby increasing childhood mortality. Morbidity and mortality from health outcomes associated with the burning of fossil fuels increases.

High challenges to adaptation and low challenges to mitigation. Inequality; a road divided

This is a world with high inequalities within and between countries, leading to low social cohesion and regular social conflict and unrest. Power is concentrated in an internationally connected and well-educated political and business elite class who contribute to the knowledge and capital-intensive sectors of the global economy. A fragmented collection of lower-income and poorly educated societies works in labour-intensive, low-tech sectors. These vulnerable groups are largely outside the mainstream globalized economic system, with little representation in national and global institutions that emphasize international competitiveness. Economic growth is medium in industrialized and middle-income countries and slow in low-income countries. Generally high oil and gas prices have large swings, with changes in demand and small disruptions of supply. Businesses hedge against price fluctuations, partly through diversifying their energy sources, including carbon-intensive fuels like coal and unconventional oil, as well as low-carbon energy sources. Challenges to mitigation are low because the pool of expertise and technologies can be brought to bear rapidly if there is a strong push toward lower emissions. Challenges to adaptation are high because of the relatively high inequality and low levels of development, along with limited access to effective institutions for coping with economic or environmental stresses.

Health in this world differs from worlds with high challenges to adaptation and mitigation in the magnitude and extent of GHG emissions. Low challenges to mitigation make some aspects of this world similar to worlds with low challenges to adaptation and mitigation, with burdens of NCDs increasing with more children surviving to adulthood, although changes in dietary patterns and reductions in air pollution from the burning of fossil fuels could reduce the burden of some chronic diseases. At the same time, the burden of climate-sensitive infectious diseases and health outcomes from extreme events and environmental changes (e.g. under-nutrition) increases with decreased funding for public health and health-care organizations and institutions. Limited cooperation across sectors increases co-harms from adaptation options implemented in other sectors.

28.6   Scenario Matrix Architecture

The realization that emission scenarios for radiative forcing levels are relatively independent of underlying socio-economic assumptions allows more flexible approaches to develop scenarios that explore the extent to which adaptation and mitigation could reduce projected risks and to estimate the costs of action and inaction (Ebi et al., 2014; Kriegler et al., 2014; O'Neill et al., 2014; van Vuuren et al., 2014). A matrix architecture will be used for the new scenarios, where the axes of the scenario matrix are the RCPs and the SSPs, noting that not all combinations are plausible. A combination of an SSP and a radiative forcing level defines a family of macroscale scenarios and provides a framework within which regional or local studies of adaptation and vulnerability can build their own narratives. Because an RCP provides only a limited specification of mitigation policy characteristics, with scarce information on adaptation policies, a third axis is being developed. These shared climate policy assumptions include additional information on mitigation and adaptation policies; for example, the global and sectoral coverage of GHG reduction regimes and the aggressiveness of adaptation in different world regions.

Using a matrix allows for several modes of analysis, including within one column to explore the implications of increasingly stringent mitigation within any one SSP, and within one row to explore the implications of various SSPs within one level of radiative forcing. There also can be analyses within any one cell (or collection of cells) to examine the implications of using different climate, mitigation or impact models within any SSP/RCP combination.

28.7   Discussion

The new scenario process, although complex, provides flexibility to answer a broader range of policy-relevant questions, such as how different levels of climate change could affect development pathways and what might be the associated health risks. For example, what health consequences might occur in a world striving to achieve sustainable development if there are large increases in the frequency and intensity of extreme weather events? Would the health impacts of climate change be manageable if climate change proceeded relatively slowly and was at the low end of projections?

Achieving useful and relevant health scenarios is the responsibility of the health sector.

The SSPs will provide limited global quantitative projections that can be used as boundary conditions within which the health sector can develop detailed scenarios, enriched with quantitative and qualitative information on the evolution of public health in different pathways (Ebi, 2013). The descriptions here provide a start to begin this process.

References

Ebi, K.L. (2013) Health in the new scenarios for climate change research. *International Journal of Environmental Research and Public Health* 10, 1 doi:10.3390/ijerph100x000x.

Ebi, K.L., Hallegatte, S., Kram, T., Arnell, N.W., Carter, T.R., Edmonds, J., *et al.* (2014) A new scenario framework for climate change research: background, process, and future directions. *Climatic Change*, doi:10.1007/s10584-013-0912-3.

IPCC (Intergovernmental Panel on Climate Change) (ed.) (2012) *Managing the Risks of Extreme Events and Disasters to Advance Climate Change Adaptation. Special Report of the Intergovernmental Panel on Climate Change.* Cambridge University Press, Cambridge, UK.

Klein, R.J.T., Huq, S., Denton, F., Downing, T.E., Richels, R.G., Robinson, J.B., *et al.* (2007) Interrelationships between adaptation and mitigation. In: Parry, M., Canziani, O., Palutikof, J., van der Linden, P.J. and Hanson, C. (eds) *Climate Change 2007: Impacts, Adaptation and Vulnerability. Contribution of Working Group II to the Fourth Assessment Report of the Intergovernmental Panel on Climate Change.* Cambridge University Press, Cambridge, UK, pp. 745–777.

Kriegler, E., Edmonds, J., Hallegatte, S., Ebi, K.L., Kram, T., Riahl, K., *et al.* (2014) A new scenario framework for climate change research: the concept of shared policy assumptions. *Climatic Change*, doi:10.1007/s10584-013-0971-5.

Moss, R., Babiker, M., Brinkman, S., Calvo, E., Carter, T., Edmonds, J., *et al.* (2008) *Towards New Scenarios for Analysis of Emissions, Climate Change, Impacts, and Response Strategies.* Intergovernmental Panel on Climate Change, Geneva, Switzerland.

Moss, R.H., Edmonds, J.A., Hibbard, K.A., Manning, M.R., Rose, S.K., van Vuuren, D.P., *et al.* (2010) The next generation of scenarios for climate change research and assessment. *Nature* 463, 747–756.

Nakicenovic, N., Alcamo, J., Davis, G., de Vries, B., Fenhann, J., Gaffin, S., *et al.* (2000) *IPCC Special Report on Emissions Scenarios (SRES).* Cambridge University Press, Cambridge, UK.

O'Neill, B.C., Kriegler, E., Riahl, K., Ebi, K.L., Hallegatte, S., Carter, T.R., *et al.* (2014) A new scenario framework for climate change research: the concept of shared socio-economic pathways. *Climatic Change*, doi:10.1007/s10584-013-0905-2.

Taylor, K.E., Stouffer, R.J. and Meehl, G.A. (2011) An overview of CMIP5 and the experiment design. *Bulletin of the American Meteorological Society*, doi:10.1175/BAMS-D-11-00094.1.

van Vuuren, D.P., Edmonds, J., Kainuma, M., Riahi, K., Thomson, A., Hibbard, K., *et al.* (2011) The representative concentration pathways: an overview. *Climatic Change*, doi:10.1007/s10584-011-0148-z.

van Vuuren, D.P., Kok, M., Girod, B., Lucas, P. and De Vries, B. (2012) Scenarios in global environmental assessments: key characteristics and lessons for future use. *Global Environmental Change* 22, 884–895.

van Vuuren, D.P., Kriegler, E., O'Neill, B.C., Ebi, K.L., Riahl, K., Carter, T.R., *et al.* (2014) A new scenario framework for climate change research: scenario matrix architecture. *Climatic Change*, doi:10.1007/s10584-013-0906-1.

Winkler, H., Baumert, K., Blanchard, O., Burch, S. and Robinson, J. (2007) What factors influence mitigative capacity? *Energy Policy* 35, 692–703.

29 Health Activism and the Challenge of Climate Change

Colin D. Butler[1] and Sue Wareham[2]
[1]Faculty of Health, The University of Canberra, Australia, National Centre for Epidemiology and Population Health, The Australian National University, Canberra, Australia, and Benevolent Organisation for Development, Health & Insight (BODHI); [2]Medical Association for Prevention of War (Australia)

The improvement of medicine will eventually prolong human life, but the improvement of social conditions can achieve this result more rapidly and more successfully.

(Rudolf Virchow, 1879)

29.1 Introduction

The direct relationship between patient and doctor or other health-care professional has traditionally been the essence of clinical practice. It is what most health-care practitioners train to do and how they spend most of their working lives. However, it is a role that is increasingly recognized as having limits, for it can often be one of patching up or simply reacting to the damage wrought by a host of factors that affect human health, be they social, economic or environmental. Recently, health activism has been introduced as a subject in some health courses. Many important advances in human health and welfare have occurred through better understanding of and education about the links between our health and external factors. These advances have occurred in association with improved living conditions for people, and may in future establish health care as a basic human right, observed as well as pledged.

Public health practitioners working with others,[1] such as family members and friends of ill people, have played many vital advocacy roles. They have addressed issues such as the health problems of aboriginal communities, road trauma, alcohol abuse, domestic violence, asbestos exposure, gun control legislation and the advertising of tobacco, junk food and other harmful substances. Unsurprisingly, success has varied by time, surrounding culture and location. Sometimes, health reformers have led public opinion; at other times, they have been swept along with unstoppable cultural change. Health activism by trained health workers often lags far behind that of lay communities, such as when concerns about HIV arose among gay men in the USA in the 1980s. In that case, activism was led mainly by the gay community (Hilliard et al., 2007).

Health activists can be particularly visible in contexts where they are desperately needed. The recent imprisonment of the Indian community health activist, Dr Binayak Sen (Sen, 2011), for allegedly providing care to people involved on both sides of a vicious conflict, is rendered all the more terrible because of the atrocious health conditions of millions of rural poor in the northern states of India. On the other hand, in China, the work of individual activists is rarely known; yet its society has mobilized successfully on a vast scale to lower the burden of many diseases, such as schistosomiasis. Mao Tse Tung, despite numerous other actions which proved catastrophic to public health in China, was personally inspirational and supportive of the long campaign to reduce the effects of this parasitic infection (Utzinger *et al.*, 2005).

Overall, health in China is now much better than in India. It is disturbing, however, that some health activists in China are oppressed, including Zhao Lianhai, the father of an infant whose kidneys were damaged by the inadvertent consumption of the toxic substance melamine, used to adulterate many processed foods, including infant formula, in order to increase profits. Zhao organized a website ('Kidney Stone Babies') for parents of similarly affected children (as many as 300,000), but was found guilty of 'inciting social disorder' and in late 2010 was sentenced to 2.5 years in prison, a year after his arrest (Reuters Health, 2010).

In Western countries, health activists have had remarkable success in limiting cigarette smoking. In 2004, Ireland became the first nation to implement a comprehensive smoking ban in indoor workplaces, including restaurants and bars. Contrary to expectations, the ban was both popular and enforceable. Pleasingly, other countries followed (Koh *et al.*, 2007). The right to clean air was being affirmed. In China, steps are now being taken to curtail that nation's once pervasive tobacco addiction, especially prevalent among males (Anon., 2011). Smoking in China is likely to be particularly harmful, due to the country's air pollution – both indoor in many villages and outdoors in most areas, including its vast cities (Tie *et al.*, 2009).

29.2 The Great London Smog

As our natural environment has come under greater strain, its links with human health have been increasingly recognized. For centuries, writers have complained of poor air quality in London, the original 'Big Smoke' (McNeill, 2000). In 1950, air pollution in most British cities – due largely to the combustion of coal, then considered essential for heating and industry – was so marked that, in combination with high rates of smoking, virtually every man had a chronic wheeze by the age of 50 (Berridge, 2007) (see Fig. 29.1).

Fig. 29.1. A trace of London's great smog remains in its profusion of chimneys, still plentiful more than 60 years since their last use. (Photo: C.D. Butler.)

In December 1952, London was again[2] afflicted by a heavy, unshifting smog, reducing visibility to metres and bringing the city to a standstill. Policemen had to guide buses through the haze. Hospital admissions for respiratory illnesses rose, as did mortality, the latter by an estimated 3500–4000 excess deaths for the month. At first, there was a perhaps surprising lack of recognition of the severity of these health impacts (Davis et al., 2002). However, 4 years later, growing agitation led to the passing of the Clean Air Act, though even parliamentarians remained fearful of the reception of new restrictions on behaviour (Brimblecombe, 2006).

29.3 Health Activism in the 19th Century

While the doctors of London in 1952 did not seem to react with alacrity in recognizing the adverse impact of smoke on health, many health pioneers before and since have seen far beyond their individual patients to the social and environmental conditions that predisposed to their illnesses (see Table 29.1). About a century before the great London smog, the German pathologist, Rudolf Virchow (1821–1902), broke ground with his revolutionary understanding of the social and physical bases of ill health. He established that cells were the fundamental unit of life, and elucidated the pathology of thrombosis, leukaemia, amyloid bodies and other conditions. But he was equally innovative in recognizing and advocating for social changes as a means to fundamentally improve many health conditions. Virchow also strongly opposed the German statesman Otto von Bismarck's 1865 proposal to divert funds to a military budget (Eisenberg, 1984).

Another early advocate for the role of primary prevention and public health was the 19th century New York doctor, John H. Griscom. Griscom advocated measures such as public hygiene, medical care for the destitute and prison and tenant reforms (Bloom, 2001). He was an early proponent of public health being a speciality, and advocated the collection of vital statistics, as did his British contemporary, William Farr (Hamlin, 1995). Griscom also played a significant role in the formation of the American Medical Association. His driving force was ethical and moral concern, combined with a strong commitment to scientific principles.

For some, the direct witnessing of large-scale environmental destruction in more recent times has led to a life in politics. Senator Bob Brown, the first leader of the Australian Greens, switched careers from general practitioner to parliamentarian in response to massive deforestation of pristine wilderness in his adopted state of Tasmania, and other assaults on the health of our planet. Through his long political career, Brown remained engaged with the health community, including with medical students.

29.4 Medical Journals

Many medical journals have been important vehicles for promoting new, reforming and humanitarian ideas. Eminent among them has been The Lancet, founded by surgeon, Thomas Wakley, in 1823. In its early days, this journal highlighted such issues as the rights of prisoners to medical care and conditions under which the poor worked (Kandela, 1998). In more recent times, it has continued to address causes considered by some to be too 'political' for professional involvement, such as opposition to apartheid in South Africa and the plight of the Palestinians. Correspondence published in The Lancet led to the formation of the Medical Association for Prevention of War in the UK in 1951, and a former Lancet editor, Ian Munro, was the first president of Physicians for Human Rights (UK), which was founded in 1989 (Kandela, 1998). The Lancet has also been ahead of most of its peers in recognizing environmental issues including climate change (Epstein and Sharp, 1993; Maskell et al., 1993) and many aspects of sustainable development (King, 1990; Anon., 1992). It was crucial in the campaign against its own parent, the conglomerate Reed Elsevier, which for a while conducted a lucrative sideline in the arms trade, marketing cluster bombs and torture equipment (Anon., 2005).

Similar to The Lancet in the UK, the New England Journal of Medicine in 1962 played a key role in alerting the profession and the wider community to the impossibility of providing medical care in the event of a nuclear war, including by its publication of four articles on

Table 29.1. Selected health activists since 1790. There are many others, including those involved in nuclear disarmament campaigning. Health workers can be proud of this record, yet much remains unfinished.

Name	Nationality	Period	Major contribution
Philippe Pinel	French	Late 18th century	Psychiatric reform
James Kay	British	Early 19th century	Public health reform, Manchester
Thomas Wakley[b]	British	Early 19th century	Founder of *The Lancet*, London
Edwin Chadwick	British	Mid-19th century	Sanitation
Florence Nightingale	British	Mid-19th century	Nursing reform
David Livingstone	British	19th century	Anti-slavery
William Farr[b]	British	Mid-19th century	Statistics, social activism
Lemuel Shattuck[b]	American	Mid-19th century	Sanitary reform
John H. Griscom[b]	American	Mid-19th century	Sanitary reform
Rudolf Virchow[b]	German	Late 19th century	A founder of social medicine
Oswaldo Cruz	Brazilian	Early 20th century	Public health
Natan Vigdorchik	Russian	Early 20th century	Social security, public health
Ronald Ross	British-Indian	Early 20th century	Malaria control
John Boyd Orr	British	Mid-20th century	Founding Director General, FAO
Rene Dubos	French-American	Mid-20th century	Ecology and health
Frantz Fanon	French Caribbean	Mid-20th century	Third World rights, decolonization
Albert Schweitzer[b]	Swiss	Mid-20th century	Human rights, international health
Jonathan Mann	American	Late 20th century	HIV/AIDS, human rights
Trevor Beard	Australian	Late 20th century	Hydatids, salt
John Goldsmith	American	Late 20th century	Environmental epidemiology
Fred Hollows	New Zealander	Late 20th century	Ophthalmology
Frank Fenner	Australian	Late 20th century	Ecology and health
Paul Epstein	American	20th–21st century	Climate change, sustainability
Catherine and Reginald Hamlin	Australian	Contemporary/ 20th century	Obstetric fistula repair and prevention
Maurice King	British	Contemporary	Primary health care, sustainability
Binayak Sen[b]	Indian	Contemporary	Human and health rights for the poor
Rani and Abhay Bang	Indian	Contemporary	Rural health care
Tushar Joshi	Indian	Contemporary	Asbestos reduction
Wan Yanhai, Tian Xi[a]	Chinese	Contemporary	HIV/AIDS activism
Zhao Lianhai[a,b]	Chinese	Contemporary	Health promotion (melamine danger)
Bob Brown[b]	Australian	Contemporary	Environmental health
Gerry Koffman	Canadian	Contemporary	Illegal organ trading
Gro Harlem Brundtland	Norwegian	Contemporary	Sustainable development
May Haddad	Lebanese	Contemporary	Public health
Tim Lang	British	Contemporary	Ecology, food, health
Mechai Viravaidya[a]	Thai	Contemporary	HIV/AIDS, family planning ('Mr Condom')
Zackie Achmat[a]	South African	Contemporary	HIV/AIDS
Waris Dirie[a]	Somali	Contemporary	Female genital mutilation

[a]Not a trained health worker; [b]mentioned or discussed in the chapter.

this topic in one issue (1962). It has since played a key role in the nuclear weapons issue on numerous occasions. It may not be coincidental that the headquarters of this famous journal are located in Shattuck Street (Boston,

USA), named for another pioneering public health figure, the legislator Lemuel Shattuck (Fair, 1952; Cassedy, 1975).

Numerous other journals have played an active role in promoting and providing avenues for research and opinion on the social and environmental factors that improve health. A very few journals have been shown to be systematically biased toward interests inimical to health, including the tobacco industry (Butler et al., 2007).

29.5 Medical Campaign to Reduce the Threat of Nuclear Weapons

In the latter part of the 20th century, an increasing suite of ecological threats joined the old threats of poverty, dirty water and undernutrition. These included the deterioration of land quality, river systems, biodiversity and fish stocks in many parts of the globe (Corvalán et al., 2005). However, the last half-century has also seen growing concern at a hidden yet powerful threat to global health – the development and possible use of nuclear weapons (Haines, 2009).

It is worth examining the medical profession's response to the nuclear weapons threat for any clues that might be applicable to our advocacy for climate change action. This is done, however, in full knowledge that, first, the nuclear weapons problem is still very much with us, and, second, the medical movement for nuclear weapons abolition is not held up as a model that other similar movements must emulate. Like all human endeavours, it has its strengths and its weaknesses, from which we can learn.

The challenges to health care posed by nuclear weapons, the world's most destructive devices, were obvious from the very start of the nuclear weapons age in 1945 (Rhodes, 1986). As early as 1951, medical practitioners in the UK formed the Medical Association for Prevention of War. Its focus was on the destructive capacity of modern warfare generally, but in particular the newly developed weapons of mass destruction. Physicians for Social Responsibility was formed 10 years later in the USA (Sidel and Levy, 2009). Both organizations were later to become part of the global International Physicians for the Prevention of Nuclear War.

In the early days of the movement against nuclear weapons, the agenda was primarily a cessation of nuclear weapons testing in the atmosphere, which was seen as a public health measure in its own right. There were serious concerns about the health effects of the radioactive

contamination that resulted from the tests. An early champion of the movement to cease nuclear tests was the distinguished physician, theologian and humanitarian, Albert Schweitzer, later awarded the Nobel Prize for Peace. On 23 April 1957, he read his Declaration of Conscience, urging an end to nuclear testing, over Radio Oslo. Broadcast in approximately 50 nations and reported in countless newspapers, his appeal had a significant impact on global opinion (Wittner, 1993).

The finding that radioactive strontium from atmospheric tests had been incorporated into children's deciduous teeth in the USA proved a powerful argument in the medical campaign and hastened the banning of atmospheric testing, which was formalized in the 1963 Partial Test Ban Treaty (Humphrey, 2009). While this was a critical milestone, it generated a certain degree of complacency and quiescence, which lasted until concerns were reignited from the late 1970s. At that time, a number of factors combined to strengthen the opposition to these weapons, including a deterioration in East–West relations, new nuclear missile deployments and official talk of actually fighting (and 'winning') a nuclear war, which led to the formation in 1980 of International Physicians for the Prevention of Nuclear War (IPPNW). Australian paediatrician, Dr Helen Caldicott, played a powerful and central role, in the USA, in Australia and globally, in this renewed struggle to prevent nuclear war, dubbed at the time 'the final epidemic' (Abrams et al., 1981).

IPPNW played a pivotal role in raising awareness of the dangers of nuclear weapons. The medical message was stark and simple: in the event of nuclear war, our profession would have very little to offer the survivors. Medical evidence to back up the message was plentiful, including the 1983 British Medical Association report, The Medical Effects of Nuclear War, which concluded that the country's health services could not cope with a single Hiroshima-sized bomb attack on the UK (British Medical Association. Board of Science and Education, 1983). Articles in the general science literature warned of 'nuclear winter', a global haze triggered by multiple nuclear explosions that might envelope the atmosphere for at least a year and kill almost all life (Turco et al., 1983). In May 1983, the World Health Organization spoke up, stating that 'nuclear weapons constitute the greatest immediate threat to the health and welfare of mankind, and that physicians have both the right and the duty to draw attention in the strongest possible terms to the catastrophic results that would follow from any use of nuclear weapons'.

A further element added to the group's success in capturing the attention of both the public and governments: its unity across the enormous political divide of the Cold War. The co-founders of IPPNW were two eminent cardiologists, an American, Dr Bernard Lown, and a Russian, Dr Eugene Chazov, who was physician to the Soviet political leadership. It was cardiology that brought them together, but a mutual concern to avert nuclear catastrophe that drove their groundbreaking alliance. The movement they founded proclaimed loudly and clearly that health trumps politics. We are all in this together. Five years after its formation, IPPNW was awarded the 1985 Nobel Prize for Peace for its role in educating the world on the health effects of nuclear weapons.

However, the strength of the medical message did not, of course, penetrate every corridor of power. General Lee Butler, head of the US Strategic Air Command in the early 1990s, provided a frightening insight into the depths of ignorance at the highest levels regarding what would actually happen in a nuclear war. Butler states that during that time, the US President had '... only a superficial understanding of what would happen in a nuclear war' and that 'Congress knew even less'. In referring to briefings of senior officials, Butler said, 'Generally no-one at the briefings wanted to ask questions because they didn't want to embarrass themselves' (http://www.publications.parliament.uk/pa/cm200809/cmselect/cmfaff/memo/1176/ucm3702.htm). Butler also paid tribute to the essential work by non-government organizations done over many years to raise awareness about the danger hanging over humanity.

Fortunately for us all, the degree of ignorance described by Butler was not universal among leaders. Mikhail Gorbachev, Soviet president in the late 1980s, credited IPPNW with influencing his decision that nuclear weapons must be abolished. In his book, *Perestroika*, he wrote:

> It is impossible to ignore what these people are saying. What they are doing commands great respect. For what they say and what they do is prompted by accurate knowledge and a

passionate desire to warn humanity about the danger looming over it... In the light of their arguments and the strictly scientific data which they possess, there seems to be no room left for politicking. And no serious politician has the right to disregard their conclusions or neglect the ideas by which they take world public opinion a stage ahead.

(Gorbachev, 1987)

Bernard Lown gave valuable insights into the struggle to get scientific knowledge on to political agendas in his book, *Prescription for Survival*, an account of his time in the early 1980s at the helm of IPPNW. Some come as no surprise: 'An advance on any political front does not come as a gift from governing establishments. It needs to be wrested by an unrelenting, well-organized struggle ... by a public clamouring for change.' Some of his observations, however, are surprising:

> I became aware of an astonishing fact. IPPNW could penetrate the Iron Curtain far more readily than it could enter the free halls of power of a democratic society. Leaders in the highest echelons of the Soviet Union were ready to meet and converse as well as listen. Such consistent access was denied to us in the West. Not only were we ignored by the political establishment in the United States and other NATO countries, but the western media shut us out as well.

(Lown, 2008)

In June 1982, an hour-long unscripted discussion between three US and three Russian members of IPPNW was broadcast – twice – on Moscow television, reaching an estimated 100 million people. Support for a cause can be found in unexpected quarters.

While nuclear weapons are still with us, and the struggle for their abolition continues, there is evidence that the widespread revulsion against weapons with such horrific effects goes a long way to explaining why they have not been used again since 1945 (Wittner, 1993). Global opinion would not tolerate it.

29.6 Climate Change, Security and Health Workers

In relation to climate change, as with nuclear weapons, the accurate knowledge to which Gorbachev referred is the key contribution our profession has to offer. Organizations such as the

International Society of Doctors for the Environment and its affiliates in at least 17 countries (http://www.isde.org/) have contributed. By the time of the UN Climate Change Conference held in Copenhagen in December 2009, the Climate and Health Council (http://www.climateand-health.org/) had gathered over 10,000 signatures from health professionals pledging to protect health through active engagement to limit the causes of human-caused climate change. Along with other health organizations, this group organized a 2011 meeting hosted by the British Medical Association, which examined the links between climate change, health and security. This brought together delegates not only from health care and climate science, but also from the military and politics (http://climat-echange.bmj.com/) (Charlesworth *et al.*, 2011; Jarvis *et al.*, 2011) The suggestion that climate change and armed conflict will become increasingly linked seems true, unless urgent preventive steps are taken (Haines, 2009).

29.7   Barriers and Solutions

Climate change, like nuclear war, threatens not only the health of individuals and communities but also that of whole nations. Perhaps even civilization as we know it is imperilled. In the face of such challenges, health professionals can, like the rest of society, retreat into feelings of fear and hopelessness.

For health professionals who undertake advocacy on these issues, another problem awaits. Coming from a world of medicine and science, where evidence is taken seriously and acted upon, they enter a world where scientific evidence may not be taken seriously but treated with contempt. Not every leader is a Gorbachev. Instead of decisions based on knowledge painstakingly collected to preserve public health and well-being, policies may be set to preserve corporate profits and political careers. In too many places, respect for scientific knowledge is not even a prerequisite for holding political office.

Another major problem we face is that human behaviour – which must change both at individual and governmental levels – is based not so much on what we know but on what we feel; that is, our emotions. Many people avoid confronting unpleasant information, dismiss evidence that makes them feel uncomfortable, and retreat into denial. Hence, dissemination of medical knowledge on the impacts of climate change is not sufficient. It is a core task and absolutely essential, but it is not enough. We must change behaviours not just by presenting information but also by inspiring hope and a desire for change.

To bring about change, it is helpful to start with a statement of values with which most people agree. This is convenient for health professionals, because good health is highly valued by practically all members of the public. Our advocacy must reinforce health as a fundamental value that must be upheld. This need not and should not apply only to current generations but also to future generations, who will have to live with the effects of today's climate policies. Most people would agree with the principle of intergenerational equity, or, more simply, not leaving a mess of a planet for our children and grandchildren.

Ed Maibach, of the Centre for Climate Change Communication at George Mason University in Virginia, USA, states that Americans are only dimly aware of the health implications of climate change, and that encouraging them to consider this as a health issue is a useful approach. He writes that 'the public health community has an important perspective to share about climate change, a perspective that makes the problem more personally relevant, significant and understandable to members of the public' (Maibach *et al.*, 2010).

Hope is vital to elicit positive emotions in those we seek to influence. We must present not only the problem but also solutions. Fiona Armstrong, convenor of the Climate and Health Alliance in Australia, reinforces the message that 'positive messages appeal' and urges us to 'paint a picture of an appealing, positive future possible through climate action' (Armstrong, 2012).

While the role of health professionals in bringing about climate change action is unique and compelling, it is vital to recognize the importance of other disciplines and perspectives. As with nuclear weapons, there are many experts in many fields with whom to join forces, and an even greater number who are not experts but are worried none the less and prepared to take action. Climate change can be framed not only as a health issue but also as an environmental, human rights, development, security and parenting issue. That is a lot of potential partners to bring strength and diversity to the calls for action.

Ultimately, we cannot know whether our efforts to prevent catastrophic climate change will be successful. The only certainty is failure if we do not try. Our hope lies in each one of us, and in the global movement of organizations and individuals working for this common cause. Our collective wisdom, learning, advocacy skills and integrity are vast, but these skills must be used to the full. Humanity's future is not preordained, but will depend on the choices we all make.

Bernard Lown once said, in relation to nuclear weapons, that if we are to prevail, 'we must never delegate in the presence of challenge, and never whisper in the presence of wrong'. Similarly with climate change, raising the alarm about its dire health impacts cannot be delegated. While there are many partners in other fields with whom to join forces, it is up to us, health professionals, to educate, advocate and agitate for a healthier planet. Our message must be loud and clear, and combined with a determination to challenge the political inertia, the vested interests, the deniers of science and every other obstacle that prevents meaningful action.

Notes

[1] Such as The Health of Towns Association (1844–1849), formed in Britain.
[2] For example, in 1873, some Londoners accidentally walked into the Thames, unable to see the river.

References

Abrams, H.L., Caldicott, H.M., Chazov, E.I., Constable, J.D., Feld, B.T., Finch, S.C., *et al.* (eds) (1981) *The Final Epidemic: Physicians and Scientists on Nuclear War.* University of Chicago Press, Chicago, Illinois.

Anon. (1992) Horse manure after Rio. *The Lancet* 339, 1515–1516.

Anon. (2005) Reed Elsevier and the arms trade. *The Lancet* 366, 868.

Anon. (2011) China's unhealthy relations with big tobacco. *The Lancet* 377, 180.

Armstrong, F. (2012) How climate change gets lost in translation. *Climate Spectator* (http://www.businessspectator.com.au/article/2012/3/21/science-environment/how-climate-change-gets-lost-translation, accessed 4 April 2014).

Berridge, V. (2007) Public health activism: lessons from history? *BMJ* 335, 1310–1312.

Bloom, M. (2001) Primary prevention and public health: an historical note on Dr. John Hoskins Griscom. *The Journal of Primary Prevention* 21, 305–308.

Brimblecombe, P. (2006) The clean air act after 50 years. *Weather* 61, 311–314.

British Medical Association. Board of Science and Education (1983) *The Medical Effects of Nuclear War.* J. Wiley, London.

Butler, C.D., Castleden, W., Ruff, T.A., Westberg, N.G. and Corra, L. (2007) A call for publishers to declare their conflicts of interest. *Journal of the Royal Society of Medicine* 100, 355.

Cassedy, J. (1975) The roots of American sanitary reform 1843–47: seven letters from John H. Griscom to Lemuel Shattuck. *Journal of the History of Medicine and Allied Sciences* 30, 136–147.

Charlesworth, A., Gray, A., Pencheon, D. and Stern, N. (2011) Assessing the health benefits of tackling climate change. *BMJ* 343, doi: http://dx.doi.org/10.1136/bmj.d6520.

Corvalán, C., Hales, S., McMichael, A.J., Butler C.D., Campbell-Lendrum D., Confalonieri U., *et al.* (2005) *Ecosystems and Human Well-Being. Health Synthesis.* World Health Organization, Geneva, Switzerland.

Davis, D., Bell, M. and Fletcher, T. (2002) A look back at the London smog of 1952 and the half century since. *Environmental Health Perspectives* 110, A734–A735.

Eisenberg, L. (1984) Rudolf Ludwig Karl Virchow, where are you now that we need you? *The American Journal of Medicine* 77, 524–532.

Epstein, P.R. and Sharp, D. (1993) Medicine in a warmer world. *The Lancet* 342, 1003–1004.

Fair, G.M. (1952) Shattuck, Chadwick, and the engineer in public health. *Public Health Reports* 67, 443–449.

Gorbachev, M. (1987) *Perestroika*. Collins, London, p. 154.

Haines, A. (2009) Redefining security (reprinted from Medicine and War, 1992). *Medicine, Conflict and Survival* 25, 282–285.

Hamlin, C. (1995) Could you starve to death in England in 1839? The Chadwick-Farr controversy and the loss of the 'social' in public health. *American Journal of Public Health* 85, 856–866.

Hilliard, N., Jenkins, R., Pashayan, N. and Powles, J. (2007) Informal knowledge transfer in the period before formal health education programmes: case studies of mass media coverage of HIV and SIDS in England and Wales. *BMC Public Health* 7, 293, doi:10.1186/1471-2458-7-293.

Humphrey, J. (2009) The development of the physicians' peace movements. *Medicine, Conflict and Survival* 25, 87–89.

Jarvis, L., Montgomery, H., Morisetti, N. and Gilmore, I. (2011) Climate change, ill health, and conflict. *BMJ* 342, 777–778.

Kandela, P. (1998) Medical journals and human rights. *The Lancet* 352, 7–12.

King, M. (1990) Health is a sustainable state. *The Lancet* 336, 664–667.

Koh, H.K., Joossens, L.X. and Connolly, G.N. (2007) Making smoking history worldwide. *New England Journal of Medicine* 356, 1496–1498.

Lown, B. (2008) *Prescription for Survival: A Doctor's Journey to End Nuclear Madness*. Berrett-Koehler Publishers, San Francisco, California, pp. 382–383.

McNeill, J.R. (2000) *Something New Under the Sun: An Environmental History of the Twentieth-Century World*. W.W. Norton, New York.

Maibach, E., Nisbet, M., Baldwin, P., Akerlof, K. and Diao, G. (2010) Reframing climate change as a public health issue: an exploratory study of public reactions. *BMC Public Health* 10, 299, doi:10.1186/1471-2458-10-299.

Maskell, K., Mintzer, I.M. and Callander, B.A. (1993) Basic science of climate change. *The Lancet* 342, 1027–1031.

Reuters Health (2010) China court sentences melamine milk activist to jail (http://www.reuters.com/article/2010/11/10/oukwd-uk-china-melamine-idAFTRE6A914Y20101110, accessed 4 April 2014).

Rhodes, R. (1986) *The Making of the Atomic Bomb*. Simon and Schuster, New York.

Sen, B. (2011) Securing the right to health for all in India. *The Lancet* 377, 532–533.

Sidel, V.W. and Levy, B.S. (2009) Homage to Janus: looking backward and forward. *Medicine, Conflict and Survival* 25, 271–274.

Tie, X., Wu, D. and Brasseur, G. (2009) Lung cancer mortality and exposure to atmospheric aerosol particles in Guangzhou, China. *Atmospheric Environment* 43, 2375–2377.

Turco, R.P., Toon, O.B., Ackerman, T.P., Pollack, J.B. and Sagan, C. (1983) Nuclear winter: global consequences of multiple nuclear explosions. *Science* 222, 1283–1292.

Utzinger, J., Xiao-Nong, Z., Ming-Gang, C. and Bergquist, R. (2005) Conquering schistosomiasis in China: the long march. *Acta Tropica* 96, 69–96.

Virchow, R. (1879) *Gesammelte Abhandlungen aus dem Gebiet der Öffentlichen Medicin und der Seuchenlehre*. Vol 1. Hirschwald, Berlin. Cited by: Waitzkin, H. (2006) One and a half centuries of forgetting and rediscovering: Virchow's lasting contributions to social medicine. *Social Medicine* 1, 5–10.

Wittner, L.S. (1993) *The Struggle Against the Bomb. Vol 2 Resisting the Bomb*. Stanford University Press, Stanford, California, p. 31.

30 Climate Change and Health: From Adaptation Towards a Solution

Colin D. Butler,[1,2] Robin Stott,[3] Kathryn Bowen,[2] Matthew Haigh,[4] Byron Smith,[5] Mark Braidwood[6] and Devin C. Bowles[2]

[1]*Faculty of Health, The University of Canberra, Australia, and Benevolent Organisation for Development, Health & Insight (BODHI);* [2]*National Centre for Epidemiology and Population Health, The Australian National University, Canberra, Australia;* [3]*Climate and Health Council, UK, and 15 Egerton Drive, London, UK;* [4]*School of Oriental and African Studies, University of London, UK;* [5]*School of Divinity, University of Edinburgh, Scotland, UK, and 398 Bronte Road, Waverley, New South Wales, Australia;* [6]*School of Public Health & Community Medicine, The University of NSW, Sydney, New South Wales, Australia*

> The main threats to sustained human existence now come from people, not from nature.
>
> (Rees, 2013)

30.1 Constrained Choices

Earlier chapters of this book have surveyed evidence for the primary, secondary and tertiary effects to health from climate change, observable now and foreseeable in coming decades. These health effects interact with broader society, which is collectively responsible for many of the underlying social and environmental determinants of health. Though many authors describe mechanisms to cushion or prevent specific adverse health effects, this chapter seeks to identify, describe and analyse the solutions that might apply on a broader scale.

The opening chapter highlighted numerous problems that society faces *additional* to climate change, now and in future. These problems of the Anthropocene, also characterized as the approach to critical 'planetary boundaries', interact with others, such as falling phosphate reserves, weapons of mass destruction and the increasing number of displaced people. These interrelated concerns, created partly by the behaviour and actions of the most fortunate billion people to live in the last century, will, unless rapidly solved, constrain future opportunities and choices, even *without* climate change.

Rising energy costs, the economic crisis and unburnable carbon: a seemingly intractable triangle

The rising cost of energy, still overwhelmingly sourced from fossil fuels,[1] is related intimately to the issue of climate change in three main ways. In 2004, the price of oil was barely US$40/barrel. By 2008, its price had more than tripled, approaching US$150/barrel (Murray and King, 2012). Though it has since fallen by about 20% (30% in the USA, due its aggressive expansion into unconventional carbon sources such as shale oil), the energy price remains persistently and historically high. This high price has been a powerful financial motivator for the development of renewable energy technology, more significant to date than the modest carbon premium, whether due to a carbon tax or the emission trading schemes which are gradually being introduced.

Second, the high energy cost is related closely to the ongoing financial crisis, as it drives up the cost of living. In many regions (e.g. Europe and Egypt), domestic purchasing power is cut, as funds are diverted to import energy (Murray and King, 2012). The economic crisis slows the transition to renewable energy, because populations experiencing economic contraction and declining living standards are less willing to pay for the required infrastructure, which in some ways is more expensive than a decade earlier.[2]

The rising energy price reflects less accessible fossil fuel sources. Although the concept of 'peak oil' is too simplistic, fossil fuels that remain unburned are of lower quality and costlier to exploit than those available a century ago. Even so, most remaining supplies[3] have far higher 'net energy' than modern renewables (Murphy and Hall, 2011). There has been almost a stampede to locate and extract these fuels, perhaps because their accessibility will continue to decline. But, while supplies are claimed as plentiful, their rate of extraction is not sufficient to lower prices much. Furthermore, ample unburned fossil fuels remain to raise average global temperature (and its associated adverse effects) well beyond the 2°C 'guard rail' (Smith et al., 2009), perhaps triggering runaway warming and even the end of civilization (Hansen, 2013).

Recognizing this, terms such as 'unburnable carbon' and the 'carbon bubble' have been coined (Leggett, 2011), to generate investor awareness of the long-term risk of fossil fuel reliance. Fossil fuel investors face a dilemma. Their investments risk becoming worthless if sufficient renewable momentum accrues, or temporarily retain value in a world that becomes locked into a trajectory toward dangerous climate change. A tenfold increase in the 'net energy' of renewables, generating good financial returns to companies who currently profit from fossil fuels, suggests an escape from this dilemma, and could evolve. But the pace of this energy transition is slowed greatly by conventional (non-ecological) economics, by social denial, by technological obstacles and by organized forces who profit from 'business as usual' (Oreskes and Conway, 2010).

30.2   Climate Change Mitigation

Since climate change emerged as a serious issue following the first Rio de Janeiro Earth Summit in 1992, which saw the formation of the UN Framework Convention on Climate Change (UNFCCC), the stated strategies to deal with it have focused on mitigation and adaptation. Mitigation is analogous in health to primary prevention (e.g. a vaccine or smoking cessation); adaptation can be likened to health interventions following an adverse health event (e.g. cardiac surgery).

The philosophy underlying mitigation, sometimes called the Precautionary Principle, has many vernacular analogies such as 'look before you leap'. Yet a counterview is 'he who hesitates is lost'. In other words, some argue, progress requires risk-taking, even if this leads to the troubling consequences outlined in this book, especially the tertiary health consequences.

Climate change mitigation relies on ways to reduce fossil fuel emissions, including energy efficiency, new carbon-sparing forms of energy capture and technological leapfrogging.[4] Value changes promoted by education, peer behaviour, environmental stewardship and voluntary simplicity are also vital, as extolled by some public health leaders in the UK (Hanlon and Carlisle, 2010). Land-use change contributes to climate change in ways additional to the release of greenhouse gases (GHGs), such as by changing transpiration and albedo[5] (Nicholls and Larsen, 2011). Modifying land use in ways that will ameliorate climate change (e.g. greater carbon sequestration in soils) is also a form of climate change mitigation, as is reduced deforestation. In turn, some of these changes can be encouraged by new economic signals that will shift civilization away from fossil fuels, to the so-called 'Green Economy' (Kosoy et al., 2012), based on principles of ecological economics (Daly, 1996).

Investing in carbon abatement

Primary strategies to implement these proposals have been a carbon tax or a carbon emissions trading scheme. In 2008, a major review by leading British economist, Nicholas Stern, expressed hope that major institutional investors might soon embrace 'long-term investing', as part of a market-based solution to respond to and help abate climate change. To date, events seem unpromising. For example, investing institutions do not apply or expect an equivalent standard of verification for carbon pollution reduction data as for audited financial records. This erodes investor confidence. Without significant change within business culture, the customary reliance of investment portfolio managers on standardized data also usually precludes data such as probability estimates of climate events from the IPCC. Moreover, untested financial products such as catastrophe bonds and climate-related exchange-traded contracts are not in the standard purview of fiduciary investor-held risk-adjusted portfolios.

The chief source of corporate carbon emissions data for investors is an investor-sponsored collective known as the Carbon Disclosure Project. On the surface, it looks impressive and encouraging. Thousands of the world's largest companies, including 82% of the Global 500[6] in 2010, report self-assessed climate change-related risks, opportunities, strategies and GHG emissions to this project. Using desk research, questionnaires and other data gathered from 94 interviews with key informants in North America, Australia and Europe, Haigh (2013) concluded that while investors had regularly collected carbon data from the Carbon Disclosure Project, none had actually used such data in their relevant decisions (Haigh, 2013). Others have noted a lack of substance among reporting firms, many of which represented their efforts at being ecologically responsive only by means of photographs, advertisements, products and self reports.[7] This has been called the social construction of 'carbon', 'sustainability' and 'environmentalism' (Milne *et al.*, 2009). Haigh also argues that ways investors have used environmental data can be thought of similarly. He suggests that such data collection sustains and reassures investors of a potentially overstated and imaginary ideal of environmental responsiveness (Haigh, 2013).

The gap between a postulated 'carbon-reduced' world and rapid-fire financial trading markets is underlined by the tiny fraction of investments in the UN-endorsed carbon emissions rights and project-based mechanisms that lie in private hands. Despite controlling almost US$10 billion invested in mechanisms to lower carbon in 2007, this accounted for barely one-thousandth of the US$74 trillion then under private management (World Bank, 2008). Optimists suggest that it is early days, and that 'low-carbon imaginary' will blossom with better-parsed data, booming carbon markets and governmental policies that promote green investing. But even the reality of increasingly available, verified, high-quality carbon data can be questioned.

30.3   Climate Change Adaptation

In the 1990s, the dominant discourse of those concerned with climate change focused on the need to avoid 'dangerous' warming. In recent years, attention has shifted from climate mitigation to adaptation. Substantial adaptation is essential, not only because significant climate change has already occurred but also because additional effects are inevitable, even if the world were to practice aggressive mitigation. In this respect, adaptation is also precautionary. However, as the evidence of climate change has strengthened and as carbon-sparing, energy-generating technologies have improved and reduced in cost, there is a strengthening view that civilization need not avoid this risk but can, instead, adapt to it. Adaptation is necessary, but should not be at the expense of mitigation.

In many developed and developing countries, mitigation efforts such as emission trading schemes are focused, politically and structurally, at the national level. This has contributed to state and local governments reducing their mitigation focus, instead switching to adaptation as a legitimate priority. They have, in effect, experienced a power shift prompting climate change adaptation. Although relevant across all scales (Adger *et al.*, 2005), adaptation necessarily occurs in practice more locally, providing an opportunity for progressive and realistic local governments. For example, the C40 Mayors Group have placed a high priority on addressing climate change (adaptation and mitigation). Also, the organization ICLEI (International Council for Local Environmental Initiatives), now called Local Governments for Sustainability, has a mandate to address local sustainability issues for its membership of 1200 local governments. Compounding delay on adaptation and mitigation is debate on key concepts such as 'attribution' and 'additionality' (see Box 30.1).

Increasing adaptation funding may create opportunities for the health community to reduce underlying health inequities, thus partly offsetting an effect of climate change, which is widely anticipated to worsen existing global health inequalities. To advance this goal, several elements are needed: relevant policies – including for health, poverty reduction and development – should identify where adaptation can be integrated. Also needed are greater transparency in decision making, genuine efforts to eliminate (or at least substantially reduce) the corruption that corrodes so much aid funding and capacity building in countries that have endured persistent hardship. These goals, if achieved, could generate 'climate-resilient development'. However, few Millennium Development Goals have been met, in a world barely touched by climate change in comparison to what is predicted. Can the additional challenges of climate change be met while the essential building blocks for development remain so frail?

Could adaptation even be a pretence, a justification by rich polluting populations to defer mitigation perpetually? Have dreams of adaptation slowed mitigation, in the same way that the chimera of clean coal (carbon sequestration) has delayed solar technology? Beyond a critical point of climate change, adaptation seems certain to be inadequate.

30.4 Values Shift

Common to many strategies for mitigation and adaptation is the need for a shift in values. While some groups may express extreme techno-optimism, in which no social changes are required because technology solves everything, such positions are increasingly rare. Collectively, the global shift in cultural, political and behavioural change might be called an eco-noösphere – a global understanding that we are one civilization on one planet (Vernadsky, 1945). Increasingly, leaders in science, development and the media are calling for such changes.

Values can also shift concerning family size. Replacement or below-replacement fertility is now widespread, not only in parts of Europe but also in China and Iran, which has lowered its fertility rate from an average of 6.5 in 1986 to 1.8 in 2008, through policies enabling 80% of women to get secondary education and easy access to family planning (Stott, 2012). Sometimes, though, family planning has been top-down and coercive, leading to forced sterilization.

Although there have been occasional statements in favour of environmental protection by religious leaders, and the name of the new Pope, Francis, suggests environmental sympathy, the contribution of the world's great faiths to these debates has been limited (Jenkins and Chapple, 2011). In some cases, faith-based movements have contributed actively to denial of climate change.[9] However, others argue more optimistically that a new 'dark green'[10] religion is evolving (Taylor, 2010).

Box 30.1

Attribution attempts to address the question, 'to what degree can the causation of events be distinctly credited ("attributed") to climate change?' The scientific view is shifting from one in which no individual weather event is a fingerprint of climate change to one in which *all* extreme events are influenced by the greater energy in the climate system, analogous to how loaded dice increase the risk of a payout (Trenberth, 2011). However, in the world of policy and diplomacy, such liberal interpretations seem unlikely, as they would open the door to successful litigation by developing countries against powerful and wealthy polluting nations.

A second contested issue is 'additionality'. Many developing countries hold rich nations primarily responsible for climate change. They resist calls to decarbonize, seeing this as ecological imperialism that places their own material development at risk. In fact, a four-degree world may lead many low-income countries to entrapment in devastating poverty. At the same time, lack of transparency and the wish to appear generous leads some rich countries to announce adaptation funds as if they are *additional* to official development assistance.

The Green Climate Fund will oversee the global allocation of adaptation funding for developing countries, to be split evenly with mitigation funding, resulting in US$50 billion annually for each by 2020. The World Bank reported in 2011 that fossil fuel subsidies exceeded US$500 billion, about ten times the cost of Hurricane Sandy, an event probably worsened by fossil fuel subsidies.[8]

Expanding selves

Patterns of environmental awareness and consumption are linked intimately with people's sense of themselves. The past few decades have witnessed a shift in many people's self-identity, from producers to entitled consumers (Hamilton, 2010), abetted by public relations utilizing advertisers, creating messages that products will enhance their purchaser, making them more attractive mates, better parents and even more fully realized individuals. For those seeking to forge or find themselves in the mall, the prospect of reduced consumption is an affront, which dwarfs the amorphous shadow of climate change.

Those with a narrow, egocentric concept of self tend to act with less regard for the environment than those who conceptualize themselves as linked with others, especially those whose self-concept incorporates aspects of nature (Hamilton, 2010). Hard economic times may prove useful in this regard. As the economy shudders and people find it harder to buy goods that would define them, they may turn to other sources of identity, such as community, family, religion, voluntary activities and connection with nature.

Medical professionals have a potentially powerful role to play in helping people grow their sense of self. As they tend to be near the top of the economic and social hierarchy, their example is potent. The social impact of bike racks, rather than parking places, being reserved for doctors outside of hospitals could be considerable, as the definition of success moves from fast cars toward toned calves.

Reducing intergenerational inequality

Intergenerational conflict is also relevant. The philosopher, Stephen Gardiner's book, *A Perfect Moral Storm* (Gardiner, 2011), identifies strong reasons each generation finds to defer action. If younger people awaken more broadly to the constriction of possibilities for flourishing that their parents have overseen, a generational consciousness could arise that scapegoats (not without validity) baby boomers, politicizing the interests of the young against the old. Despite a

weakening of the ancient respect for age, much of the present economic system remains geared strongly in favour of the old.

Contraction and convergence

The 'contraction and convergence' movement appeals to those attracted by fairness, rationality and incrementalism, by creating pollution budgets which are divided between over and underconsumers. In this model, high consumers voluntarily legislate to reduce their pollution (e.g. carbon emissions) as underconsumers increase theirs, so that per capita consumption converges on an equitable level (Meyer, 2000), which limits total environmental harm (http://www.gci.org.uk/).

Unfortunately, the success of contraction and convergence is constrained by several factors. To date, there is little evidence that those who are the most affluent have reduced their carbon or other environmental footprints (Rosa and Dietz, 2012). Another form of voluntary redistribution, the Tobin tax, also is a good idea, but remains theoretical. The Pareto Principle, espoused by Italian dictator Benito Mussolini's favourite economist (Ackerman and Heinzerling, 2004), holds that scarcity is best solved not by redistribution but by lifting the consumption of the poor, that is convergence *without* contraction. Pareto's strategy fails in a resource-constrained world, but humanity does not yet seem to have gained sufficient selflessness for contraction and convergence to be plausible.

Responses by the health sector

One variant of contraction and convergence holds promise. This is of health and other forms of co-benefits. This strategy appeals to the sense of justice and fairness held by those who are more fortunate (Stott, 2012), including with regard to global meat consumption. People who ingest high levels of red meat (especially from digastric animals such as cattle and sheep) have an enhanced risk of bowel cancer and heart disease, and also contribute disproportionately to GHG emissions. Therefore, reduced ingestion of red meat by some people offers to improve health and slow climate change. On the other hand, an increased intake of red meat will benefit more than a billion people who are chronically iron deficient, though other ways to improve their iron status, such as treatment of parasites, will also help (McMichael *et al.*, 2007). Two more win–win strategies are increased active transport and better housing. Such strategies increase fitness, reduce energy use and lessen social isolation, thus potentially improving social capital (Haines *et al.*, 2009). Being a climate activist may also deliver co-benefits, including to mental health.

Various health organizations have been founded which attempt to promote this cultural shift, including by health workers. These include the Climate and Health Council. Early steps are also being taken to introduce the topic of climate change and sustainability in health courses, including in some medical schools.

30.5 Muddling Through with Technology, Assisted by a Transition in Subsidies

The high energy price appears here to stay. Scarcity may drive civilization toward shared global frugality, but an intensifying fortress world, in which anxious populations try to preserve high consumption, even though surrounded by want, is more likely. Such prospects may seem blasé to those used to living within the fortress, but conditions within currently prosperous regions could deteriorate. For example, the high unemployment in Europe may contribute to a return of extremism that will reduce general well-being. Already, some look back wistfully to a decade ago, when low energy prices and comparative prosperity offered greater opportunity to prepare for and to avert now looming problems.

But might sufficient technological breakthroughs occur to force the global renewable energy price below that of fossil fuels? If so, this could be accelerated by the use of more subsidies (Victor, 2009). There is hope. Germany generates a quarter of its electricity from solar, wind and biomass, hosting a third of the world's installed solar capacity. Most of this is owned by individuals, cooperatives and communities, marginalizing the country's once all-powerful utilities (Davidson, 2013).

30.6 Balancing Fear and Hope: Devising a Social Vaccine to Protect Civilization

Many trends in population are cyclic, driven by the waxing and waning of social memory and behaviour (Butler, 2005), in a way that can be likened to physical immunity. Following epidemics, surviving populations have sufficient antibodies to inhibit repeat infection until enough immunologically naive people accrue, due to waning immunity and the maturation of a new generation. After World War II, strong collective action led to a sufficiently vigorous social contract to inhibit large-scale war, fortified by numerous institutions, including the United Nations. Almost 70 years later, the 'social immunity' generated by the two World Wars is still fairly powerful, though some of these institutions are weakening. Following the Depression and World War II, the birth of the UK National Health Service liberated social forces to consolidate reduced inequality for several decades. Decolonization became unstoppable. However, gradually, many forms of inequality have reappeared, including in most formerly Communist nations.

Today, global civilization is at risk. Numerous civilizations have collapsed in the past, driven by tight ecosocially connected systems (Redman, 1999). Could a social vaccine (Baum *et al.*, 2009) be developed to help protect against collapse? Such a vaccine would stimulate civilization to protect itself, due to collective self-interest. This would work by 'immunizing' enough people to generate sufficient self-corrective action. Something like this was employed by the peace movement during the Cold War (Chivian, 2013).

Others, however, point out that too blunt a warning has a paralysing, counterproductive effect (Moser and Dilling, 2007), as an excessive antigen in a physical vaccine can cause harm. But a placebo social vaccine may be more dangerous, by inducing complacency. A balance between these two extremes is thus needed. Failure to alter planetary consciousness is likely to lead to an enforced and unpleasant change, following a time of turmoil that may make the Dark Ages seem desirable.[11] Instead, it is vital to educate a sufficient number of people to call for, develop and support the radical new policies that are required. Such policies include acceptance that resources are limited, green (ecological) economic systems that explicitly price negative externalities, and a revival of fairness of opportunity. But, of course, education is not enough; people need to change behaviour.

30.7   Last Words

No single solution is likely. Former US Vice President Al Gore has called for 'silver buckshot'. This extends not just to technologies but also to social strategies and subsidies such as feed-in tariffs (Stott, 2012). The preceding chapters in this book indicate the complexity of the web of interconnections in the eco-noösphere, which links GHG emissions and ill health. This complexity, which resists human understanding, can lend the challenge of climate change an aura of intractability. Yet the high degree of interconnection can aid solutions. Webs have key supporting strands on which they rely. Individual efforts, when well targeted, have the potential to yield disproportionately powerful results.

One such key strand is charismatic individuals who serve as catalysts and sources of inspiration. There are many: from the late Wangari Maathai to Australian Senator Bob Brown.

Robert May, the former president of the Royal Society, once told of a game in which cooperation led to a better group outcome (such as a repeated prisoner's dilemma game). In some settings, groups learned quickly to cooperate. But in other settings, some players seemed to derive perverse satisfaction from seeing others lose, even if at a personal cost.

Following a sufficiently severe global crisis, might a more cooperative global community emerge? For example, could a limited nuclear war or a sharp increase in the rate of sea level rise finally stimulate effective global action?[12] Maybe, but even better would be if the collective imagination of these possibilities could avoid such a crisis, before its worst manifestations. Increasingly, this seems unlikely, yet it is vital not to lose all hope.

Perhaps the most realistic solution lies in adequate high-level understanding of the risks of 'business as usual' (including over-reliance on adaptation). Numerous speeches by UN Secretary General, Ban Ki Moon, give hope. However, to date, such leadership is largely symbolic. Increasing discussion of the Green Economy (or 'GDP Plus') seems rhetorical while fossil fuel subsidies remain so high.

Greater use of the Internet, powered by clean energy, is an obvious way to increase opportunities, knowledge and social interaction and stimulation, especially in rural areas. This will also slow rural to urban migration and lead to a decline in birth rates in many low-income countries and locations, such as Niger, Somalia and rural India. Greater gender equity combined with other forms of human rights will also slow fertility, an important adaptation and a future mitigation strategy (Bryant *et al.*, 2009) Perhaps the 'Great Transition' (Raskin *et al.*, 2002) foreseen by some optimists will not emerge, but at least a 'muddle through world' may (Butler, 2010).

Notes

[1] That is to say, energy used by humans for transport, housing, electricity. The vast majority of energy used by humans is, of course, from the sun, for light, photosynthesis, warmth, and so on.

[2] Even though the cost of many advanced renewable technologies has fallen, more 'mundane' components, such as building materials, roads and the cost of supporting construction crews in remote locations, have not. Expanding the supply of fossil fuels requires similar basic infrastructure, but the energy in most fossil fuels is denser than in renewables, hence favoured by an economic system which ignores both climate change and the fact that fossil fuels are a limited resource.

[3] Other than tar sands, which have an energy return on energy investment as low as two.

[4] Such as from economies powered mainly by livestock to those fuelled by the sun, bypassing fossil fuels.

[5] Earth's 'shininess', or capacity to reflect sunlight.

[6] The Global 500 are the world's largest corporations, in terms of revenue, ranked each year by Standard and Poor's.

[7] There are claims that BP spent more on advertising its recently abandoned 'Beyond Petroleum' campaign than it did on investing in ways to move beyond fossil fuels.

[8] http://www.abc.net.au/environment/articles/2013/04/10/3733204.htm.

[9] For example, http://www.cornwallalliance.org/alert/cornwall-alliance-releases-an-evangelical-declaration-on-global-warming/.

[10] Beliefs that nature is sacred nature, entailing an ethical obligation to try to protect other species. Some supporters of the Green movement may fit in this category.

[11] Perhaps the most plausible dark scenario is of a fortress (triage) world in which diminishing populations maintain, at least for a time, reasonable order and well-being amidst an increasing sea of disorder, violence and coercion.

[12] A limited nuclear war may, in fact, end the taboo, triggering more. A sharp rise in sea level rise may trigger emergency Arctic cooling via geoengineering, with even worse consequences, such as to the global hydrological cycle.

References

Ackerman, F. and Heinzerling, L. (2004) *Priceless: On Knowing the Price of Everything and the Value of Nothing*. New Press, New York.

Adger, N.W., Arnell, N.W. and Tompkins, E.L. (2005) Successful adaptation to climate change across scales. *Global Environmental Change* 15, 77–86.

Baum, F., Narayan, R., Sanders, D., Patel, V. and Quizhpe, A. (2009) Social vaccines to resist and change unhealthy social and economic structures: a useful metaphor for health promotion. *Health Promotion International* 24, 428–433.

Bryant, L., Carver, L., Butler, C.D. and Anage, A. (2009) Climate change and family planning: least developed countries define the agenda. *Bulletin of the World Health Organization* 87, 852–857.

Butler, C.D. (2005) Peering into the fog: ecologic change, human affairs and the future. *EcoHealth* 2, 17–21.

Butler, C.D. (2010) The climate crisis, global health, and the medical response. *World Medical Journal* 56, 56–58.

Chivian, E. (2013) Global environmental threats: why they are hard to see and how a medical model may contribute to their understanding. *Cardiovascular Diagnosis and Therapy* 3, 93–104.

Daly, H. (1996) *Beyond Growth*. Beacon Press, Boston, Massachusetts.

Davidson, O.G. (2013) The Sustainable Economy. (http://www.ecobuddhism.org/solutions/sustainable_economy/gcee, accessed 3 April 2014).

Gardiner, S. (2011) *A Perfect Moral Storm*. Oxford University Press, Oxford, UK.

Haigh, M. (2013) Deconstructing myth: low-carbon sustainability. *Social Semiotics* 23, 47–66.

Haines, A., McMichael, A.J., Smith, K.R., Roberts, I., Woodcock, J., Markandya, A., *et al.* (2009) Public health effects of strategies to reduce greenhouse-gas emissions: overview and implications for policy makers. *The Lancet* 374, 2104–2114.

Hamilton, C. (2010) *Requiem for a Species. Why We Resist the Truth About Climate Change*. Allen and Unwin, Sydney, Australia.

Hanlon, P. and Carlisle, S. (2010) Re-orienting public health: rhetoric, challenges and possibilities for sustainability. *Critical Public Health* 20, 299–309.

Hansen, J.E. (2013) Climate urgency. *Simulation and Gaming* 44, 232–243.

Jenkins, W. and Chapple, C.K. (2011) Religion and environment. *Annual Review of Environment and Resources* 36, 441–463.

Kosoy, N., Brown, P.G., Bosselmann, K., Duraiappah, A., Mackey, B., Martinez-Alier, J., *et al.* (2012) Pillars for a flourishing Earth: planetary boundaries, economic growth delusion and green economy. *Current Opinion in Environmental Sustainability* 4, 74–79.

Leggett, J. (2011) Unburnable carbon – are the world's financial markets carrying a carbon bubble? (http://www.carbontracker.org/carbonbubble, accessed 3 April 2014).

McMichael, A.J., Powles, J., Butler, C.D. and Uauy, R. (2007) Food, livestock production, energy, climate change and health. *The Lancet* 370, 1253–1263.

Meyer, A. (2000) *Contraction and Convergence. The Global Solution to Climate Change*. Green Books for the Schumacher Society, Foxhole, Devon, UK.

Milne, M.J., Tregidga, H. and Walton, S. (2009) Words not actions! The ideological role of sustainable development reporting. *Accounting, Auditing and Accountability Journal* 22, 1211–1257.

Moser, S. and Dilling, L. (2007) *Creating a Climate for Change: Communicating Climate Change and Facilitating Social Change.* Cambridge University Press, Cambridge, UK.

Murphy, D.J. and Hall, C.A.S. (2011) Energy return on investment, peak oil, and the end of economic growth. *Annals of the New York Academy of Sciences* 1219, 52–72.

Murray, J. and King, D. (2012) Climate policy: oil's tipping point has passed. *Nature* 481, 433–435.

Nicholls, N. and Larsen, S. (2011) Impact of drought on temperature extremes in Melbourne, Australia. *Australian Meteorological and Oceanographic Journal* 61, 113–116.

Oreskes, N. and Conway, E.M. (2010) *Merchants of Doubt: How a Handful of Scientists Obscured the Truth on Issues from Tobacco Smoke to Global Warming.* Bloomsbury Press, New York.

Raskin, P., Gallopin, G., Gutman, P., Hammond, A., Kates, R. and Swart, R. (2002) *Great Transition: The Promise and Lure of the Times Ahead.* Stockholm Environment Institute, Boston, Massachusetts.

Redman, C.L. (1999) *Human Impact on Ancient Environments.* University of Arizona Press, Tucson, Arizona.

Rees, M. (2013) Denial of catastrophic risks. *Science* 339, 1123.

Rosa, E.A. and Dietz, T. (2012) Human drivers of national greenhouse-gas emissions. *Nature Climate Change* 2, 581–586.

Smith, J.B., Schneider, S.H., Oppenheimer, M., Yohe, G.W., Hare, W., Mastrandrea, M.D., *et al.* (2009) Assessing dangerous climate change through an update of the Intergovernmental Panel on Climate Change (IPCC) 'reasons for concern'. *Proceedings of the National Academy of Sciences* 106, 4133–4137.

Stott, R. (2012) Contract and convergence: the best possible solution to the twin problems of climate change and inequity. *BMJ* 344, e1765, doi: http://dx.doi.org/10.1136/bmj.e1765.

Taylor, B. (2010) *Dark Green Religion: Nature Spirituality and the Planetary Future.* University of California Press, Berkeley, California.

Trenberth, K.E. (2011) Attribution of climate variations and trends to human influences and natural variability. *WIREs Climate Change* 2, 925–930.

Vernadsky, V. (1945) The biosphere and the noösphere. *American Scientist* 33, 1–12.

Victor, D. (2009) The politics of fossil-fuel subsidies. (http://papers.ssrn.com/sol3/papers.cfm?abstract_id=1520984, accessed 3 April 2014).

World Bank (2008) State and Trends of the Carbon Market 2008 (https://wbcarbonfinance.org/Router.cfm?Page=DocLib&ht=25621&dtype=25622&dl=0, accessed 5 April 2014).

Index